Microbore Column Chromatography

CHROMATOGRAPHIC SCIENCE

A Series of Monographs

Editor: JACK CAZES
Sanki Laboratories, Inc.
Sharon Hill, Pennsylvania

Additional Volumes in Preparation

Microbore Column Chromatography

A UNIFIED APPROACH TO CHROMATOGRAPHY

edited by

Frank J. Yang
LEE SCIENTIFIC, INC.
SALT LAKE CITY, UTAH

CRC Press
Taylor & Francis Group
Boca Raton London New York

CRC Press is an imprint of the
Taylor & Francis Group, an **informa** business

First published 1989 by Marcel Dekker, Inc

Published 2019 by CRC Press
Taylor & Francis Group
6000 Broken Sound Parkway NW, Suite 300
Boca Raton, FL 33487-2742

© 1989 by Taylor & Francis Group, LLC
CRC Press is an imprint of Taylor & Francis Group, an Informa business

First issued in paperback 2019

No claim to original U.S. Government works

ISBN-13: 978-0-367-45127-1 (pbk)
ISBN-13: 978-0-8247-7989-4 (hbk)

Visit the Taylor & Francis Web site at
http://www.taylorandfrancis.com

and the CRC Press Web site at
http://www.crcpress.com

Library of Congress Cataloging-in-Publication Data

Microbore Column chromatography : a unified approach
   to chromatography / edited by Frank J. Yang.
       p.   cm. -- (Chromatographic science ; v. 45)
     Includes index.
     ISBN 0-8247-7989-4
     1. Chromatographic analysis. I. Yang, Frank J.   II. Title:
   Column chromatography. III. Series.
   QD79.C4M53 1988
   543'.0894--dc19   88-20276
                                                              CIP

Preface

Microbore column chromatography has been the major focus of much recent research and development effort in column technology, instrumentation, and detector technology, and in applications to chemical separation and characterization. The apparent trend toward microbore column chromatography is due to its development as a unified approach to chromatography that encompasses both instrumentation and applications of capillary gas chromatography (GC), supercritical fluid chromatography (SFC), and microbore high-performance liquid chromatography (micro-HPLC). The emergence of small-diameter open-tubular columns and microparticulate-fused silica-packed columns has resulted in many common features—such as the use of the same injector, column, detector, and system components—in the practice of GC, SFC, and HPLC.

Microbore column chromatography is growing rapidly; its technology and practice are no longer the exclusive domain of a handful of experts. Rather, it is becoming a routinely applied methodology in many industrial laboratories. It seemed timely to bring together in book form the practice and application of microbore column chromatography using either packed or open-tubular columns. The chapters on micro-HPLC and SFC will provide readers with an in-depth understanding of the subject as well as its underlying thought processes to stimulate new ideas for future study.

The use of micro-HPLC has achieved tremendous growth in column technology, gradient elution techniques, optical detection systems, and applications to solve analytical problems. A vast amount of information, including state-of-the-art detector technology, contributed by many leading experts, has been included in this book. SFC as a fast-growing technology has undoubtedly become the method of choice for applications in the analysis of thermally labile, nonvolatile, and high-molecular-weight compounds. Chapters on some important subjects such as SFC instrumentation and applications were contributed to this volume, by pioneers and experts in their fields. These chapters should serve as a good reference source for the practice and applications of micro-HPLC SFC techniques in solving analytical problems.

I am grateful to the chapter authors, without whom this book could not have been completed. I am also indebted to Dr. Keith D. Bartle for proofreading the first chapter. I take special pleasure in thanking Dr. Milton L. Lee and Mr. Hal Rosen for their encouragement and friendship, and Ms. Angela Sobieszyk for typing manuscripts. Finally, I am grateful for the support and understanding of my family, who encouraged me in its completion.

Frank J. Yang

Contents

Contributors

VERN V. BERRY   President, SepCon Separations Consultants, Boston, Massachusetts, and Assistant Professor, Chemistry Department, Salem State College, Salem, Massachusetts

THOMAS L. CHESTER   Section Head, Corporate Research Division, The Procter & Gamble Company, Cincinnati, Ohio

HERNAN J. CORTES   Project Leader, Analytical Laboratories, Dow Chemical Company, Midland, Michigan

C. DEWAELE   Research Associate, Laboratory of Organic Chemistry, State University of Ghent, Ghent, Belgium

JENNIFER C. GLUCKMAN*   Department of Chemistry, Indiana University, Bloomington, Indiana

HERBERT H. HILL, JR.   Professor, Department of Chemistry, Washington State University, Pullman, Washington

---

*Current affiliation: Research Scientist, Central Research, Pfizer, Inc., Groton, Connecticut.

KIYOKATSU JINNO Associate Professor, Materials Science, Toyohashi University of Technology, Toyohashi, Japan

DOUGLAS W. LATER Vice President of Operation, Lee Scientific, Inc. Salt Lake City, Utah

MILTON L. LEE Professor, Department of Chemistry, Brigham Young University, Provo, Utah

KARIN E. MARKIDES Assistant Professor, Department of Chemistry, Brigham Young University, Provo, Utah

MILOS V. NOVOTNY Professor, Department of Chemistry, Indiana University, Bloomington, Indiana

HERBERT E. SCHWARTZ Research Chemist, Santa Clara Analytical Division, Applied Biosystems, Inc. (formerly Brownlee Labs), Santa Clara, California

CHRISTOPHER B. SHUMATE Research Assistant, Department of Chemistry, Washington State University, Pullman, Washington

RICHARD D. SMITH Group Leader/Staff Scientist, Chemical Sciences Department, Batelle Pacific Northwest Laboratory, Richland, Washington

MAURICE VERZELE Professor, Laboratory of Organic Chemistry, State University of Ghent, Ghent, Belgium

BOB W. WRIGHT Senior Research Scientist, Chemical Sciences Department, Batelle Pacific Northwest Laboratory, Richland, Washington

FRANK J. YANG / Lee Scientific, Inc., Salt Lake City, Utah

EDWARD S. YEUNG Professor, Department of Chemistry and Ames Laboratory, Iowa State University, Ames, Iowa

Microbore Column Chromatography

1

Microbore Column Chromatography: A Unified Approach to Chromatography

FRANK J. YANG   Vice President, Lee Scientific, Inc., Salt Lake City, Utah

INTRODUCTION

Microbore column chromatography is a unified approach to chromatography. It can be depicted with the aid of the chromatography triangle shown in Figure 1. Microbore column chromatography offers high-resolution separation methodology within the scope of capillary gas chromatography (GC), supercritical fluid chromatography (SFC), and micro-high-performance liquid chromatography (HPLC). It covers the sample application domain of GC, SFC, and column liquid chromatography. Capillary GC, SFC, and micro-HPLC each has its own range of applications, instrumentation requirements, practical constraints, and technological uniqueness. However, the trend in the development of a unified microbore column chromatographic approach is apparent with the instrumentation development that allows the utilization of the same capillary column or columns of the same small capillary dimensions. Although developments in capillary GC, SFC, and micro-HPLC instrumentation and methodologies have taken place at different times, the separation techniques, column technology, and fundamental principles are parallel or identical in many aspects. In his book *Dynamics in Chromatography*, Giddings [1] argued that the divergence in thought mode between GC and LC is arbitrary, artificial, and counterproductive. As the column

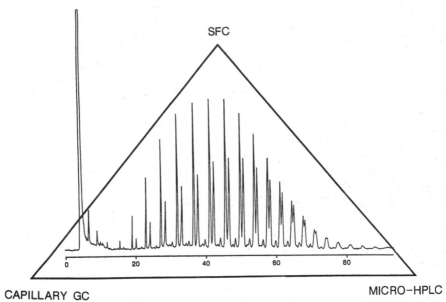

CAPILLARY GC   MICRO–HPLC

Fig. 1 Chromatography triangle of capillary GC, capillary SFC,
and micro-HPLC. The figure depicted the high resolution capability
of the microbore column chromatography. A unified approach to
chromatography for the application sample domain of GC, SFC, and
HPLC. Chromatogram of the separation of a polyglycol with an
average molecular weight of 1800 (chromatogram courtesy of B. E.
Richter, Lee Scientific).

diameter becomes smaller, the practice of GC, SFC, and HPLC becomes more similar. Table 1 compares some functional aspects of
capillary GC, capillary SFC, and micro-HPLC as practiced today.
It points out many common features and the potential for the practice of a unified microbore column chromatography, that is, the use
of the same injector, detector, column, and system components for
applications in capillary GC, SFC, or micro-HPLC. Current column
technology and detector advances will undoubtedly make microbore
column chromatography a unified approach for the generation of
total chromatographic information for an unknown sample.

    Microbore column chromatography is growing rapidly. Its de-
velopment and advances in instrumentation, column technology, and
applications are vast. Detailed discussions of various aspects of
microbore column chromatography are given by many contributing
authors in the subsequent chapters. A brief introduction of state-
of-the-art capillary GC, SFC, and micro-HPLC is given in the following sections.

TABLE 1 Comparison of Some Key System and Operation Parameters for Capillary GC, SFC, and Micro-HPLC

	Capillary GC	SFC	Micro-HPLC
Column types	Open-tubular, packed capillary	Open-tubular, packed capillary, 1 mm i.d. LC packed column	Open-tubular, packed capillary, 1 mm i.d. LC packed column
Key system components	Column oven	High pressure pump Column oven	High pressure pump Column oven
Mobile phase	High pressure gases	Supercritical fluids	Liquid solvents
Sample injection	Split/splitless (syringe)	Split (valve)	Split (valve)
	Cold on-column (syringe)	Direct (valve)	Direct (valve)
	Temperature-programmable (syringe)	Time-controlled sampling (valve)	Time-controlled sampling (valve)
	Split (valve)		
	Time-controlled sampling (valve)		
Detectors	FID, TSD, FPD, PID, ECD, MED, MS, FTIR	FID, TSD, FPD, ECD, UV, fluorescence, MS, FTIR	TSD, FPD, UV, fluorescence, MS, FTIR
Samples	Volatiles	Volatiles and nonvolatiles	Volatiles and nonvolatiles
	Thermally stable	Thermally stable and labile	Thermally stable and labile
	Low molecular weights	Low and high molecular weight	Low and high molecular weight
	Nonionic	Nonionic	Ionic and Nonionic

3

CAPILLARY GC

Capillary GC utilizes both open-tubular and packed capillary columns. As an example of its versatility, fused silica columns packed with conventional GC packing material are being used in high-efficiency separation of congeners in fruit brandies [2]. There is great potential for the use of short capillary columns packed with small particles (≤10 μm) for high-speed analysis and resolution of complex samples. A capillary-packed column has several advantages over the conventional 2-mm-inner diameter (i.d.) GC-packed column, not only in column efficiency and speed of analysis but also in enhancement in trace detection. Because of the reduced solute peak dilution in a capillary diameter-packed column, the minimum detectable concentration of the solute zones is greatly enhanced when a concentration-dependent detector such as TCD or ECD is employed.

Open-tubular capillary GC, proposed in 1957 by Golay [3], is enormously important in the practice of modern gas chromatography. Its rapid growth is evidenced by the exponential increase in the number of journal publications and routine laboratory applications in recent years. Advances in open-tubular capillary GC technology, practice, and application have been reviewed in many recent books [4−7]. Open-tubular capillary GC methodology has replaced many routine packed column practices, particularly, for new methodologies developed for the analysis of complex samples.

Open-tubular capillary GC has been developed to its full potential following the invention of fused silica open-tubular columns [8], the development of reliable capillary GC sampling techniques, commercialization of capillary gas chromatographs, and an intense education and training process. The major reasons for the interest and application of open-tubular capillary GC are:

1. Unsurpassed resolving power for the rapid separation of complex samples. A one million theoretical plate can be obtained in an analysis time of 30 minutes using a 20 m open-tubular column with a diameter of 25 μm. The column can resolve, with unit resolution, as many as 550 solute components in an analysis time of 30 minutes.

2. Excellent reproducibility in peak area and peak retention. Recent advances in capillary GC sampling techniques such as cold on-column, temperature-programmable sampling, and direct valve injection achieve better than 1 to 2% relative standard deviation (RSD) peak area reproducibility and 0.05% RSD in peak retention time reproducibility. These sampling techniques when employed with an autosampler provide the best reliable system for quantitative analysis and routine application in quality control.

3. Ease of method development. Because of the "brute force" resolving power of the open-tubular column, method development becomes relatively easy. The choice of the selectivity of the stationary phase becomes less critical when compared to packed column GC.

4. Advantage of rapid analysis. With the inherent high resolving power of small-diameter open-tubular columns, analysis can be very rapid for samples containing only a few components. An open-tubular column of 30-μm diameter can generate more than 500 theoretical plates per second. Open-tubular column GC is the method of choice in terms of speed of analysis for the analysis of volatile and thermally stable compounds.

5. Advances in fused silica column technology. Highly efficient, chemically inert, thermally stable, and nonextractable cross-linked open-tubular columns have recently become commercially available for routine applications. The technological breakthrough in capillary column deactivation and the stationary-phase cross-linking process has brought acceptance and application of fused silica capillary columns to many laboratories.

6. Extended GC application range to high-molecular weight samples and to trace analyses. The open-tubular column GC technique offers great advantages in trace analysis due to the sharpness of the elution peaks and the consequent enhanced detectability. Subfemtogram sample masses can be detected with an ECD using 50-μm-i.d. columns. Recent developments in high-temperature stationary phases and the aluminum-clad fused silica capillary columns [9,10] have greatly extended GC applications to large molecules such as polystyrene (MW = 1200) and polydimethylsiloxane ps 340 (MW = 1700).

7. Low flow rates facilitate ease of interfacing open-tubular column GC with mass spectrometers. Fourier transform infrared (FTIR), and microwave and other plasma emission (MED) detectors. Capillary GC-MS and capillary GC-FTIR are now very important identification tools. There is no doubt that the success in capillary GC-MS and capillary GC-FTIR interfaces ensure widespread acceptance of capillary GC techniques in many industrial laboratories. Microwave plasma emission detectors are on the edge of becoming real analytical tools. The low flow rate used in open-tubular capillary GC makes it relatively easy to interface directly to the microwave plasma emission detector. Successes with capillary GC-MED interfaces will produce many more

applications of capillary GC due to the multielement selective
detection capability for the environmental, petroleum, petrochemical, and pharmaceutical industries.

In comparison with SFC and micro-HPLC, open-tubular column GC
is always the method of choice for unknown sample screening if the
sample is: (a) thermally stable, (b) volatile, and (c) contains many
components. The choice of the column in terms of column length,
diameter, stationary phase, film thickness and selectivity, etc., depends upon the complexity of the sample and the desired speed of
analysis. One needs to know the easiest separation for a mixture of
components of different chemical classes. Such a separation can be
easily achieved by using the very high resolution available from the
capillary column. A good example of the application of small-diameter
long open-tubular columns for the separation of complex samples can
be found in the routine application of a 50 m × 100 μm SE-30 open-
tubular column in gasoline PONA (paraffin—olefin—naphtha—aromatics)
analysis where more than 250 components are separated in 70 minutes.
    For rapid analysis, a short-length small-diameter column is the
best choice. Figure 2 shows the rapid separation of a mixture of
some long-chain fatty acid methylesters on a 10 m × 10 μm fused
silica open-tubular column coated with 0.1 μm 90% biscyanopropyl
vinyl methylsilicone [11].
    Proper choice of the selectivity of the column stationary phase
is essential for fine tuning the separation of one or more difficult
pairs in the sample. This choice should be based on consideration
of the intermolecular forces and interactions such as dispersion,
dipole—dipole, and dipole-induced dipole between the solute and the
stationary-phase molecules [12]. The dispersion interaction between
solute and stationary-phase molecules accounts for separations based
on differences in solute boiling points and sizes. The dipole—dipole
interaction accounts for the separation of polar compounds on a
polar stationary-phase column. The dipole-induced dipole interaction can be used for the separation of dipolar solute molecules
which act as electron donors or electron-receptors. Figures 3a and
3b compare the separation of mutagenic aminophenanthrene isomers
on (a) 50% phenylmethylpolysiloxane, and (b) 25% biphenyl methylpolysiloxane open-tubular columns (12 m × 310 μm i.d.) under
identical chromatographic conditions. The 25% biphenyl methylpolysiloxane column clearly showed better selectivity in the separation
of the polar aminophenanthrene isomers based on the molecular
dipole-induced dipole interactions.
    Open-tubular column GC is capable of achieving even more dif-
ficult separation of geometric isomers having differences in length to
breadth ratios. Figure 4 [12] shows an example of the separation of
five-ring polycyclic aromatic hydrocarbons on a 25% biphenylcarboxylate

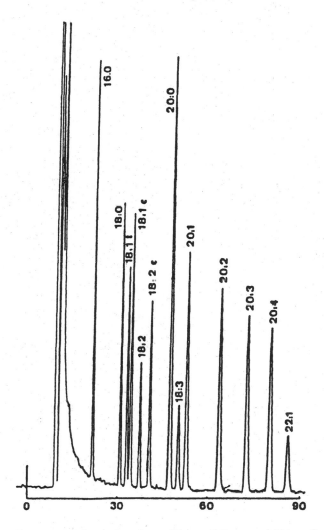

Fig. 2　High-speed capillary GC separation of a long-chain fatty acid methylesters on a 10 m × 100 μm i.d. column coated with a 0.1 μm film of the 90% CP phase, column temperature was 180°C. Hydrogen carrier gas inlet pressure was at 60 psig.　(Reproduced from Ref. 11 with permission of the publisher, copyright Dr. Alfred Heuthig.)

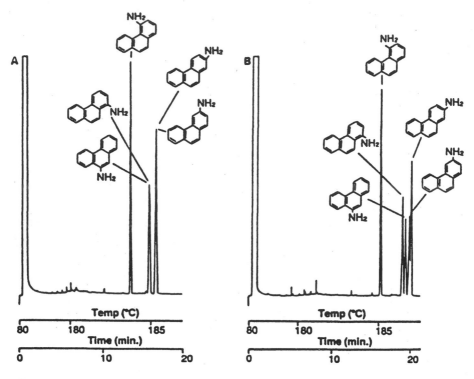

Fig. 3   Comparison of capillary GC column selectivity for the separation of mutagenic aminophenanthrene isomers on (A) 5% phenyl methylpolysiloxane and (B) 25% biphenyl methylpolysiloxane (df = 0.25 µm) capillary columns of 12 m × 310 µm i.d.   Column temperature programmed from 80°C to 180°C at 20°C/min and from 180°C to 220°C at 0.5°C/min after an isothermal period of 1 minute.   (Reproduced from Ref. 12 with permission of the publisher.)

ester liquid crystalline polysiloxane stationary phase [13] open-tubular column of 12 m × 200 µm i.d.   The rod-like isomer (peak 12) is able to interact more strongly with the parallel arranged mesomorphic side chain of the polysiloxane smectic liquid crystal phase than are the other disk-like or spherical (peak 1) isomers. Other examples of the application of the same smectic liquid geometric isomers in an environmental priority pollutant sample

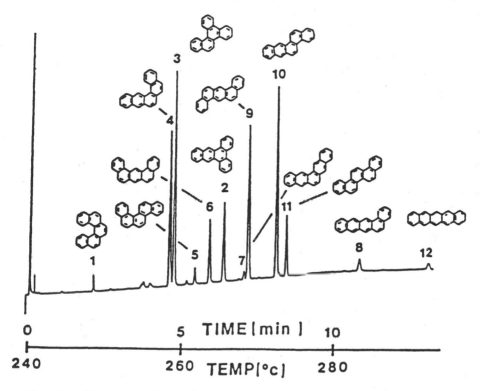

Fig. 4   The separation of annealed five-ring polycyclic aromatic hydrocarbons on a 12 m × 200 μm i.d. open-tubular column coated with 25% biphenylcarboxylate ester liquid crystalline polysiloxane stationary phase (df = 0.15 μm).   Column temperature programmed from 24°C to 280°C at 5°C/min.   (Reproduced from Ref. 12 with permission of the publisher.)

containing 2, 3, 7, 8-class congeners of polychlorodibenzodioxins and dibenzofurans can be found in Ref. 14.

Advances in open-tubular column technology have also led to successes in the separation of enantiomers.   Fused silica open-tubular columns coated with D-Chirasil-Val or L-Chirasil-Val are successfully employed for the separation of many enantiomers and sequential isomers of dipeptides containing glycine and leucine. Figure 5 shows rapid resolution of four stereoisomers of N-TFA-ala-phe-OME on a 25 m × 300 μm i.d. fused silica open-tubular column coated with D-Chirasil-Val [15].

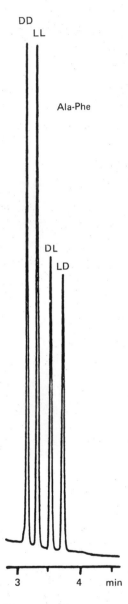

Fig. 5 Fast open-tubular GC separation of stereoisomers (D-D/L-L, and D-L/L-D) of N-TFA-Ala-Phe-OME on a 25 m × 300 μm i.d. column coated with D-Chirasil-Val. Column was at 200°C isothermal. (Reproduced from Ref. 15 with permission of the publisher.)

Some of the success and rapid growth of capillary GC is due to the recent advances in column technology. The state-of-the-art procedure for the preparation of fused silica open-tubular columns has been reviewed [12,16,17,83]. As capillary SFC follows the growth of capillary GC, new column technology in areas such as specialty phases, high polarity phases, and selective phases for biological sample applications is expected to emerge in the near future. Since the growth of microbore column chromatography depends upon the development of small-diameter columns, future trends in open-tubular column technology are likely to follow the development of small-diameter columns for high-speed and high-resolution capillary GC, capillary SFC, and open-tubular column HPLC.

SFC

Although capillary GC is the method of choice for rapid chromatography and sample screening, its application is limited to volatile and thermally stable compounds. The separation of thermally labile, nonvolatile, and high-molecular-weight samples is best carried out by SFC or HPLC. SFC is preferable to conventional HPLC due to the fast analysis and a wide range of compatible GC and LC detectors. It is preferred over HPLC for samples which can be analyzed or eluted in supercritical fluid mobile phases. It is also the method of choice for the initial screening of unknown samples which may contain both volatile and nonvolatile, high molecular weight components, thermally labile compounds, and compounds not detectable by ultraviolet (UV) or fluorescence detectors.

Many gases listed in Table 2 exhibit excellent elution strength when compressed to high densities. Giddings [18] stated that "one of the most interesting features of ultrahigh pressure gas chromatography would be its convergence with classical liquid chromatography. A liquid is ordinarily about 1000 times denser than a gas; at 1000 atmospheres, however, gas molecules crowd together with a liquid-like density. At such densities, intermolecular forces become very large, and are undoubtedly capable of extracting big molecules from the stationary phase." The pioneering work on supercritical fluid chromatography was first reported by Klesper et al. [19]. Their work demonstrated the enhanced solubility of some porphyrins in dichlorodifluoromethane and monochlorodifluoromethane at pressures up to 136 atm on a 77 m length column packed with 33% polyethylene on 60–80 mesh Chromosorb N. Later work by Giddings et al. [20–22] further revealed the great elution potential of dense gas chromatography at pressures up to 2000 atm for complex and nonvolatile

TABLE 2   Supercritical Temperature and
Pressure for Some Selected Gases

Gases	Tc (°C)	Pc (atm)
CO_2	31.3	72.9
NH_3	132.5	112.5
n-Butane	152.0	37.5
n-Pentane	196.6	33.3
N_2O	36.5	71.7
Xe	16.6	58.4
SF_6	45.5	37.1
CHF_3	25.9	46.9
$CHClF_2$	96.0	49.1
CCl_2F_2	111.8	40.7
$CClF_3$	28.8	39.0
CH_3OH	238.7	79.9
C_2H_5OH	242.3	63.0
$1-C_3H_7OH$	262.8	51.0
$2-C_3H_7OH$	235.4	47.0
$CH_2 = CH_2$	217.1	55.0

solutes with molecular weights as high as 400,000.   Giddings stressed
the advantages of employing the solvation power of high-density
gases over liquids especially in the control of solvent strength by
rapidly and easily changeable gas pressure, in speed of analysis,
and in detection with the sensitive and selective GC detectors such
as FID and NPD.   Further developments in packed column SFC were
reported by Sie et al. [23–25], who demonstrated the separation of
polynuclear aromatic hydrocarbons using CO_2, n-pentane, and isopropanol mobile fluids at temperatures up to 245°C.
    Progress in the development of SFC of the late 1960s was over-
shadowed by the advent of HPLC and by numerous technical difficulties in pump technology and in the lack of high-performance
SFC columns.   Recent advances in the development of high-pressure

pulseless syringe pumps and in the availability of the HPLC-packed and capillary GC open-tubular columns revived interest in supercritical fluid chromatography. Gere et al. [26,27] and Greibrokk et al. [28] modified a reciprocating HPLC pump for packed column SFC applications. An SFC system based on the work of Gere et al. [26,27] was first made commercially available by Hewlett-Packard Company [26,27,29] in 1982. Unfortunately, the packed column SFC system was later made obsolete due to the lack of density programming and FID detection features. The system also suffered from many difficulties such as flow rate instability, pulsation noise, and poor retention time reproducibility.

Following the work of Novotny et al. [30], dedicated SFC pumping systems utilizing pulseless high-pressure syringe pumps were made available in 1985 from several commercial sources. The application of SFC for the separation of thermally labile compounds, nonvolatile solutes, and high-molecular-weight samples is now a widely utilized methodology.

The rapid growth of SFC applications has also been made possible by the following major factors: (a) the invention of the open-tubular column SFC by Novotny et al. [30] in 1981; (b) the availability of high-performance open-tubular columns with non-extractable stationary phases; (c) various advances in SFC instrumentation, detectors, autosamplers, and sampling systems; and (d) the need for the analysis of many industrial samples that cannot be analyzed by GC and/or HPLC techniques.

SFC today has centered on the application of both open-tubular columns and microbore columns of 1-mm-i.d. packed with 5 and 10 μm particles [31]. Although still in its early developmental stage, packed capillary columns are expected to play a very important role in the future of packed column SFC applications. The advantages of the fused silica capillary-packed columns over 1 mm i.d. micro-packed columns for SFC applications are: (a) the inherent high permeability of the fused silica capillary-packed columns allowing the use of longer columns for better resolution; (b) low column mobile fluid flow rates allow direct interfacing to GC detectors, MS, FTIR, etc.; (c) small elution peak volumes enhance the ease of detection in conjunction with a concentration-dependent detector; and (d) small elution peak width or volume allows high resolution and high concentration sample transferring or heart cutting in multi-dimensional SFC-HPLC or SFC-GC techniques.

Micropacked columns of 1-mm i.d. currently used in SFC were developed for micro-HPLC applications. Unfortunately, these columns do not have the chemical inertness required for elution of polar compounds by a supercritical CO_2 mobile phase. The silanol groups on the surface of the silica particle cannot be masked with the supercritical CO_2 and the result is severe tailing and poor

resolution of polar solutes. Recently, micropacked columns packed
with silica that have been deactivated with polymethylsiloxane gums
[17,32,82] have become commercially available. The columns were
reported to have good chemical inertness and high efficiency.

Capillary SFC first reported by Novotny et al. in 1981 [30] has
undoubtedly spurred a significant new interest and rapid growth in
the development of SFC instrumentation, column technology, application techniques, and methodology. Capillary SFC offers the following advantages over packed column SFC: (a) low pressure drop
per unit length allowing long columns to be employed for achieving
high resolution; (b) large peak capacity for the analysis of very
complex samples; (c) low surface adsorption which maintains good
peak shapes and minimizes sample loss; (d) column permeability
allows rapid pressure programming; and (e) ease of interface to GC
and MS detectors. Capillary SFC, however, does not match the
performance of packed column SFC in speed of analysis, in sample
capacity, and in the forgiveness of the injection of dirty samples.

Following the rapid advances in SFC instrumentation and detection systems (see Chaps. 8, 9, and 10), many key SFC applications
have been identified. The practice and application of SFC in the
analysis of industrial samples is reviewed in Chapter 11. Some
other novel applications are given below:

Simulated Distillation of Heavy Petroleum Fraction and Crude Oil

Capillary SFC, when operated at the mild column temperature of 100°C
in conjunction with an FID, provides the best method for simulated
distillation. A typical SFC chromatogram of a crude oil sample
eluted from a short nonpolar column (7 m × 50 μm 50% n-octyl
methylpolysiloxane) under pressure-programming conditions is given
in Figure 6. Excellent sample recovery and minimum baseline drift
were observed. Schwartz et al. [33] and Raynie et al. [34] demonstrated the application of capillary SFC for the simulated distillation
of the heavy petroleum fractions with a true boiling point up to
1400°F. A complete sample recovery of the DISTRACT fraction from
a 10 m × 50 μm DB-5 column was obtained with a CO_2 mobile phase
at 100°C with pressure programming. A computer-reconstructed
area slice chromatogram with the boiling point distribution corre-
spondent to the accumulative percentage eluted is shown in Figure 7.
In comparison to the standard GC ASTM D2887 method, which limits
the maximum upper boiling point to 1000°F as indicated in the figure
by the dotted line, the capillary SFC technique elutes 40% more
area for the heavy components of the sample. In addition, the data
obtained from the capillary SFC method demonstrates more accurately
the true boiling points than that from the ASTM D2887 method
shown in Table 3 [34].

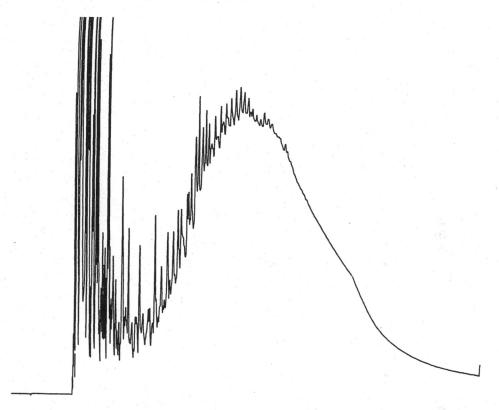

Fig. 6 Supercritical fluid chromatography of crude oil for the application in simulated distillation. A 7 m × 50 µm i.d. open-tubular column coated with 25% octyl methylpolysiloxane was employed. Column temperature was at 100°C. FID detection with CO_2 mobile phase programmed from 75 atm to 100 atm at 5 atm/min after an initial 5 minute hold at 75 atm, then to 350 atm at 10 atm/min.

Group Fractionation of Gasoline and Middle Distillate Fuels

Several methods have been used for many years by the petroleum industry for the determination of saturates, olefins, and aromatic hydrocarbons in petroleum products. The fluorescence indication absorption (FIA) method, ASTM D1319 [35], and LC methods [36–40] suffer from many problems such as poor resolution, long analysis time, poor precision and reproducibility, and the lack of sensitive and uniform response detectors such as the FID. Supercritical fluid chromatography, using CO_2 and SF_6 as mobile phases

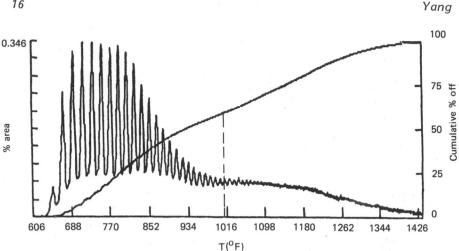

Fig. 7 Boiling point distribution (% off, right-hand scale) and area plots (left-hand scale) of an atmospheric residium of a petroleum sample. (Reproduced from Ref. 33 with permission of the publisher.)

Table 3 Comparison of the Deviation of Simulated Distillation Boiling Points of Some PAHs from True Boiling Points

Compounds	Deviations from true boiling points (°C)		True boiling (°C)
	ASTM D2887	SFC	
Naphthalene	−11	7	218
Phenanthrene	−35	6	339
Pyrene	−48	8	395
Chrysene	−60	−4	447

Source: From Ref. 34.

presents the best method for obtaining good resolution and uniform response with FID detection. Many reports have been published [41—44] in recent years concerning the use of packed column SFC for petroleum hydrocarbon group separations. Campbell et al. [45] recently proposed the use of a column-switching technique utilizing a silica column and an Ag^+-loaded strong cation-exchange column packed with Nucleosil 5-μm sulfonic acid silica particles for obtaining baseline resolution of saturates, olefins, and aromatics in a single run. Figure 8 shows an SFC hydrocarbon group separation of a gasoline sample using a two-column switching system. The gasoline sample was first injected onto the silica column where aromatics were selectively retained at 50°C using a 10% CO_2/90% SF6 mobile phase. The olefins, eluted with saturates from the silica column, were then retained in the Ag^+-loaded ion-exchange column. The unretained saturates fraction was eluted from the Ag^+-loaded ion-exchange column and was detected by FID. After the complete elution of the saturates, the Ag^+-loaded ion-exchange column was backflushed and the olefin fraction was detected by the FID. Finally, the silica column was temperature programmed from 50°C to 150°C to elute aromatics from the silica column and detected as the last group of peaks in Figure 8. The method is suggested as having advantages over a single silica column method in the following ways: (a) it achieves complete resolution of saturates and olefins; (b) better reproducibility; and (c) less need for frequent regeneration of silica activity.

Thermally Labile Compounds

Some azo compounds and agricultural chemicals such as alkyl azo dyes and carbamate pesticides are good examples that demonstrate the usefulness of the SFC technique in the analysis of thermally labile compounds. Figure 9 is a chromatogram of a group of azo compounds [46] eluted from a capillary column. Excellent resolution and detection with the FID were observed. The SFC technique has received considerable attention in the analysis of environmental samples. It offers not only high resolution columns for the analysis of the environmental samples in complex matrices (volatiles and nonvolatiles), but also sensitive and selective detectors such as FID, NPD, and FPD, for both universal and selective detection. A rapid separation of a mixture of thermally labile carbamate and acid pesticides using a short open-tubular column (1.5 m × 25 μm) is shown in Figure 10 [47]. The analysis time was 130 seconds. The conventional HPLC technique requires nearly 20 minutes with gradient elution and the use of a postcolumn reaction for fluorescence detection.

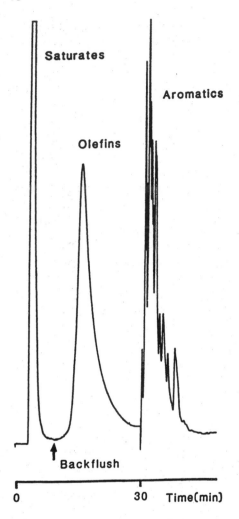

Fig. 8   Group separation of a gasoline sample using a silica and a Ag^+-loaded ion-exchange 1 mm i.d. packed columns in a two-column switching system. (Courtesy of M. L. Lee, Chemistry Department, Brigham Young University.)

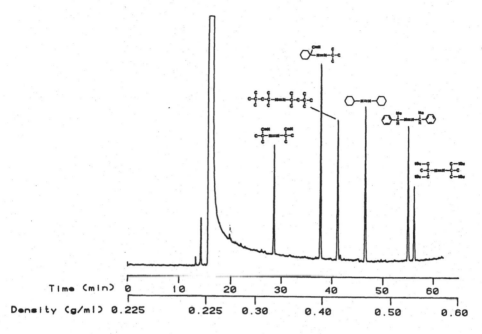

Fig. 9   Capillary SFC separation of thermally labile azo compounds. (Reproduced from Ref. 46 with permission of the publisher.)

Polymers and Oligomers

Gel permeation chromatography (GPC) is the primary method for the analysis of polymers and oligomers today. It is a popular methodology for the determination of molecular weight distributions and the average molecular weights of polymers. However, there are many difficulties often encountered because many polymers and oligomers do not contain chromophores for UV detection. The GPC method also suffers in its limited resolving power for differentiating small sample differences. The capillary SFC technique offers a significant breakthrough for the separation and characterization of polymers and oligomers. The resolving power of capillary SFC is demonstrated in Figure 11 where a very complex mixture of polyglycol oligomer was chromatographed using a 25 m × 50 µm i.d. capillary column and FID detection [48]. Excellent resolution of three

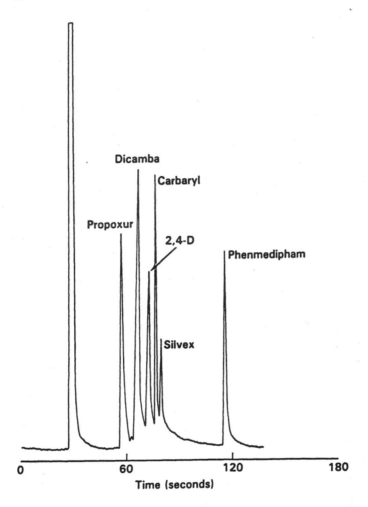

Fig. 10   Rapid SFC separation of a thermally labile pesticide mixture on a 1.5 m × 25 μm i.d. column coated with 5% phenyl polymethylsiloxane (0.15 μm film thickness).   Column was at 100°C isothermal temperature.   Column mobile phase CO_2 pressure was initially at 135 atm and was programmed to 300 atm at 50 atm/min. (Reproduced from Ref. 47 with permission of the publisher.)

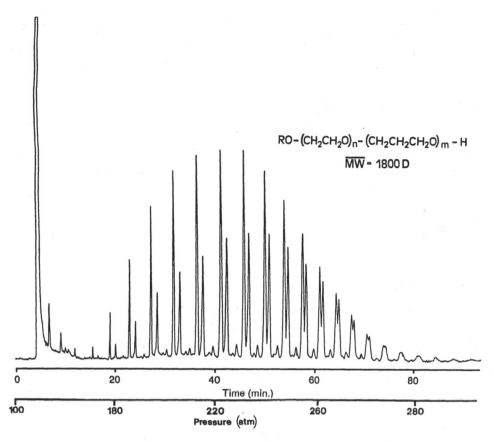

Fig. 11   High-resolution SFC chromatography of a polyglycol
with an average molecular weight of 1800 on a 25 m × 50 μm i.d.
fused silica column coated with methylpolysiloxane stationary phase
(0.5 μm film).   Mobile phase was CO_2 from 100 atm to 165 atm at
40 atm/min with a 10 minute hold at 100 atm, then to 267 atm at
2 atm/min, then to 320 atm at 1 atm/min.   FID detector at 350°C,
column temperature at 160°C.   (Courtesy of B. E. Richter, Lee
Scientific.)

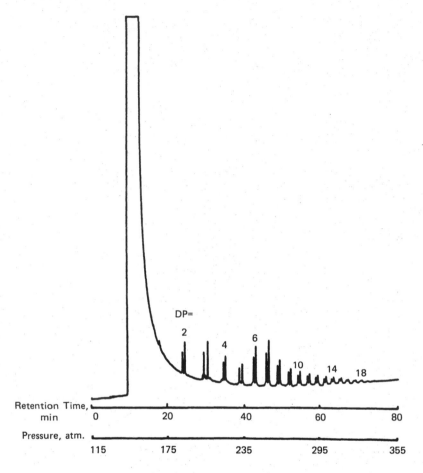

Fig. 12 SFC separation of the α and β anomers of silylated maltrin 100 on a 10 m × 50 μm i.d. fused silica column coated with DB-1. Column temperature was at 89°C. FID detection. (Reproduced from Ref. 50 with permission of the publisher.)

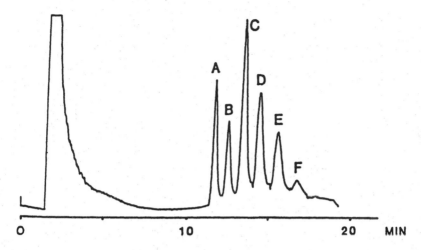

Fig. 13 SFC separation of the thermoplasma tetraether standard on a 10 m × 75 μm i.d. fused silica column coated with methyl-phenylpolysiloxane. CO_2 mobile phase programmed from 150 atm to 300 atm at 15 atm/min. Column temperature at 120°C. The identities of the peaks were given in Ref. 56. (Reproduced from Ref. 53 with permission of the publisher.)

distinct oligomer distributions is observed. Chester and Innis [49, 50] applied capillary SFC (10 m × 50 μm DB-1) for the separation of a series of glucose polymers. Figure 12 shows the excellent separation and FID detection of silylated glucose polysaccharides into α and β anomers. It is important to note that the last peak on the chromatogram, DP = 18, has a molecular weight of 6986. The polysaccharide, DP = 18, has a molecular weight of 2934 before derivatization and contains 56 hydroxyl groups.

Lipids

The separation and characterization of lipids is a major problem faced by many researchers in the life science industries. Separation and detection of phospholipids is especially difficult. Conventional HPLC and GPC methods do not provide adequate sensitivity [51,52], and thus require a large amount of sample for the analysis (e.g., 5 μg). Recently, a capillary SFC technique has been successfully applied in the separation of polar lipids of archaebacteria [53]. Figure 13 shows a chromatogram [53] of the six thermoplasma-glycerol tetraethers that are eluted from a 10 m × 75 μm i.d. 50% methylphenylpolysiloxane column with supercritical CO_2 mobile phase.

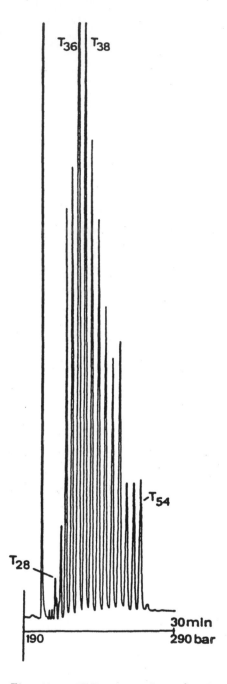

Fig. 14 SFC separation of palm kernal oil on a 10 m × 100 μm SE-54 column at 170°C. Mobile phase CO_2 programmed from 190 bar to 290 bar in 30 minutes. (Reproduced from Ref. 55 with permission of the publisher.)

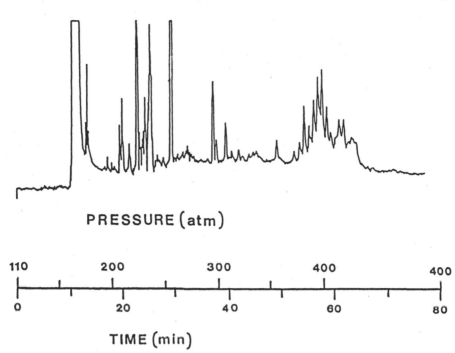

PRESSURE (atm)

| 110 | 200 | 300 | 400 | 400 |

| 0 | 20 | 40 | 60 | 80 |

TIME (min)

Fig. 15   SFC separation of a permethylated human meconium extract on a 10 m × 50 μm, DB-5 fused silica column at 120°C.   Mobile
phase was CO_2.   Column pressure programmed from 100 atm to 400
atm at 5 atm/min after an initial 5 minute hold.   (Courtesy of V. N.
Reinhold, Harvard School of Public Health, Boston, MA.)

The application to a complex mixture of mono-, di-, and triglycerides
was also demonstrated by Chester [54].   The application and advantage of using capillary SFC for the analysis of thermally labile lipids
were further reported by Proot et al. [55].   Figure 14 demonstrates
the separation of palm kernel oil containing glycerides with 28 to 56
carbons using a 10 m × 100 μm i.d. SE-54 capillary column.   Capillary SFC offers great potential for the analysis of lipids in biological
samples.   Kwei et al. [56] demonstrated the application of capillary
SFC to glycosphinygolipids.   Figure 15 shows the separation of a
permethylated human meconium extract.   There is a relatively high
concentration of eluting peaks with retention times corresponding to
those of globo- and gangliosides.
    There are many other applications of the SFC technique to bio-
logical, environmental, petroleum, petrochemical, pharmaceutical,
food, flavors, and polymer samples.   However, more effort is needed

in the refinement and expansion of SFC for more effective applications. Enhancement in the elution of polar compounds and high molecular samples by increasing operating limits of the pump pressure, by providing new columns, and by using more polar mobile fluids are important steps for the future growth of SFC. More effort needs to be undertaken in developing more sensitive and selective detectors for enhancing the application range of SFC.

MICRO-HPLC

Recent advances in micro-HPLC center on the miniaturization of the chromatographic system for applications with very small diameter open-tubular columns, 1-mm-i.d. micropacked columns, or fused silica capillary-packed columns. Open-tubular column HPLC has been investigated by a number of researchers [57–68], Knox and Gilbert [65], Yang [68], Guiochon [67]. Pretorius and Smuts [66] have also discussed open-tubular column LC from the viewpoint of theoretical optimization. As indicated by both theory and experimental results, the open-tubular column diameter needs to be reduced to the size of a packed column particle diameter to match the speed of analysis of conventional packed column HPLC. For example, a 7.5-µm-i.d. open-tubular column is required to match the speed of the analysis of a column packed with 5 µm particles [68].

The small-diameter open-tubular column can be deactivated and coated in the same way as that utilized for the open-tubular GC and SFC columns. However, the practice of open-tubular LC requires the use of very small injector and detector volumes [65,68]. In addition, the very small flow rate and high pressure drop per unit length of the small diameter column impose very difficult problems for the design of a high-pressure hydraulic pumping system that can deliver sub-µl/min flow rates and can also be used for gradient elution. Interest in the development of open-tubular column HPLC has significantly reduced due to the introduction of the fused silica capillary-packed column [69,70].

The development and application of a 1-mm-i.d. micropacked column for HPLC was first reported by Scott and Kucera [71]. They employed a modified, commercially available liquid chromatograph and demonstrated its utility for solvent economy and high resolution. Their work spurred interest and resulted in the recent commercialization of the micro-HPLC instrumentation by many LC instrument and column manufacturers. Unfortunately, interest in 1-mm-i.d. micropacked columns soon faded because users could not realize any significant improvements and advantages over the conventional 4.6-mm and 2-mm-i.d. packed columns except some saving in consumption of the mobile fluid.

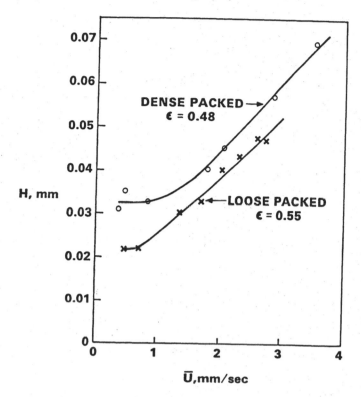

Fig. 16   Comparison of column efficiency in terms of plate height
(H) between dense-packed and loose-packed fused silica packed columns of 50 cm × 320 μm i.d. Packing material was 10 μm ODS silica
particles. Mobile phase was 50:50 % ACN: H_2O.

While the economic and environmental aspects of the reduction of
solvent consumption are advantageous, other important benefits, as
given below, can only be realized with the use of the small diameter
($\leqslant 0.6$ mm i.d.) fused silica capillary-packed columns [69,70].

1. A 100—600-times savings in the cost of packing material.
2. Long columns (up to 2 m) can be prepared [70] because of
    the high-packed bed porosity ($\geqslant 0.5$, see Fig. 16) and low
    flow rate resistance factor ($\geqslant 400$).
3. A smooth and chemically inert fused silica inner surface min-
    imizes the interaction between solute molecules and the col-
    umn wall.

4. Good optical transparency of the fused silica column allows
    visual observation of the column-packed bed as well as on-
    column adsorbance and/or fluorescence detection.
5. The fused silica column walls can be easily modified [72] to
    achieve high column efficiency and selectivity.
6. Minimum sample requirements allow analysis in cases of limited
    availability.
7. Low mobile phase flow rates facilitate direct interfacing of the
    column with GC flame base detectors and mass spectrometers.
8. Small column diameters and the corresponding small dilution
    of solute zones enhance solute detection and quantitative
    accuracy.
9. Small elution peak width allows high-resolution and high-con-
    centration sample transferring and heart-cutting in multi-
    dimensional micro-HPLC—SFC and micro-HPLC—capillary GC
    techniques (see Chap. 7).

A detailed packing procedure for the preparation of the fused
silica capillary packed column has been given [69,70,72]. Other packing techniques which yield successful fused silica capillary-packed
columns have also been published [73—75]. High efficiency fused
silica columns packed with 3-, 5-, and 10-µm particles can be prepared easily. As the particle sizes are reduced to below 3 µm, special
techniques [72] for modifying the column wall are found to be very
important in order to obtain optimum performance. As shown in Figure 17, a 35 cm × 320 µm i.d. fused silica capillary-packed column
with the column wall modified with Carbowax 20 m and then packed
with 2-µm-o.d.s. particles produces 257,000 plates per meter. The

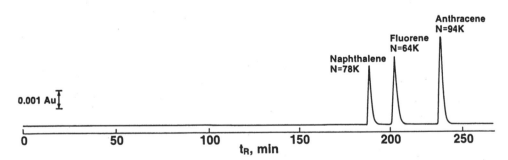

Fig. 17 Performance of a 35 cm × 320 µm i.d. fused silica column
packed with 2 µm ODS particles. Mobile phase was 50:50% ACN:H_2O
at 2.6 µl/min. UV detection at 254 mm with 0.5 µl flow cell volume.
0.1 µl time-controlled on-column injection.

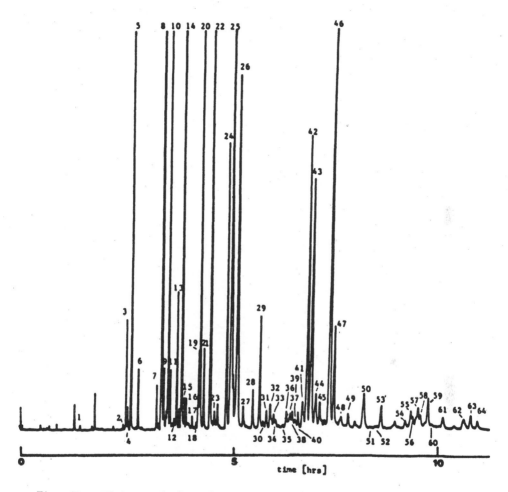

Fig. 18   High-resolution chromatogram of large polyaromatic compounds extracted from carbon black on a 180 cm × 200 μm i.d. fused silica column packed with 3-μm C_{18} spherisorb. Chromatographic conditions and peak identification were given in Ref. 76. (Reproduced from Ref. 76 with permission of the publisher.)

unmodified fused silica column packed with the same 2-μm-o.d.s. particles generated 71,000 plates per meter. Application of the ultra-high resolving power of long fused silica packed capillary columns has been demonstrated by Hirose et al. [76] and Novotny et al. [77] and are shown in Figures 18 and 19.

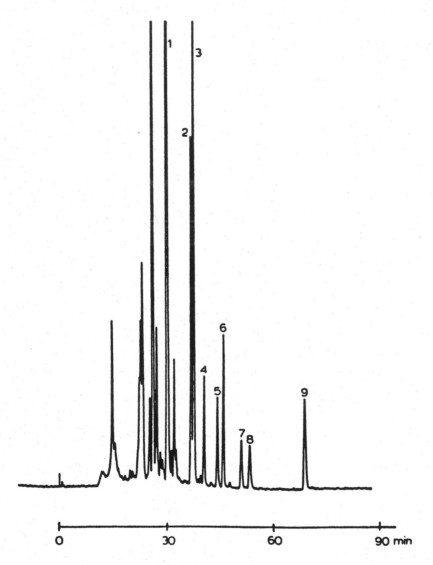

Fig. 19   High-resolution chromatogram of bile acid standards on a
1 m × 240 μm i.d. fused silica capillary packed column of 5 μm spherisorb ODS particles.   Chromatographic condition and peak names were
given in Ref. 77.   (Reproduced from Ref. 77 with permission of the
publisher.)

Many other significant advances in micro-HPLC instrumentation have also been made. Detailed discussions of some new frontiers of the micro-HPLC technology such as gradient elution, detectors, and column-packing materials can be found in the subsequent chapters. Reviews on micro-HPLC can be found in Refs. 78, 79, and 80.

FUTURE TRENDS: A UNIFIED CHROMATOGRAPHIC APPROACH

Advances in microbore column chromatography technology have been made on all three major fronts, namely capillary GC, SFC, and micro-HPLC. However, it is becoming apparent that all three chromatography techniques can be performed using the same column and instrumentation. The trend in the development of the microbore column chromatography technology is that there is rapid convergence to a unified chromatographic approach. For example, the time-controlled pneumatic switching valve designed for micro-HPLC applications is being adapted as a standard injection technique for SFC. The application of these valves and fast switching techniques have also been investigated [81] for capillary GC sample injection. The developments in tubular columns and capillary-packed column technologies have benefited all fronts in microbore column chromatography. The trend in miniaturization of the chromatographs, especially in the reduction of the detector cell volumes and response times strongly implies that future chromatography will be performed using the same injector, column, detector, pump, and oven systems. The only difference is the choice of mobile phases and the control of the physical parameters of the mobile fluid according to the application needs.

A simple chromatograph that can be used for capillary GC, SFC, and micro-HPLC has the following six major components:

1. A high-pressure pumping system that can deliver mobile phase at pressures up to 1000 atm. The pump control allows pressure, density, and solvent composition programming. The pump can also deliver mobile fluid either under pressure or flow control. The pump should also be equipped with a stream selection valve for automatic selection of the mobile fluid for the separation.

2. A high-pressure thermostatted injection valve which can be time-controlled for the injection of a wide range of sample volumes (e.g., 2-nl to 1-μl). The advantages of the time-controlled sampling technique over the conventional direct valve or split injections are: (a) sample volume as small as 2-nl can be injected on the column (no splitting of sample at the inlet of the column is required); (b) saving in mobile fluid consumption (no inlet stream splitting and waste of

mobile phase occurs); and (c) fast pneumatic switching valve and short injection time eliminate peak tailing and ensure excellent quantitative reproducibility.

3. A high temperature oven equipped for both open-tubular and capillary-packed columns. A maximum column oven temperature of 500°C is desirable so that the application potential of the new high-temperature stationary phases and the aluminum-clad fused silica capillary columns [9,10] can be fully explored. The temperature stability should be better than ±0.1°C and the thermal gradients within the column envelope should be very small (≤±0.5°C) so that low thermal mass fused silica columns can be employed.

4. Microbore columns of small diameter are the heart of the separation process. Fused silica capillary columns, either in open-tubular or in packed column form, have brought enormous growth to the practice and application of micropore column chromatography in both industrial application laboratories and research institutes. Future growth is expected in the development of special phases for dedicated applications such as crude oil-simulated distillation, and gasoline PONA group separations. The need for life science applications should also spur more interest in the development of microbore columns that are specially prepared for the separation of biological samples.

5. Detectors promise to be one of the most exciting areas in the development of microbore column chromatography. One of the major driving forces in the microbore column chromatography is that of low mobile fluid flow rates and the resulting ease of direct interfacing of the column to both GC and LC detectors. Among the various detectors that have great potential for microbore column chromatography are the FID and other selective detectors such as UV, fluorescence, chemiluminescent, MS, FTIR, ECD, NPD, and microwave emission. Capillary GC-MS is routine and is widely utilized in industrial laboratories. SFC- and micro-HPLC-MS links are easier than the conventional HPLC-MS interfaces. SFC-MS is particularly attractive due to the volatility of the SFC mobile phase and the ease of operation in both chemical ionization and electron impact mass spectrometry of non-volatile and thermally labile compounds. The development of sensitive and selective detectors is a key to the growth of microbore column chromatography for industrial and life science applications.

6. Sample preparation combined with sample clean-up, preconcentration, and injection is of particular importance in microbore column chromatography. The small-diameter fused

silica column cannot tolerate the injection of samples containing particulates. On-line coupling of micro-HPLC to capillary GC, SFC to capillary GC, micro-HPLC to SFC, etc., offers great potential for sample clean-up, prefractionation, and preconcentration on the micro-HPLC or SFC column prior to the high resolution stage of the microbore column chromatography. On-line couplings of a supercritical fluid extractor with microbore column chromatographic system allows selective fractions of extractable sample components to be transferred to a microbore column for separation. On-line supercritical fluid extraction is a particularly important avenue in selective extraction and injection of components of interest without introducing interferences from the sample matrix into the column such as that which occurs when using liquid solvent extraction methods. Significant developments in computer-controlled robotic sample preparation and extraction via supercritical fluid systems are also expected in the near future.

Finally, it is difficult to predict just how fast microbore column chromatography will grow and how soon a microbore column chromatograph will be commercially available for applications in capillary GC, SFC, and micro-HPLC. When such a chromatograph is available, it will be a joy for method and application development. The same unknown sample should be able to be chromatographed in any one of the three chromatographic techniques by means of microbore column chromatography.

REFERENCES

1. J. C. Giddings, *Dynamics of Chromatography*, Marcel Dekker, New York, 1965.
2. S. Schindler and K. Wasserfallen, *J. High Resolut. Chromatogr. Chromatogr. Commun.* 10, 371 (1987).
3. M. J. E. Golay, *Gas Chromatography*, V. J. Coates, H. J. Noebels, and I. S. Fagerson, ed., Academic Press, New York, 1958, p. 1.
4. M. L. Lee, F. J. Yang, and K. D. Bartle, *Open-Tubular Column Gas Chromatography: Theory and Practice*, John Wiley & Sons, New York, 1984.
5. W. Jenning, *Gas Chromatography with Glass Capillary Columns*, Second Edition, Academic Press, New York, 1980.
6. R. R. Freeman, ed., *High Resolution Gas Chromatography*, Second Edition, Hewlett-Packard, Avondale, PA, 1981.
7. W. G. Jennings, ed., *Applications of Glass Capillary Gas Chromatography*, Marcel Dekker, New York, 1981.

8.   R. Dandeneau and E. H. Zerenner, *J. High Resolut. Chroma-togr. Chromatogr. Commun. 2*, 351 (1979).
9.   S. R. Lipsky and M. L. Duffy, *J. High Resolut. Chromatogr. Chromatogr. Commun. 9*, 376 (1986).
10. S. R. Lipsky and M. L. Duffy, *J. High Resolut. Chromatogr. Chromatogr. Commun. 9*, 725 (1988).
11. M. Proot and P. Sandra, *J. High Resolut. Chromtogr. Chromatogr. Commun 9*, 618 (1986).
12. B. A. Jones, K. E. Markides, J. S. Bradshaw, and M. L. Lee, Chromatogr. Forum, *May—June*, 38 (1986).
13. K. E. Markides, M. Nishioka, B. J. Tarbet, J. S. Bradshaw, and M. L. Lee, *Anal. Chem. 57*, 1296 (1985).
14. M. Swerev and K. Ballschmitter, *J. High Resolut. Chromatogr. Chromatogr. Commun. 10*, 544 (1987).
15. B. Koppenhoefer, H. Allmendiger, and E. Bayer, *J. High Resolut. Chromatogr. Chromatogr. Commun. 10*, 325 (1987).
16. A. Bemgard, L. Blomberg, M. Lymann, S. Claude, and R. Tabacchi, *J. High Resolut. Chromatogr. Chromatogr. Commun. 10*, 302 (1987).
17. C. L. Woolley, K. E. Markides, M. L. Lee, and K. D. Bartle, *J. High Resolut. Chromatogr. Chromatogr. Commun. 9*, 506 (1986).
18. J. C. Giddings, *Gas Chromatography*, A. Gold Up, ed., Elsevier Publishing Co., Amsterdam, 1964.
19. E. Klesper, A. M. Corwein, and D. A. Turner, *J. Org. Chem. 27*, 700 (1962).
20. M. N. Myers and J. C. Giddings, *Separation Sci. 1*, 761 (1966).
21. L. McLaren, M. N. Myers, and J. C. Giddings, *J. Chromatogr. 159*, 197 (1968).
22. J. C. Giddings, M. N. Myers, L. McLaren, and R. A. Keller, *Science 162*, 67 (1968).
23. S. T. Sie, W. Van Beersam, and G. W. A. Rijnders, *Separation Sci. 1*, 459 (1966).
24. S. T. Sie and G. W. A. Rijnders, *Separation Sci. 2*, 729 (1967).
25. S. T. Sie and G. W. A. Rijnders, *Separation Sci. 2*, 755 (1967).
26. D. R. Gere, R. Board, and D. McManigill, *Anal. Chem. 54*, 736 (1982).
27. D. R. Gere, *Science 222*, 253 (1983).
28. T. Greibrokk, A. L. Blilie, E. J. Johnson, and E. Lundanes, *Anal. Chem. 56*, 2681 (1984).
29. H. Lauer, D. McManigill, and R. Board, *Anal. Chem. 55*, 1370 (1983).
30. M. Novotny, S. R. Springston, P. A. Peaden, J. C. Fjeldsted, and M. L. Lee, *Anal. Chem. 53*, 407a (1981).
31. Y. Hirata, F. Nakata, and M. Kawasaki, *J. High Resolut. Chromatogr. Chromatogr. Commun. 9*, 633 (1986).

32. C. L. Woolley, R. C. Kong, B. E. Richter, and M. L. Lee, *J. High Resolut. Chromatogr. Chromatogr. Commun.* 7, 329 (1984).
33. H. E. Schwartz, J. W. Higgins, and R. G. Brownlee, *LC/GC 4*, 639 (1986).
34. D. E. Raynie, K. E. Markides, and M. L. Lee, *Anal. Chem.* in press.
35. *Manual on Hydrocarbon Analysis*, 3rd ed., American Society for Testing and Materials, Philadelphia, 1977.
36. J. C. Suatoni and R. E. Swab, *J. Chromatogr. Sci. 13*, 361 (1975).
37. T. V. Alfredson, *J. Chromatogr. 218*, 715 (1981).
38. R. L. Miller, L. S. Ettre, and N. G. Johansen, *J. Chromatogr. 259*, 393 (1983).
39. R. L. Miller, L. S. Ettre, and N. G. Johansen, *J. Chromatogr. 264*, 19 (1983).
40. D. J. Cookson, C. J. Rix, I. M. Shaw, and B. E. Smith, *J. Chromatogr. 312*, 237 (1984).
41. T. A. Norris and M. G. Rawdon, *Anal. Chem. 56*, 1767 (1984).
42. E. Lundanes and T. Greibrokk, *J. Chromatogr. 349*, 439 (1985).
43. E. Lundanes, B. Iverson, and T. Greibrokk, *J. Chromatogr. 366*, 391 (1986).
44. H. E. Schwartz and R. G. Brownlee, *J. Chromatogr. 353*. 77 (1986).
45. R. M. Campbell, N. M. Djordjevic, K. E. Markides, and M. L. Lee, *Anal. Chem. 60*, 356 (1988).
46. J. C. Fjeldsted, R. C. Kong, and M. L. Lee, *J. Chromatogr. 279*, 449 (1983).
47. B. W. Wright and R. D. Smith, *J. High Resolut. Chromatogr. Chromatogr. Commun. 9*, 73 (1986).
48. B. E. Richter, *Chromatogr. Forum* Nov.–Dec, 52 (1986).
49. T. L. Chester and D. P. Innis, *J. High Resolut. Chromatogr. Chromatogr. Commun. 9*, 178 (1986).
50. T. L. Chester and D. P. Innis, *J. High Resolut. Chromatogr. Chromatogr. Commun. 9*, 209 (1986).
51. R. H. McCluer and F. B. Jungalwala, *J. Chromatogr. Sci. 10*, 7 (1979).
52. R. L. Briand, S. Harold, and K. G. Blass, *J. Chromatogr. 223*, 277 (1981).
53. S. J. Deluca, K. J. Voorhees, T. A. Langworth, and G. Holzer, *J. High Resolut. Chromatogr. Chromatogr. Commun. 9*, 182 (1986).
54. T. L. Chester, *J. Chromatogr. 299*, 424 (1984).
55. M. Proot, P. Sandra, and E. Geeraert, *J. High Resolut. Chromatogr. Chromatogr. Commun. 9*, 189 (1986).

56. J. Kuei, G. R. Her, and V. N. Reinhold, *Anal. Biochem.* submitted for publication.

57. K. Hibi, D. Ishii, I. Fujishim, T. Takeuchi, and T. Nakanishi, *J. High Resolut. Chromatogr. Chromatogr. Commun. 1*, 21 (1978).

58. T. Tsuda, K. Hibi, T. Nakanishi, T. Takeuchi, and D. Ishii, *J. Chromatogr. 158*, 227 (1978).

59. K. Hibi, T. Tsuda, T. Takeuchi, T. Nakanishi, and D. Ishii, *J. Chromatogr. 175*, 104 (1979).

60. D. Ishii, T. Tsuda, and T. Takeuchi, *J. Chromaogr. 185*, 73 (1979).

61. D. Ishii, T. Tsuda, and T. Takeuchi, *J. Chromatogr. 199*, 249 (1980).

62. D. Ishii and T. Takeuchi, *J. Chromatogr. Sci. 18*, 462 (1980).

63. F. J. Yang, *J. High Resolut. Chromatogr. Chromatogr. Commun. 3*, 589 (1980).

64. F. J. Yang, *J. High Resolut. Chromatogr. Chromatogr. Commun. 4*, 83 (1981).

65. J. H. Knox and M. T. Gilbert, *J. Chromatogr. 186*, 405 (1979).

66. V. Pretorius and T. W. Smuts, *Anal. Chem. 38*, 272 (1966).

67. G. Guidochon, *J. Chromatogr. 185*, 3 (1979).

68. F. J. Yang, *J. Chromatogr. Sci. 20*, 241 (1982).

69. F. J. Yang, *J. Chromatogr. 236*, 265 (1982).

70. F. J. Yang, U.S. Patent 4,483,773 (1984).

71. R. P. W. Scott and P. Kucera, *J. Chromatogr. 169*, 51 (1979).

72. F. J. Yang, U.S. Patent pending (1988).

73. T. Tokeuchi and D. Ishii, *J. Chromatogr. 238*, 409 (1982).

74. Y. Hirata and K. Jinno, *J. High Resolut. Chromatogr. Chromatogr. Commun. 4*, 571 (1983).

75. J. Gluckman, A. Hirose, V. L. McGuffin, and M. Novotny, *Chromatographia 17*, 303 (1983).

76. A. Hirose, D. Wiesler, and M. Novotny, *Chromatographia 18* 239 (1984).

77. M. Novotny, K. Karlsson, M. Konish, and M. Alasandro, *J. Chromatogr. 292*, 159 (1984).

78. F. J. Yang, *J. High Resolut. Chromatogr. Chromatogr. Commun. 6*, 348 (1983).

79. S. Ahuja, ed., *Ultrahigh Resolution Chromatography*, American Chemists Society, Washington, D.C., 1984.

80. P. Kucera, ed., *Microcolumn High-Performance Liquid Chromatography*, Elsevier, New York, 1984.

81. A. VanEs, J. Janssen, R. Bally, C. Cramers, and J. Rijks, *J. High Resolut. Chromatogr. Chromatogr. Commun. 10*, 273 (1987).

82. G. Schomburg, *LC/GC 6*, 36 (1988).

83. B. J. Tarbet, J. S. Bradshaw, K. E. Markides, M. L. Lee, and B. A. Jones, *LC/GC 6*, 232 (1988).

2

Packing Materials and Packing Techniques for Micro–HPLC Columns

MAURICE VERZELE and C. DEWAELE / State University of Ghent, Ghent, Belgium

Only three parameters of high-performance liquid chromatography (HPLC) columns can be miniaturized to lead to micro-HPLC: the internal diameter (i.d.), the length (L), and the mean diameter (dp) of the column packing material. This chapter about packing materials is mainly concerned with the dp factor.

At present there are no packing materials exclusively suited for micro-HPLC columns. The same materials are used for all HPLC sizes. There is a tendency to use coarser particles in columns of larger internal diameter, for preparative-scale HPLC.

Therefore it seems possible that smaller than usual particle materials would be indicated for micro-HPLC. The question is, however, what are smaller than usual particles and which ones should be used?

PARTICLE SIZE THROUGH CHROMATOGRAPHIC HISTORY

An interesting aspect of chromatographic history is the evolution of the size of the particles used. Zechmeister and von Cholnoky [1] gave a table of some adsorption materials with their particle sizes. This is reproduced in Table 1. The very small size of the particles

TABLE 1   Mean Particle Size of Some Adsorbents Used in the 1930s

Adsorbent	Particle size (μm)
Aluminum oxide (Merck, standardized after Brockmann)	7
Aluminum oxide, made in-house	2
Acid clay from Java (Ca—Al—Mg silicate)	10
Calcium carbonate, precipitated (Merck)	1.5
Calcium carbonate (commercial chalk)	1.2
Calcium hydroxide, prepared in-house	2.5
Gypsum (hydrated) (after Karrer and Weber)	10.5
Magnesium oxide	1.5
Fuller's earth (bleaching clay)	3
Floridin (a fuller's earth)	1.5—7
Florisil XXF	1.5—6

Source: From Ref. 1.

mentioned in Table 1 is striking. On average, they are even smaller than the particles currently used for chromatography. Permeability problems must have been very prominent in those early days of chromatography. Comparatively soon, therefore, the use of vacuum and/or pressure appeared, as well as the admixing of "inert" coarser grain particles [2].

The absence of silica gel from Table 1 is noteworthy. Of the adsorbents listed in the table, only aluminum oxide is still in use. Silica gel, as such or in chemically derivatized form, is now the favored packing material. More than 90% of all separations reported in 1983 were carried out on silica gel [3].

Because of the permeability problems, the particle size of the column packing material increased. This has not been discussed clearly in the literature. We believe that another reason could be that most chromatography at that time was on a heavily loaded preparative scale, where the difference between smaller and larger particles is not so clear. There was thus apparently no compelling reason to accept the permeability problems of the earlier smaller particles. Finally, the particle size of chromatographic packing materials even reached 50 to 200 μm. Indeed, silica gel of that size is still much in use today for chromatography [4—7]. More specifically, when large-scale preparative separations are considered, the coarse particles automatically come to the foreground.

The revival of analytical LC demanded particles with better mass transfer characteristics. A short period during which pellicular materials were used is noted at the beginning of the 1970s. The return to fully porous microparticulate material is mainly due to Majors [8]. For about 10 years, 10-μm irregular-shaped silica gel particles were the most popular HPLC column packing material. Now, 5-μm particles, preferably spherical in shape, have joined the 10-μm particles mentioned above. Still smaller particles of 3-μm size are also commercially available, but their use is only slowly being accepted by the chromatographic community. Particles of 2-1-μm size have been mentioned in the literature [9,10]; they were developed at this laboratory. They could be the materials of the future for micro-HPLC. Although silica gel is still the favorite material, there is strong interest in alternatives. Intensive efforts are focused on styrene-divinylbenzene copolymers. At the moment these do not produce the efficiency and chromatographic versatility of silica gel, but the progress made in recent years is such that it seems reasonable to postulate the arrival soon of competitive synthetic resin packing materials.

CONTEMPORARY PACKING MATERIALS FOR MICRO-HPLC

The only material to be considered is silica gel and perhaps to a lesser extent styrene-divinylbenzene resin.

What Is Silica Gel?

Silica gel is the basis of many industrial catalysts and is used widely as a filling material in plastics, paints, and food. It has been discussed very often in the literature. A fairly recent book by Unger [11] describes silica gel in depth. Silica gel is usually produced by precipitating sodium silicate solution with acid, followed by washing the gel and drying. Preparation recipes have been published [12—14]. Milling, sieving, agglomeration, etc., give various forms of commercial silica gel. One particular extremely finely divided form is obtained by burning silicon tetrachloride in a hydrogen flame (Cab—O—Sil, Aerosil). Silica gel is produced on a large scale by nearly 50 multinational chemical companies, according to the Directory of World Chemical Producers [15]. Silica gel is mainly characterized by its surface chemistry, specific surface area, and mean pore diameter. The pore volume, apparent density, particle size and shape, trace elements, and pH of the

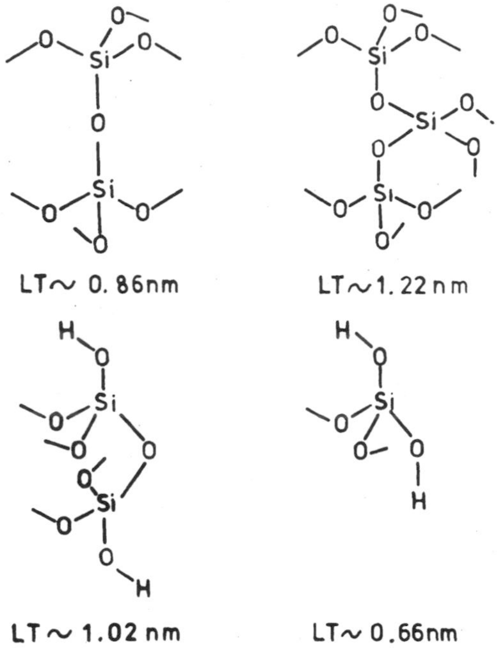

Fig. 1   Structures of silica gel with different layer thicknesses (LTs).

surface are also important for chromatographic purposes. We recently published a discussion of the quality factors of silica gel [16]. The most characteristic property of silica gel is its very large specific *surface area*, which can reach 1000 m^2/g. For chromatographic purposes the specific surface area is between 100 and 400 m^2/g. With such figures, calculation leads to the conclusion that the wall thickness of the porous silica gel structure is of the order of nanometers. This then leads to the conclusion that most silicon atoms are surface atoms and that, not surprisingly, many bear a silanol function. Indeed, the silicon–oxygen bond at the surface can be a siloxane bond but it can also be a silanol function. We present silica gel as an open reticulated foam with a structure as shown in Fig. 1.

The porosity of silica gel derives from the formation of the polycondensate in a large excess of water, which is later removed. This leads to the large specific surface area. The surface area is determined by so-called BET measurements [17]. We have described a simple instrument, which can be made by any glass-blower, to obtain BET figures [18].

According to the *mean pore size*, a distinction is made between microporous silica gel (mean pore size <3 nm), mesoporous silica gel (mean pore size between 3 and 50 nm), and macroporous silica gel (mean pore size above 50 nm). This is not so relevant to chromatography. Micropores should be avoided since they will hold large molecules too strongly, resulting in tailing and/or band broadening. For normal chromatographic purposes the best mean pore size is 5 to 10 nm. For special cases and for size exclusion chromatography (SEC) a larger pore size, e.g., 30 nm, may be desirable. In the past, pore size determination was rather difficult and needed special instrumentation—mercury intrusion technology [19]. An example was described by Kirkland [20]. Recently, mainly through contributions by Halász and co-workers [21,22], it has become possible to determine mean pore size from SEC measurements with suitable reference compounds.

The *surface chemistry* of silica gel is characterized by the presence of two functions: siloxane bridges with a valence bond angle of 140° and silanol functions. Silica gel has been compared to a carpet of silanol functions. The properties of silica gel are determined mainly by these silanol functions [23—27]. Two types of silanol functions are usually considered: free ones (isolated from one another) and hydrogen-bridged ones (close together). Erard and Kováts [28] and Berendsen and de Galan [29] mention a silanol group concentration of about $4.8/nm^2$. This corresponds to about one silanol group per six silicon atoms.

Most silica gels contain rather large amounts of *trace elements* [30—32]. These can be simply adsorbed or incorporated in the silica gel structure. Trace elements change the chemical and catalytic properties of silica gel profoundly. Iron is such a common silica gel impurity. It can contribute to adsorption and even form change transfer complexes [33]. Zechmeister reported difficulties with iron-containing silica gel and gave a method for removing the iron from Florisil by tenfold treatment with boiling 3 N HCl. Although acid treatment, also of silica gel, was mentioned repeatedly [34—37], a more detailed study was reported only recently [38]. Percolation of silica gel columns with hydrochloric acid solutions is not efficient enough. The only way to get rid of the trace metals is by repeated boiling in 1—3 N HCl and extensive washing in between. Contrary to assertions in the literature, we have found that this treatment has no effect or only a small effect on the specific surface area of most silica gels. It is extremely difficult to remove excess acid from silica gel by washing. Hydrochloric acid, for instance, is very strongly adsorbed [39]. This adsorbed acid can cause catalysis of chemical reactions of the chromatographed compounds [40,41]. The acid condition of silica gel and its effects can be demonstrated by chromatographing colored compounds, such

as dimethylaminoazobenzene, that change color in contact with acid.
More recently, Engelhardt and Muller studied the importance of
silica gel pH [42]. They showed that adding suitable acidic or
basic compounds to the eluting solvent can control the influence of
adsorbed acid.

The *apparent density* and the *pore volume* of silica gel are
closely related. The *apparent density*, as we call it, is determined
by tapping an amount of silica gel in a graduated cylinder until the
volume does not change further. The value is not very different
from the *packing density* [4,7,43] and for normal silica gel is about
0.4 to 0.5 g/ml. Since such silica gel is obtained from solutes of
about 10% (which should lead to a density of 0.1 g/ml), it follows
that the process of drying the precipitated silicic acid is accompanied
by considerable contraction. The true density of silica gel is also
of interest. For quartz it is 2.65 g/ml and earlier figures for
amorphous silica gel are around 2.2 g/ml [33,44]. More recent figures are lower; Strubert mentions 1.9 [45] and we find 1.8 [16].

The apparent density is related to the specific pore volume.
With an apparent density of 0.5 g/ml and a true density of the
silica gel skeleton of 1.9, the true volume taken up by the silica
gel per milliliter of material is only 0.26 ml. The total free volume
in such a silica gel is therefore 0.74 ml/ml or 1.48 ml/g.

A calculation similar to that above for silica gel with apparent
density 0.2 leads to a silica gel volume per milliliter of particles of
about 0.1 ml and to a total free volume of 0.9 ml/ml or 4.5 ml/g.
The difference in total free volume for these two silica gels is very
great. The apparent density of silica gel is important, because
silica gels for which this figure is below 0.40 g/ml will have poor
strength and will therefore not pack efficiently with the high-pressure techniques now in use.

Conditioning, Drying, and Heating of Silica Gel

Silica gel can absorb up to its own weight of water without apparently becoming wet. In equilibrium with a normal laboratory atmosphere, silica gel will contain 5 to 15% adsorbed water. This has a
strong effect on its adsorption characteristics. Heating silica gel
in a drying oven will restore or increase its adsorption power and/
or activity. For alumina a similar situation exists. As long ago as
1941 Brockmann and Schodder [46] added known amounts of water
to alumina (0, 3, 6, 10, and up to 15%) to obtain the five Brockmann activity classes of alumina. Such grading for silica gel has
not received the same attention. The activity of silica gel diminishes
very rapidly on addition of even small amounts of water. Running
an undried solvent through a silica gel column will result in changes

of its adsorption characteristics—shortening the retention times and changing the α values. Conversely, a dry solvent will remove water from an incompletely dried silica gel with equally frustrating results. Changes in retention times with adsorption chromatography on silica gel are a serious drawback of this technique. This is the basic reason for the trend toward derivatized silica gels with better repeatability.

Drying or activating of silica gel has been described extensively in the literature. Physically bonded water is removed by heat treatment. The best temperature for this is between 105 and 150°C [47]. At higher temperatures (>200°C) "chemically bonded water" is removed. This means that two suitably placed silanol functions eliminate water to form a siloxane bond.

Above 400°C silanol groups start to migrate over the surface and the silica gel skeleton starts to contract. More water will be removed. This process goes on up to very high temperatures. Thermal gravimetric analysis (TGA) of silica gels shows that about 2% water is removed between 200 and 900°C [48]. Rehydration of such silica gel is not a fast process. Opening of siloxane bridges thus formed may require boiling in relatively concentrated hydrochloric acid. Heating silica gel above 400°C and up to 1000−1200°C contracts the skeleton, reduces the specific surface area, and increases the mean pore size diameter. This has been reported in several publications, but there is little unanimity in the reported figures. Addition of a "flux" such as NaCl or Na_2CO_3 is often advocated to promote this process. In our hands, however, this was not reproducible. Moreover, very pure silica gel did not change its specific surface area or pore size structure in our experiments, even when heated up to 900°C. Perhaps many of the studies reported were carried out on silica gel that contained trace elements that acted as flux.

Specific Adsorption

A specific adsorption material would retain only one compound or one group of compounds. Silica gel is fairly selective for the separation of polar compounds, aromatic hydrocarbons, olefins, and saturated hydrocarbons. To enhance its selectivity for olefins, silica gel has been impregnated with silver nitrate [49−52]. It can also be made more specific by impregnation with other chemicals, or preferably by derivatization, as will be discussed further. One curious method of obtaining specificity is that mentioned by Dickey [53], who produced the gel in the presence of the compound of interest, which was then removed by washing or extraction. The gel would retain the imprint of the compound and thus adsorb it preferentially. Not much has been heard about this approach since

1955. What about the considerable contraction of the gel, by a
factor of five, on drying? Our consideration of specific adsorption
could be enlarged very much, but a full discussion of the manifold
possibilities is outside the scope of this chapter.

On the Hazard of Silica Gel

Crystalline silica dust is an industrial hazard because it can react
with the wall of the lung and create nonfunctional tissues. This has
been recognized by the Occupational Safety and Health Administra-
tion (OSHA) in the United States. Silicosis is a well-known disease
resulting from this hazard. Williams and Hawley [54] discussed this
topic at length and stated that "Amorphous silicas pose a lesser
hazard to health and, as such, are not included under the general
term 'silica' as used throughout this text." This is a very important remark and would seem to exclude silica gel, as we know it
in chromatography, as a health hazard.

Davison (a Grace affiliate), in a leaflet on silica gel [55], mentions that "Silica gel is chemically inert, non-toxic and non-dusting.
No special handling requirements are necessary for silica gel in its
delivered form. Silica gel is amorphous in structure and does not
have the silicosis hazard attendant to crystalline forms of silica."

A time-weighted average (TWA) maximum dust exposure limit of
20 million particles per cubic foot is currently listed in OSHA Standard 29, CRF 1910.1000, Table Z-3 for amorphous silicas. The
weight concentration TWA given by the formula in Table Z-3 should
not be used. A TWA of 10 mg/m^3 is the OSHA standard for nuisance
dusts and is an absolute maximum for any dusty material.

The American Conference of Governmental Industrial Hygienists
has recommended a threshold limit value for amorphous silica of
6 mg/m^3 for total dust and 3 mg/m^3 for respirable dust (>5 μm).

Thus we can conclude that there may be no serious health problems associated with silica gel. This is not surprising, since we all
inhale considerable amounts of sand particles during a lifetime. On
the other hand, it is obvious that dust of any kind must be avoided,
and all necessary protective measures should be taken.

Derivatized Silica Gel—Partition Chromatography

Initially, chromatography was based exclusively on adsorption. In
1941 Martin and Synge in their famous paper [56] proposed replace-
ment of the stationary adsorbent by a stationary liquid. This would
be held in place on an inert finely divided porous material. An example of this could be water or a buffer impregnated on diatomaceous
earth and eluted with carbon tetrachloride. At first, as above, a
polar liquid is mostly adsorbed on a polar support and elution is then

performed with an apolar solvent (buffer on silica gel eluted with isooctane [57]). This "straight" sequence can also be "reversed" by using a hydrophobic support (silanized silica gel or polystyrene) and eluting with a polar solvent (mixtures of alcohol and water). Such "straight phase" liquid—liquid partition chromatography works very well, as does the "reversed" form, but column stability is not good. Although carbon tetrachloride and water are immiscible, some stationary water is nevertheless dissolved and removed from the column. Gradually the phase ratio changes, which alters the chromatographic parameters. Efforts to physically bind the stationary phase more strongly were unsuccessful. This naturally led to the development of chemical bonding of polar and apolar liquids. Many contributions to this field can be found in the literature; a selection is given in Ref. [58]. As time went by, bonding of chemicals to change chromatographic properties turned out to be achieved most easily and successfully with silica gel as the starting material. The degree of coverage of the silica gel surface and the stability of the chemical linkage formed between the silica gel and the liquid are the determining factors. The most stable bonds are obtained with silanes. It can be safely assumed that silanization is therefore the only reaction now used for derivatizing silica gel for chromatography.

The silanes have become very useful and versatile reagents. They range in polarity from apolar, with long alkyl chains on silicon, to polar, with amino, cyano, and alcohol functions bonded on a short alkyl chain. One of the groups attached to the silicon atom of the silane must be a reactive one that can condense with surface silanol functions of silica gel. Chlorine and alkoxy substituents are used for this purpose. The reaction is therefore:

$$\text{Si—OH} + \text{R}_3\text{SiCl or R}_3\text{SiOEt} \longrightarrow \text{Si—OSiR}_3$$

Among the large number of possible silanes, those most commonly used are listed in Table 2.

Following the reaction presented above, only one ligand group can be bonded per silanol group on the silica gel. With the last five silanes of Table 2 subsequent polymerization and multiple bonding can occur, as shown by the following equations of some possible reactions:

$$\text{Si—OH} + \text{RSiX}_3 \longrightarrow \text{Si—O—SiR} \overset{\text{X}}{\underset{\text{X}}{|}} \tag{1}$$

TABLE 2 Silanes for Silica Gel Derivatization

Substituents present on silicon atom	Trivial phase name	Polarity	Applications
1. Trimethyl (Cl)	TMS	Medium	Miscellaneous
2. Butyldimethyl (Cl)	Butyl	Medium-apolar	Proteins
3. Octyldimethyl (Cl)	Octyl	Apolar	Reversed phase
4. Octadecyldimethyl (Cl)	ODS	Very apolar	Reversed phase
5. Phenyldimethyl (Cl)	Phenyl	Apolar	Reversed phase
6. Octyl (OEt)$_3$	Octyl	Apolar	Reversed phase
7. Octadecyl (Cl)$_3$	ODS	Very apolar	Reversed phase
8. Aminopropyl (OEt)$_3$	Amino	Polar	Carbohydrates
9. Cyanopropyl (Cl)$_3$	Cyano	Polar	Straight phase
10. Phenyl (Cl)$_3$	Phenyl	Apolar	Reversed phase

$$\begin{array}{c} \wr \\ \mathrm{Si-OH} \\ \wr \\ \\ +\mathrm{RSiX}_3 \\ \\ \wr \\ \mathrm{Si-OH} \\ \wr \end{array} \longrightarrow \begin{array}{c} \wr \\ \mathrm{Si-O} \quad \mathrm{X} \\ \wr \qquad \backslash \; | \\ \qquad\quad \mathrm{SiR} \\ \wr \qquad / \\ \mathrm{Si-O} \\ \wr \end{array} \qquad (2)$$

$$\begin{array}{c} \quad /\mathrm{OH} \\ \wr / \\ \mathrm{Si} \qquad + \; \mathrm{RSiX}_3 \\ \wr\backslash \\ \quad \backslash\mathrm{OH} \\ \\ \wr \\ \mathrm{Si-OH} \\ \wr \end{array} \longrightarrow \begin{array}{c} \quad \mathrm{O} \\ \wr/ \quad \backslash \\ \mathrm{Si} \quad \mathrm{Si-R} \\ \wr\backslash \; / \\ \quad \mathrm{O} \quad | \\ \\ \wr \\ \mathrm{Si---O} \\ \wr \end{array} \qquad (3)$$

$$\mathrm{a} + \mathrm{H_2O} \text{ when washing, working up} \longrightarrow \begin{array}{c} \qquad\quad \mathrm{OH} \\ \qquad\quad | \\ \wr \\ \mathrm{Si-O-Si-R} \\ \wr \\ \qquad\quad | \\ \qquad\quad \mathrm{OH} \end{array} \quad (4)$$

a + partially hydrolyzed reagent during the reaction

$$\begin{array}{ccc} & \mathrm{R} & \mathrm{X} \\ \wr & | & | \\ \mathrm{Si-O-Si-X} & + & \mathrm{RSi-OH} \\ \wr & | & | \\ & \mathrm{X} & \mathrm{X} \end{array} \longrightarrow \begin{array}{ccc} & \mathrm{R} & \mathrm{X} \\ \wr & | & | \\ \mathrm{Si-O-Si-O-Si-R} \\ \wr & | & | \\ & \mathrm{X} & \mathrm{X} \end{array} \qquad (5)$$

The reagent can oligomerize and only then react with either silica gel or a, etc.

$$\begin{array}{ccc} & \mathrm{R} & \mathrm{X} \quad \mathrm{X} \\ \wr & | & | \quad | \\ \mathrm{Si-O-Si-X} + \mathrm{HO-Si-O-Si-R} \\ \wr & | & | \quad | \\ & \mathrm{R} & \mathrm{R} \quad \mathrm{X} \end{array} \longrightarrow$$

$$\begin{array}{ccc} \mathrm{R} & \mathrm{X} & \mathrm{X} \\ | & | & | \\ \wr \\ \mathrm{Si-O-Si-O-Si-O-Si-R} \\ \wr \\ | & | & | \\ \mathrm{R} & \mathrm{R} & \mathrm{X} \end{array} \qquad (6)$$

Another possibility is

$$2 \overset{\xi}{\underset{\xi}{Si}}-OH + RSiX3 \quad \longrightarrow \quad \overset{\xi}{\underset{\xi}{Si}}-O-\overset{\overset{R}{\mid}}{\underset{\underset{X}{\mid}}{Si}}-O-\overset{\xi}{\underset{\xi}{Si}} \qquad (7)$$

The chemistry looks complex, but in practice we believe that (2), (3), (5), and similar reactions do not occur because of steric hindrance. Monofunctional silanes lead to monomeric phases and polyfunctional silanes lead to polymeric phases. This view is often clearly taken in the literature; for a review see the contribution of Majors [58]. Monomeric phases should be more efficient but hydrolytically not very stable, whereas polymeric phases should be less efficient but more stable to hydrolysis [reaction (9) would explain this]. This matter of monomeric-polymeric phases has been discussed by us at greater length [59]. We state that the difference in chromatographic properties for allegedly monomeric or polymeric phases is minor. However, it is more difficult to synthesize a good "polymeric" phase, and this difference in synthetic accessibility is the reason for most of the negative remarks made about polymeric phases. Real "polymeric" bases need special conditions to be made. They present special selectivity for, e.g., polycyclic aromatic hydricarbons [60].

The two phases still have a number of free silanol functions after the bonding reaction (see below). Removing as many of these as possible by trimethylsilylation is called end-capping or deactivation. Still, extensively substituted and fully end-capped phases obtained from silane 4 and silane 7 in Table 2 are not exactly the same. The main difference is as shown in the following formulas:

$$\overset{\xi}{\underset{\xi}{Si}}-O-\overset{\overset{Me}{\mid}}{\underset{\underset{Me}{\mid}}{Si}}-C_{18}H_{37} \qquad\qquad \overset{\xi}{\underset{\xi}{Si}}-O-\overset{\overset{Me}{\overset{\mid}{\underset{\mid}{MeSiMe}}}\atop{\mid}\atop O\atop\mid}{\underset{\underset{O}{\mid}\atop{\mid}\atop MeSiMe\atop\mid\atop Me}{S}}-C_{18}H_{37}$$

Silane 4   Silane 7

This difference is small, but it can result in selectivity differences for some compounds. This is one of the reasons why the ODS phases of different manufacturers can exhibit different selectivities. There are many other reasons, however. Even silica gels themselves, with the same pore size and specific area, do not behave in exactly the same way in chromatography, for unknown reasons. Silanes of the type $RSiX_3$ are more readily available than the others. They are obtained in fewer synthesis steps than the monofunctional silanes with two methyl groups on silicon and are therefore much cheaper. These are the silanes to consider for preparative purposes, but for micro-HPLC the cost of the phase is of minor importance and therefore exotic, difficult, and expensive reagents can all be considered.

Coverage of Derivatized Silica Gel

The possible coverage of a silica gel depends on the number of silanol functions on its surface. This has been studied and determined many times. According to Berendsen and de Galan [29], the most plausible value for silanol group concentration, considering all arguments and contributions in the literature, must be $4.8/nm^2$. Expressed in other units, this is equal to a silanol amount of 7.8 $\mu mol/m^2$ or to 2.3 mmol/g for a 300 m^2/g silica gel. Only one in every six silicon atoms bears a silanol function in an average silica gel [16]. Erard and Kováts also mention 7.8 μmol OH groups/m^2 for Cab-O-Sil with a specific surface area of 100 m^2/g [28]. Replacing a hydrogen atom by a trimethylsilyl group introduces a considerable difference in steric shielding of the surface. At most about half or about 4 $\mu mol/m^2$ of the available silanol groups can be substituted because of this steric hindrance. Not surprisingly, this figure is not very different with a large alkyldimethylsilane because the long alkyl chain can stand vertically on the silica gel surface and does not necessarily contribute to steric shielding. Therefore, with octyl and octadecyl side chains also, a substitution of more than 3 $\mu mol/m^2$ can be reached. The vertical position of the alkyl groups would occur in the apolar solvents used for the derivatization reaction. In the dry state and in the more polar solvents of reverse-phase chromatography the chains would be collapsed on the surface and cover it completely. A traditional silica gel for chromatography has a specific surface area of 300–400 m^2/g and would therefore, when highly loaded with alkyl chains, take up 900 to 1200 $\mu mol/g$. For octadecyl chains this corresponds to about 20% bonded organic material. Today, however, many silica gels for chromatography have a lower specific surface area. They could thus be fully or highly loaded with a much smaller amount than 20% bonded organic material. This reasoning clearly shows that the usual method of

giving a percentage of bonded material is not enough to characterize
a derivatized silica gel. It is better to give the substitution degree
in micromoles per square meter. There are also several ways to determine the degree of substitution. Thermal gravimetric analysis
has the advantage of being readily obtainable but the disadvantage
that silicon-containing species can be evaporated [61] during the
pyrolysis.

Elemental analysis would be correct but is much more difficult
than might be expected. Especially for low loadings, large deviations from the real figure can be obtained. Total hydrolysis of the
sample followed by gas chromatographic (GC) analysis has also been
proposed [5,62].

Particle Shape and Size of Silica Gel
Stationary Phases

The quality factors of silica gel discussed above are equally important for analytical, preparative, and micro-HPLC packing materials. Considering the small amount of silica gel needed for micro-
HPLC, the most stringent quality requirements can be observed.

From theoretical considerations, Knox and Saleem [63] deduced
that 2 μm was the optimal particle size for HPLC. Halàsz et al./
[64] derived a minimum particle size between 1 and 3 μm. At the
same time the opinion was often expressed that 5 μm was about the
limit of practical particle sizes. Since then, however, 3-μm silica
gel has manifested itself clearly and we have further shown the possibility of using even smaller 2- and 1-μm particles [9,10].

These apparent contradictions have to do with the shape of the
particles. With irregularly shaped silica gel particles, 5 μm is indeed about the limit. At least, we have not been able to obtain
good columns with irregularly shaped silica gel particles only 3 μm
in diameter. Best results are obtained with irregular 10-μm particles.
In this context, performance or quality of results is best expressed
as reduced plate height at optimum flow rate. Furthermore, the
data are given for octadecylated silica gel and with acetonitrile/
water as the eluent on pyrene as the sample substance. In these
conditions it is possible to go down to 1.4 dp with 10-μm irregularly
shaped silica gel. With 5 μm of the same material 2.5 dp is the
best we ever reached, while for 3 μm our best result was only
around 10 dp. With spherical particles, the situation is different,
at least for the very small particles. Here the best results obtainable were with 5-μm materials, producing reduced plate heights
of only 1.6 dp. For particles with a mean diameter below 3 μm, we
obtained good results only with octadecylated materials. Still, the
possibility of using very small particles is most important for micro-
HPLC, as they have the desirable feature of not showing an early
upswing in the H/u curve as illustrated in Fig. 2.

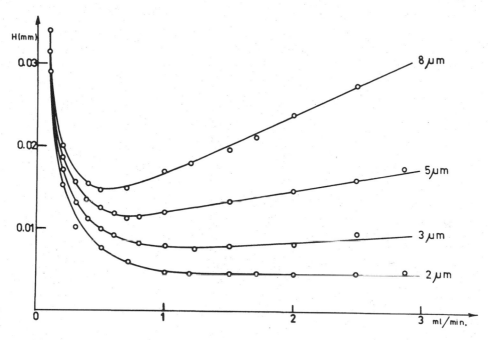

Fig. 2   Curves of H versus u for small particles. Column: 25 ×
0.46 cm for 8- and 5-μm ROSil-C18-D, 15 × 0.46 cm for 3-μm
ROSil-C18-D, and 4 × 0.46 cm for 2 μm; mobile phase: 75:25
acetonitrile/water; sample: pyrene (k' ∿ 6).

      An interesting aspect of the flat region of the H/u curve is
that the sensitivity is independent of flow rate with these very
small particle sizes, at least for flow rates lower than 6 to 8 mm/s.
This is shown in Fig. 3. The chromatogram with the 5-μm particle
column shows a decided reduction in sensitivity when the flow rate
is increased by a factor of four. This is not the case for the comparable chromatogram obtained with 2-μm particles.
      One-micrometer particles seem about the limit that can be
handled by traditional techniques of filtration, sieving, derivatization, viewing under a light microscope, etc. With these very small
particles the problem of friction heat dissipation comes to the foreground. Friction heat dissipation has been discussed theoretically
by Halàsz et al. [64], Horvàth and Lin [65], and Poppe et al.
[66,67]. Recently Guiochon and Colin [68] have also discussed
friction heat, They concluded that the internal diameter is a very
important factor. Our experience shows that under normal operating conditions (v < 3−4 ml/min or u < 6−8 mm/s) viscous heating is not the major contribution to band broadening for conventional
4- to 5-mm i.d. columns. This is shown in Fig. 2. However, when

A

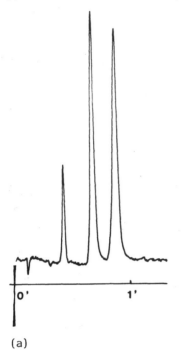

B

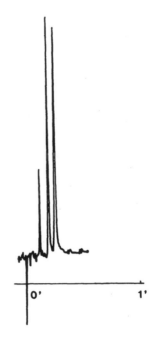

(a)

A

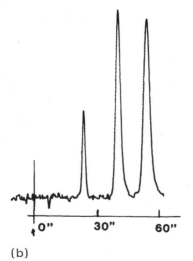

B

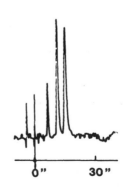

(b)

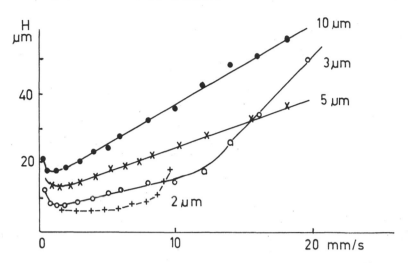

Fig. 4   Curves of H versus u for small particles for high u values.
Column: 10 × 4.6 mm filled with 2-, 3-, 5-, and 10-µm ROSiL-
C18-D.   Mobile phases and samples as in Fig. 2.

viscous solvents and high flow rates are used a deviation from the
normal H/u curve is observed (Fig. 4).   The small 2- and 3-µm
particles show peak tailing or doubling at these high flow rates.
This is not the case for 5- and 10-µm particles.   For this reason
alone, microbore or capillary columns are of the highest interest.
The flow rates mentioned above are not particularly low.   Therefore,
there is no reason not to use very small particles in HPLC, certainly
not in micro-HPLC if we can correctly pack the columns with these
particles.   Since it is rather difficult to prepare very small particles,
they must be very expensive.   As already pointed out, this is not
a drawback for micro-HPLC, where only a very small amount of
packing material is required.

The range or distribution of particle sizes for a given material
is important.   A suitable way of expressing this numerically is provided by the ratio dp_{90}/dp_{10} (dp_{90} and dp_{10} are the particle sizes

Fig. 3   Sensitivity as a function of speed and particle size.   (a)
Column: 1.5 × 0.4 cm filled with 2 µm ROSiL-C18-D; mobile phase:
75:25 acetonitrile/water; A: 0.8 ml/min; B: 3.5 ml/min; sample:
naphthalene, anthracene, pyrene.   (b) Column: 1.5 × 0.4 cm filled
with 5 µm ROSiL-C18-D; mobile phase: 75:25 acetonitrile/water; A:
0.8 ml/min; B: 3.5 ml/min.

(a)

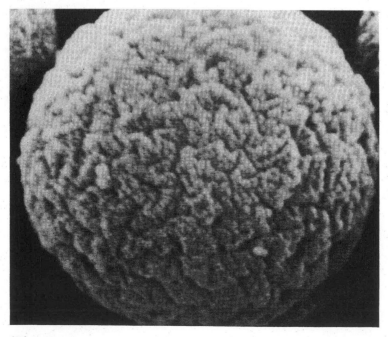

(b)

limiting 90 and 10% of the total weight of the particles). For most commercial silica gel packing materials the $dp90/dp10$ ratio is about 1.5. We have shown that this is generally adequate [31]. For micro-HPLC stringent demands can be placed on this factor, which should at least not exceed the value of 1.5.

Particle size can be measured in different ways. The easiest and most generally available methods are Coulter counting [32] and microscopic viewing [69]. Scanning electron microscopy (SEM) pictures of a 2-μm ROSil-C18-D packing material are shown in Fig. 5. Figure 5a shows a batch of 2-μm ROSil-C18-D, and Fig. 5b shows one particle. The mean particle size of this particular batch is 2.2 μm and the particle size distribution dp_{90}/dp_{10} is 1.5. When using microscopy for irregularly shaped particles, it should be borne in mind that particles present their largest surface area to the viewer [18]. For micro-HPLC, however, only spherical silica particles should be considered. This is further elaborated in the next section. As they come or eventually after prolonged use and repeated packing of the same batch, silica gels can contain relatively large amounts of fines. This will reduce the permeability. Washing by decantation is a solution that is easily implemented.

Packing Methods for Very Small Particles

Small (<20 μm), compressible, and/or swelling particles can be packed efficiently only by a slurry technique. The slurry is pressed in the column and the solids (the packing material) are filtered off in the column. Silica gel-based phases can be pressurized at very high values (up to 1000 bar). High pressure has the following effects:

Packing time is shorter.
Sedimentation is avoided and replaced by filtration. The speed of the particles is greater than their Stokes speed. This avoids segregation of particle sizes in the axial direction [70,71].
The impact speed of the particles on the forming bed is increased and this is beneficial for column stability. This impact speed is higher in lower-viscosity solvents [72]. We feel that this is a neglected factor, often of great importance for the packing of small particles. The impact speed must be high enough to overcome interparticle

---

Fig. 5   Scanning electron microscopy of 2-μm ROSiL-C18-D.
(a)   2-μm ROSil-C18-D (magnification × 1000, bar = 10 μm).
(b)   One particle of 2-μm ROSil-C18-D (magnification × 30,000).

friction. This is all the more important for smaller particles because speed alone is obviously not enough. The small mass in the expression for the kinetic energy must be compensated by a very high impact speed in the case of small particles. This explains the difference in results obtained with spherical and irregularly shaped particles of dp below 5 μm and also the increased difficulty of obtaining good packing results with smaller dp. The available evidence seems to indicate that once a particle has come to rest on a packed bed it is no longer displaced, or is displaced only after a considerable time when the packing has been loosened somewhat, e.g., by dissolution of some of the packing material. Dense packing requires that the particle forcibly displace some of the particles already in position. This will be accomplished most effectively when the approaching particle has a high kinetic energy. Frictional forces are also important, however, even more so for irregularly shaped particles than for spherical ones. We believe this theory explains many of the observed phenomena:

1. The beneficial effect of higher pressure
2. The better result for very small particles with low-viscosity solvents
3. The better result for spherical particles when the size of the particles becomes very small
4. The greater difficulty of packing very large pore packing materials

We realize that the above list is controversial. There are several reports in the literature stating that increased pressure is not relevant. This may be so for a particular packing technique. In general, however, the highest pressure that the packing material can withstand is the best. Point 4 is perhaps not so evident. In general, large pore particles have a lower density and so have a lower kinetic energy during packing. The greater difficulty mentioned above may also be related to the greater brittleness of large-pore silica gel.

Packing can be achieved downward or upward. It is not yet generally accepted which approach gives the best results. We feel that the smaller particles have a better probability of giving optimum packing by an upward technique. Both upward and downward packing should be tried in optimizing a particular column packing material. This may sound frustrating, but that is the way things are at this time [73]. At least this is so for analytical-sized columns as used today. For micro-HPLC columns the little information that is available on the subject is discussed further in this book.

Slurry packing is carried out with a solvent. Several possibilities can be distinguished:

1. The "balanced density slurry" using a mixture of solvents with the same density as the particles. This stabilizes the slurry, which cannot settle out. Tetrabromoethane usually is one of the solvents [74–80]. Drawbacks of this method are the toxicity of the bromoalkanes and their easy decomposition, leading to bromine formation. Although the balanced density concept is appealing, it is being used less and less.

2. Electrostatic dispersion of the particles in a polar medium containing ionizing species [81,82]. Adsorption columns will require very long and costly rinsing with apolar solvents to remove the polar solvents and ions. We have tried this approach several times but without significant improvement of the results over other methods.

3. The viscosity slurry method in which a viscous solvent is used to make the slurry, e.g., glycol or glycerol. Settling out of the suspension is again avoided because it is so slow in the high-viscosity medium [83]. Packing time is very long.

4. Solvents like carbon tetrachloride, chloroform, tetrahydrofuran, acetone, or methanol. Settling out of the slurries is now avoided by carrying out the packing procedure very rapidly.

5. Low-viscosity solvent packing as introduced by us [73]. To increase the impact speed of the particles during packing, solvents with the lowest possible viscosity such as pentane or ether are used ($\eta = 0.23$). Fast packing again avoids untimely settling out. Moreover, in our experience, a good packing material does not settle out as rapidly as a bad packing material, which will not lead to a good column anyway whatever the packing method.

The packing methods described above are generally applicable. Detailed procedures for packing microbore and fused silica columns are given in Refs. 84, 85, 86, 87, and 88. Table 3 compares the packing parameters for conventional, microbore, and fused silica packed capillary columns.

Microbore columns were [89,90], and even now are [86], packed at very high pressure (up to 25,000 psi) in rather viscous solvents (glycol or glycerine in methanol). The reason for this was that rather coarse particles were used. For the smaller particles which today seem more indicated, less viscous solvents and lower pressures are being used [85]. There is also a tendency to use lower slurry

TABLE 3   General Tendencies in Column Packing

Parameter	Conventional, i.d.: 4–5 mm	Microbore, i.d.: 0.5–2 mm	Fused silica, i.d.: 0.1–0.5 mm
1. Slurry solvent	See text	Viscous solvents (MeOH/glycol)	Nonviscous acetonitrile
2. Slurry concentration	5–30%	1–5%	20%
3. Packing pressure	300–1000 atm	600–1800 atm	300–500 atm
4. Down or up	Both	Down	Down
5. Column	As such or polished	Polished or glass-lined	As such or inner wall coated

concentrations (1—5%) for microbore columns. Changing the packing direction (upward or downward) can have a strong influence on column performance. Why this should be is not at all clear. No reports have been published on upward packing of microbore or fused silica columns. The quality of the column wall is sometimes a problem for conventional and microbore columns. Polishing the column wall or using glass-lined tubing is therefore often recommended. One of the advantages of fused silica tubing is the highly polished nature of the wall. Specific procedures for packing fused silica capillaries have been published [87,88]. Some are even patented [91].

Novotny and co-workers [87] described a method for packing 3- and 5-μm particles in 1 m × 0.25 mm fused silica capillaries. The acetonitrile slurry is pressure-forced into the column at 6000 psi. The slurry ratio is 3.8 to 1. Takeuchi et al. [88] described a method for shorter 0.3-mm-i.d. columns also with 3- and 5-μm materials.

A very important new technique for packed fused silica capillaries uses tubing with an inner wall elastic coating. This holds the packing structure and leads to highly efficient columns with excellent stability. 5 μm sized particles appear as the optimum in this context [96].

THE FUTURE

In our opinion the future of micro-HPLC is in the field of packed capillaries of about 50 to 300 μm internal diameter as used by Novotny and co-workers [92] and Yang [93]. An example is shown in Fig. 6. At present, columns are packed with 3- or 5-μm particles. They can produce very high plate numbers but require very long analysis times. A remarkable thing about these columns is the good permeability, even for lengths which would not be possible with a more conventional internal diameter. This must mean that these microcapillary columns are very loosely packed while still being stable because of the small internal diameter.

With 2-μm reverse-phase spherical silica gel packing material the longest normal-sized analytical column that could be run at optimum flow rate while not exceeding 250—300 bars would be around 10 cm in length with a potential plate number of 20,000. For 1-μm particles the corresponding figures would be 4 cm and possibly 13,000 plates. An example of the high efficiency per unit length for 1-μm particles is shown in Fig. 7. If these particles could be loosely packed in microcapillaries, 10 to 20 cm column length even with 1-μm particles seems possible. The plate number could be around 50,000 for an analysis time that could still be shorter than that provided by a classical analytical column of about 25 cm length. To us, this seems to be one of the possible developments of micro-HPLC worth exploring. Several attempts in this direction have been reported [94, 95]. An example of this type of micro-HPLC is shown in Fig. 8.

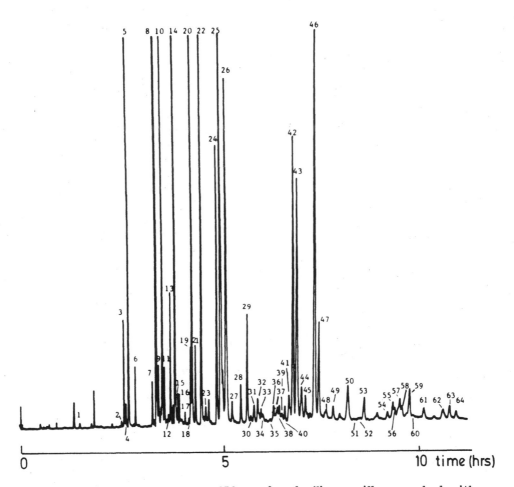

Fig. 6   Column: 1.8 m × 250 μm fused silica capillary packed with
3-μm spherical C18 silica gel; detection: fluorescence, stepwise
gradient.   (From Ref. 92.)

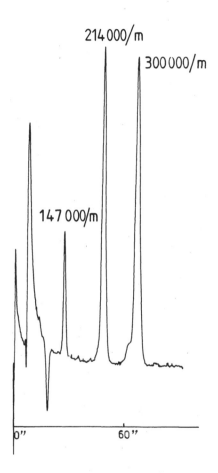

Fig. 7   Micro-HPLC on 1-μm particles. Column: 1.5 × 0.4 cm filled with 1-μm ROSil-C18-D; mobile phase: 75:25 acetonitrile/water, 0.8 ml/min, ΔP = 80 atm; detection: 254 nm (1.4-μl cell and 20-ms time constant); sample: naphthalene, anthracene, pyrene (k' ∿ 6), HEPT = 3 μm.

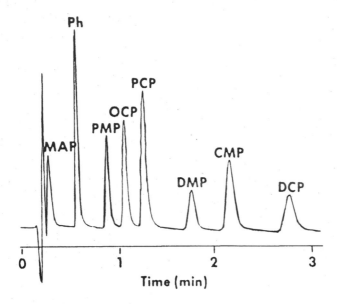

Fig. 8   Example of a high-speed packed capillary chromatogram.
Column: 75 × 0.32 mm, 3-μm MicroPak SP C18; mobile phase:
34:0.1:0.1:66 acetonitrile/acetic acid/triethylamine/water, flow rate
20 μl/min (390 atm); detection: electrochemical.   Peaks: MAP =
m-aminophenol, Ph = phenol, PMP = *p*-cresol, OCP = *o*-chlorophenol,
DMP = 2,6-dimethylphenol, CMP = 4-chloro-3-methylphenol, DCP =
2,4-dichlorophenol.   (From Ref. 94.)

ACKNOWLEDGMENTS

We thank the "Nationaal Fonds voor Wetenschappelijk Onderzoek-
NFWO," the "Instituut voor Wetenschappelijk Onderzoek in Nijverheid
en Landbouw—IWONL," and the "Ministerie voor Wetenschapsbeleid"
for financial support to the laboratory.   J. Sleeckx from Dow Chemicals is thanked for the SEM experiments.   We thank TEC-GENT for
permission to use material from our book Preparative HPLC—A
Practical Guideline (Alltech, 1986).

REFERENCES

1.   L. Zechmeister and L. von Cholnoky, Die Chromatographische
     Adsorptionmethode, Springer-Verlag, Vienna (1937).
2.   C. Giles and I. Easton, *Adv. Chromatogr.* 3, 67 (1966).

3. R. Majors, G. Barth, and C. Lockmuller, *Anal. Chem. 56*, 300R (1984).
4. H. Veening, *Crit. Rev. Anal. Chem. 5*, 165 (1975).
5. Z. Deyl, K. Macek, and J. Janak, *Liquid Column Chromatography, a Survey of Modern Techniques and Applications*, p. 1176, Elsevier, Amsterdam (1975).
6. J. Kirkland, in *Modern Practice of Liquid Chromatography* (J. Kirkland, ed.), Wiley-Interscience, New York (1971).
7. L. Snyder and J. Kirkland, *Introduction to Modern Liquid Chromatography*, p. 534, Wiley-Interscience, New York (1974).
8. R. Majors, *Anal. Chem. 44*, 1772 (1972); *J. Chromatogr. Sci. 11*, 88 (1973).
9. M. Verzele and C. Dewaele, *J. Chromatogr. 282*, 341 (1983).
10. M. Verzele and C. Dewaele, *J. Chromatogr.*, Libr. No. 32 (1985) 435. A. J. P. Martin honorary symposium, Urbino, Italy.
11. K. Unger, Porous Silica, *J. Chromatogr.*, Libr. No. 16, Elsevier, Amsterdam (1979).
12. E. Lederer and M. Lederer, *Chromatography, a Review of Principles and Applications*, p. 460, Elsevier, Amsterdam (1953).
13. *Gmelin's Handb. Anorg. Chem. 15B* (1959).
14. K. Unger, Schick-Kalb, and K. Krebs, *J. Chromatogr. 83*, 5—9 (1973).
15. *Directory of World Chemical Producers*, p. 471, Chemical Information Services, Oceanside, N.Y. (1981).
16. M. Verzele, C. Dewaele, and D. Duquet, *J. Chromatogr. 329*, 351 (1985).
17. S. Brunauer, P. Emmett, and E. Teller, *J. Am. Chem. Soc. 60*, 309 (1938).
18. M. Verzele, J. Lammens, and M. Van Roelenbosch, *J. Chromatogr. 186*, 435 (1979).
19. L. Drake and H. Ritter, *Ind. Eng. Chem. Anal. Ed. 17*, 787 (1945).
20. J. Kirkland, *J. Chromatogr. Sci. 10*, 593 (1972).
21. R. Nikolov, W. Werner, and I. Halàsz, *J. Chromatogr. Sci. 18*, (1980).
22. W. Werner and I. Halàsz, *J. Chromatogr. Sci. 18*, 277 (1980).
23. R. M. Scott, *J. Chromatogr. Sci. 11*, 129 (1973).
24. J. Hockey, *Chem. Ind. (London) 9*, 57—63 (January 1965).
25. C. Giles and I. Easton, *Adv. Chromatogr. 3*, 57—63 (1966).
26. L. Snyder and J. Ward, *J. Phys. Chem. 70*, 3941—3952 (1966).
27. R. P. Scott and P. Kucera, *J. Chromatogr. Sci. 13*, 337—346 (1975).
28. J. Erard and E. Kovàts, *Anal. Chem. 54*, 193 (1982). '

29. G. Berendsen and L. de Galan, *J. Liq. Chromatogr.* *1*(4), 403 (1978).
30. H. Cassidy, *Fundamentals of Chromatography*, Vol. X, Interscience, New York (1951).
31. C. Dewaele and M. Verzele, *J. Chromatogr.* *260*, 13 (1983).
32. T. Allen, *Particle Size Measurement*. Chapman & Hall, London (1974).
33. T. Hanai and K. Fujimura, *J. Chromatogr. Sci.* *14*, 140 (1976).
34. E. Stahl, *Dunschicht-Chromatographie*, p. 534, Springer, Berlin (1963).
35. H. Seiler and M. Seiler, *Helv. Chim. Acta* *43*, 1939–1941 (1960).
36. K. Randerath, *Dunnschicht-Chromatographie*, p. 250, Verlag Chemie GmbH, Weinheim (1962).
37. E. Herrmann, *J. Chromatogr.* *38*, 498–507 (1968).
38. M. De Potter, J. Gheysels, and M. Verzele, *J. High Resolut. Chromatogr. Chromatogr.* *3*, 151 (1979).
39. H. Kohlschutter, A. Risch, K. Unger, and K. Vogel, *Ber. Bunsenges. Phys. Chem.* *69*, 349–356 (1965).
40. A. Meyers, T. Slade, R. Smith, and E. Michetlich, *J. Am. Chem. Soc.* *44*, 2247 (1979).
41. C. Walling, *J. Am. Chem. Soc.* *72*, 1164–1168 (1950).
42. H. Engelhardt and H. Muller, *J. Chromatogr.* *218*, 395 (1981).
43. D. Ottenstein, *J. Chromatogr. Sci.* *11*, 136–144 (1973).
44. H. Halpaap, *J. Chromatogr.* *78*, 63 (1973).
45. W. Strubert, *Chromatographia* *6*, 50 (1973).
46. H. Brockmann and H. Schodder, *Chem. Ber.* *74*, 73 (1941).
47. L. Snyder and J. Kirkland, *Introduction to Modern Liquid Chromatography*, Wiley-Interscience, New York (1974).
48. M. Verzele, unpublished.
49. F. Mikes, V. Schurig, and E. Gil-Av, *J. Chromatogr.* *83*, 91–97 (1973).
50. N. Houx, S. Voerman, and W. Jongen, *J. Chromatogr.* *96*, 25–32 (1974).
51. R. Heath, J. Tumlinson, R. Doolittle, and A. Proveaux, *J. Chromatogr. Sci.* *13*, 380–382 (1975).
52. R. Aigner, H. Spitzy, and R. Frei, *Anal. Chem.* *48*, 2–7 (1976).
53. F. Dickey, *J. Phys. Chem.* *59*, 695–707 (1955).
54. C. Williams and R. Hawley, *Int. Lab.*, 67 (1975).
55. Davison Chemical Division (Grace), IC-219-581.
56. A. Martin and R. Synge, *J. Biochem.* *35*, 91 (1941).
57. M. Verzele, *Bull. Soc. Chim. Belges* *64*, 70 (1955).
58. R. Majors, *J. Chromatogr. Sci.* *18*, 488 (1980).

59. M. Verzele and P. Mussche, *J. Chromatogr. 154,* 117 (1983).
60. L. Sander, S. Wise, *J. Chromatogr. 316,* 163 (1984).
61. P. Mussche and M. Verzele, *J. Anal. Appl. Pyrolysis 4,* 273 (1983).
62. M. Verzele, P. Mussche, and P. Sandra, *J. Chromatogr. 190,* 331 (1980).
63. J. Knox and M. Saleem, *J. Chromatogr. Sci. 12,* 614 (1969).
64. I. Halàsz, R. Endele, and J. Asshauer, *J. Chromatogr. 112,* 37 (1975).
65. Cs. Horvath and H. Lin, *J. Chromatogr. 149,* 43 (1978).
66. H. Poppe, J. Kraak, J. Huber, and J. van den Berg, *Chromatographia 14,* 515 (1981).
67. II. Poppe and J. Kraak, *J. Chromatogr. 282,* 413 (1983).
68. G. Guiochon and H. Colin, in *Microcolumn High Performance Liquid Chromatography* (P. Kucera, ed.), p. 1, Elsevier, Amsterdam (1984).
69. Methods for the Determination of Particle Size of Powders, British Standard 3406, Part 4 (1963).
70. L. Snyder, *J. Chromatogr. Sci. 7,* 352 (1969).
71. S. Sie and N. van den Hoed, *J. Chromatogr. Sci. 7,* 257 (1969).
72. P. Bristow, P. Brittain, C. Riley, and B. Williamson, *J. Chromatogr. 139,* 57 (1977).
73. M. Verzele, *J. Chromatogr. 269,* 81 (1984).
74. R. Majors, *Anal. Chem. 44,* 1722 (1972).
75. L. Snyder, *J. Chromatogr. Sci. 7,* 352 (1969).
76. J. Kirkland, *Chromatographia 10,* 661 (1975).
77. J. Kirkland, *J. Chromatogr. Sci. 9,* 206 (1971).
78. W. Strubert, *Chromatographia 6,* 50 (1973).
79. H. Engelhardt, J. Asshauer, U. Neue, and N. Weigand, *Anal. Chem. 46,* 336 (1974).
80. R. Cassidy, D. Le Gay, and R. Frei, *Anal. Chem. 46,* 340 (1974).
81. J. Kirkland, *J. Chromatogr. Sci. 10,* 593 (1972).
82. M. Caude, L. Phan, B. Terlain, and J. Thomas, *J. Chromatogr. Sci. 13,* 390 (1975).
83. J. Asshauer and I. Halàsz, *J. Chromatogr. Sci. 12,* 139 (1974).
84. C. Lochmuller, in *Small Bore Liquid Chromatography Columns* (R. P. W. Scott, ed.), p. 115, Wiley-Interscience, New York (1984).
85. J. C. Kraak, *LC Magazine 3*(2), 88 (1985).
86. H. Menet, P. Gareil, M. Caude, and R. Rosset, *Chromatographia 18*(2), 73 (1984).

87. J. C. Gluckman, A. Hirose, V. L. McGuffin, and M. Novotny, *Chromatographia* *17*(6), 303 (1983).
88. T. Takeuchi, D. Ishii, and A. Nakanishi, *J. Chromatogr.* *285*, 97 (1984).
89. R. W. P. Scott and P. Kucera, *J. Chromatogr.* *125*, 251 (1976).
90. R. W. P. Scott and P. Kucera, *J. Chromatogr.* *169*, 51 (1979).
91. F. Yang, U.S. Patent 4,483,773.
92. A. Hirose, D. Wiesler, and M. Novotny, *Chromatographia 18*, 239 (1984).
93. F. Yang, *J. Chromatogr.* *236*, 263 (1982).
94. Sj. Van der Wal, *LC Magazine* *3*(6), 488 (1985).
95. T. Takeuchi and D. Ishii, *J. High Resolut. Chromatogr. Chromatogr. Commun.* *6*, 683 (1983).
96. M. Verzele, C. Dewaele, M. DeWeerdt, and S. Abbott, *J. High Resolut. Chromatogr. Chromatogr. Commun.* (in press).

3

Gradients in Microbore LC: Techniques and Applications

VERN V. BERRY / SepCon Separations Consultants, Boston, Massachusetts, and Salem State College, Salem, Massachusetts

HERBERT E. SCHWARTZ / Applied Biosystems, Inc. (formerly Brownlee Labs), Santa Clara, California

INTRODUCTION

This chapter reviews the development of high-performance liquid chromatography (LC) instrumentation for gradient elution with packed microbore columns, i.e., columns of 2 mm inside diameter (i.d.) or less. The discussion focuses on

1. Methods for generating gradients
2. Commercially available instrumentation
3. Column types (diameter, length, packing, etc.)
4. Applications

In a previous review, Schwartz and Berry [1] covered some of the principles and instrumentation for gradient elution in microbore LC. This chapter discusses flows and gradient volumes for different column sizes, the expected reproducibility of retention times and peak areas with gradients, and the measurement and causes of gradient deviations.

Gradient elution based on changing the LC solvent composition is useful in achieving good speed and detection of samples having a wide range of polarities. The theory and practice of gradient elution as applied to conventional (4 to 5 mm) columns have been reviewed in several articles and books (e.g., [2,3]).

TABLE 1 Nomenclature and Comparison of Typical Diameters, Cross-Sectional Areas, Approximate Flows, and Approximate Gradient Volumes for Common Microbore Columns

Common type and i.d. (mm)	Area (mm²)	Approximate flows (μl/min)	Approximate gradient volumes (μl)	
Conventional bore				
4.6	21.16	1000–4000	15,000	60,000
Narrow bore				
2	4	200–800	3,000	12,000
1	1	50–200	700	2,800
Micro-HPLC				
0.5	0.25	10–40	175	600
0.3	0.09	5–20	60	240
Open tubular				
0.01	0.0001	0.005–0.020	0.07	0.28

Different LC column types have been categorized by column dimensions, materials, and corresponding flows by several writers (e.g., [1]). In this chapter the term *microbore* is used to describe columns having inside diameters smaller than 2 mm. *Micro-HPLC* is used for columns in the range 0.2—0.5 mm i.d. Table 1 summarizes the nomenclature used.

For proper comparison, flow (microliters per minute) should be in proportion to column cross-sectional areas if the linear velocities are to be kept the same. Guidelines for preparative LC indicate that the volume of the gradient also should be in proportion to the cross-sectional area, provided the column lengths are the same. Comparison of gradient volumes is sometimes complicated by the greater lengths (often more than 1 m) used with microbore columns. Table 1 is a rough guide to typical gradient volumes and flows, based on a conventional column 250 mm long by 4.6 mm i.d. (250 × 4.6 mm) and a gradient 15 min long with a flow of 1 to 4 ml/min (first row of table).

From Table 1, the ratios of the areas show that flows and gradient volumes for microbore columns of 2 and 1 mm i.d. are approximately 5 and 21 times lower, respectively, than for conventional 4.6-mm-i.d. columns, assuming the same linear velocity of eluent. Micro-HPLC columns of 0.5 and 0.3 mm i.d. require flows about 85 and 235 times lower than for conventional columns, while with open-tubular columns this factor may be as large as 200,000. The open-tubular columns would use much less than 1 µl for a full gradient! Of course, solvent savings and solvent disposal costs fall in the same proportion as flows and gradient volumes.

REPRODUCIBILITY OF DATA WITH GRADIENTS

A number of factors influence the reproducibility of chromatographic data measurements, including sample load, column stability, column equilibration, solvent preparation, column temperature control, sample preparation, and data measurement methods. Variability in most of these factors can be minimized by the operator, with the result that factors related to the solvent delivery system are often the limiting ones. For quantitative work, reproducibility of retention times, peak areas, and peak heights is of primary importance. For reproducible LC measurements on the same instrument, two pump-related factors are key: (a) flow reproducibility and (b) composition reproducibility. For comparison of data between instruments, (c) flow accuracy and (d) composition accuracy become important. Accuracy reflects how close the set value is to the "true" flow or "true" accuracy.

A survey of commercial instruments by van den Berg et al. indicated that the flow accuracy and composition accuracy of current

commercial LC systems are adequate for conventional columns (i.e.,
at 1 ml/min) [4], but problems did occur with microbore systems
(i.e., at 0.01 ml/min) [5].

At low flows, flow reproducibility and flow accuracy are marginal
with unmodified equipment. However, in the view of the analyst,
reproducibility of chromatographic data is perhaps more important
than the absolute accuracy of instrumental factors, because analyses
are often performed on a single instrument and calibration standards
are run on the same instrument. In this section, the factors influ-
encing precision and accuracy are discussed. First, it is shown
how gradients can be defined experimentally. Causes of gradient
deviations are discussed next. Finally, the reproducibilities of retention times and areas that can be expected in microbore LC are
touched upon.

Measurements to Define Gradients

It should be noted that the actual gradient profile can be observed
experimentally by putting an unretained ultraviolet (UV)-absorbing
component in either eluent, for example, 0.1% acetone in acetonitrile.

Response Volume

The response volume may be defined as the volume of eluent between
the point where a step change is made in the composition of the
mobile phase (where the gradient components come together) and
where this changed composition reaches a detector (with the column
removed). This is illustrated in Fig. 1a with high-pressure mixing
of two eluents in a dynamic (stirred) mixer connected directly to
the detector.

Changes in composition seen by the detector do not appear
instantly since various components in the hydraulic system, such as
mixers, tubing, pressure transducers, and pulsation dampers, act
to delay and to modify the intended gradient shape. Gradients
deviate from the intended profile in different ways, depending on
whether devices that produce significant exponential mixing (dynamic mixers) or eddy mixing (static mixers) are in the system.
These two types of devices may require different methods for quan-
titatively locating a fixed "final" composition. Recently, a theoretical
treatment of equipment design and operating conditions permitted
Quarry et al. [6] to select gradient conditions for minimal gradient
distortion.

Response volume determination is illustrated in Fig. 1a with the
combination of high-pressure (two-pump) gradient generation with a
dynamic exponential mixer and no column. Delay volume (next section) is also illustrated (Fig. 1b) with the combination of low-pressure gradient generation with a static (packed-bed) mixer. Of
course, other combinations of pumps and mixers are possible, such

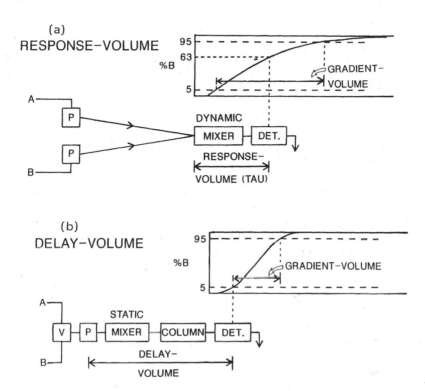

Fig. 1   Volume measurements important in gradient elution.   Examples of two common methods for generating gradients, high-pressure mixing (a) and low-pressure mixing (b), and two common devices for mixing eluents, a dynamic (stirred) mixer (a) and a static (packed-bed) mixer (b).   If the same mixers are used to generate a gradient by switching abruptly from eluent A to B, the methods for determining the gradient volumes are shown below the concentration versus time curves.   Components are: A, weak eluent; B, strong eluent; P, high-pressure pumps, V, high-speed switching valve for low-pressure gradient generation; and DET, detector.

as low-pressure gradient generation and a dynamic mixer.   The combinations in Fig. 1 illustrate the general shape of the concentration profile of eluent B in A if an instantaneous step change is made from A to B.

Note that the response volume includes the volume between the mixer (whether static or dynamic) and the detector (column removed) for either the high-pressure or low-pressure mixing system.   The low-pressure mixing system may include the pump head in the delay volume if significant mixing occurs in the pump head (i.e., the vol-

ume of the pump head is significantly larger than the "pulses" of
different eluents delivered).

For *dynamic (stirred) mixers* interposed between the gradient
generator and the column, the intended gradient may be distorted
by an exponential profile superimposed on the composition profile,
like that shown in Fig. 1a.

For quantitative measurements with dynamic mixers, the response
volume may be defined in terms of a time constant (tau) determined
as follows. The composition of final eluent (%B) leaving a dynamic
mixer varies exponentially with time (t) by Eq. (1).

$$\%B = 100 \ (1 - e^{-Ft/V})\% \tag{1}$$

Here V is the volume of the mixer, and F is flow. When the flow-
time product (Ft) equals the volume of the exponential diluter
chamber, V, a *response time*, tau, can be defined at the point
where the gradient has progressed to 63% of the final composition
[7]. Alternatively, 4 tau, may be used for dynamic mixers. This
corresponds to about 99% B. Tau may be used to compare various
dynamic (exponential) mixers.

For *static or packed-bed mixers* a symmetrical S-shaped change
will be superimposed on a step change in composition (Fig. 1b).
The step is rounded at the beginning and end because of eddy diffusion in the packed bed and (some) viscous drag in the lines.
The broadness of the S increases with larger particles because of
the increased eddy diffusion (see later section on Gradients from a
Breakthrough Curve). The broadness of the S will depend on flow
only if porous particles are used (because of mass transfer in and
out of the pores). If nonporous particles are used in the packed-
bed mixer, the broadness of the S will be independent of flow.

For quantitative measurement with static mixers, the volume at
which the actual gradient reaches some arbitrary composition, such
as 95% B, may be taken as a fixed end point for the gradient. A
volume measurement for 95% B could also be used for exponential mixers.

Delay Volume

Figure 1b shows one definition of the delay volume as the volume it
takes for a step change to pass through the column and cause a
measurable change in the detector (e.g., a 5% change, Fig. 1b).
In this case, the delay volume increases with increased column and
extracolumn volume, Long (1 m) columns for high plate counts in
microbore LC have the disadvantage that the large delay volume
makes separations very long.

Sometimes the delay volume is determined in the same way as
the response volume as defined above, i.e., with the column re-

moved. A step change in eluent is made and the delay volume determined from when a certain %B (e.g., 63%) reaches the detector.

Depending on the design of the system and the approach to
generating gradients, the delay volume can vary considerably.
For high sample throughput it is desirable to have a small delay
volume, or, more important, a small delay time, since, by using a
diverter valve just before the injector for "venting," large delay
volumes can be reduced to low delay times if the pumping is initially faster. In some approaches (e.g., with "breakthrough" gradients
for the submilliliter gradient volumes that are useful for micro-HPLC
columns), the delay volume must be dealt with specifically by using
the venting approach since it can make the run two or more times
longer, as shown by Cotter [8].

Gradient Volume

For instruments that permit direct generation of gradients by changing pump speeds, the gradient volume may be taken as the product
of the flow rate and the time over which the gradient is run.

For systems in which the gradient is generated indirectly, the
gradient volume may be determined experimentally by adding a UV-
absorbing component to one eluent.

A clearly defined initial composition may be taken as the intercept of the linear portion of the actual profile with the 0% B line,
or (often easier to define) the intercept of the actual profile with
the 5% B line (as shown in Figs. 1a and 1b). A clearly defined
final composition may be taken as the intercept of the actual profile
with the 95% B volume, for either dynamic or static mixers. The
gradient volume, then, is the difference between the initial 5% B
intercept volume and the final 95% B intercept volume [9,10]. This
measurement is sensitive to slow changes in the last portion of the
gradient but still does not require waiting for a "final" concentration
to be reached, which can take a long time with a dynamic exponential
mixer.

Figure 1 illustrates the determination of gradient volume that
can be calculated for a step change using a dynamic mixer (Fig. 1a)
or a static mixer (Fig. 1b).

Causes of Gradient Deviations

The actual flow and gradient profile may deviate from the instrument
set values for several reasons:

Delay Volume and Response Volume

As mentioned above, the mixers, tubing, column, etc., between the
initial point at which the gradient components come together and the

head of the column, can result in unwanted distortion of the gradient. For the same mixer volume, dynamic mixers distort the gradient more than static mixers. While the delay volume can be measured and subtracted from the chromatogram, it can greatly reduce problem solving throughput by lengthening the analysis by the additional time for both the gradient and the re-equilibrium to the starting conditions.

Mechanical/Electronic Imperfections

Gradient imperfections can be caused by electronic or mechanical imperfections, for example, in the operation of check valves, stepping-motor drives, gears, or pistons. Among the most frequent sources of mechanical failure are the check valves used in conventional LC pumps. In particular, accumulation of particles (mold fibrils, bacteria, buffer salts, ion pair agents, dirt, etc.) leads to systematic and random malfunctions of check valves as well as changes in efficiency, resolution, and column permeability [11,12]. Leakage can also result from formation of air bubbles due to the slight vacuum in the pump head on the suction stroke of the pump. Operating conventional pumps at low flows increases check valve problems.

Eluent Volume Contractions, Eluent Compressibility, and System Compliance

These effects can act to distort the true flow and expected gradient shape, depending on how pumping and gradient generation are achieved. The reproducibility of a gradient may be high even if important flow or gradient deviations exist.

Volume contraction resulting from the mixing of two or more eluents can affect the actual flow [13,14]. Pumping systems that use low-pressure mixing minimize flow deviations due to this effect (e.g., Fig. 1b).

Compressibility differences between different eluents can also affect flow. This affect can be considerable, depending on the flow, pressure, temperature, type of pump, and method for gradient generation (e.g., high- versus low-pressure mixing). Lower flows can cause greater deviations. For example, Rainin Instrument Company pointed out that with a two-piston pump (piston displacement, 100 μl; refill time, 100 μl/min) delivering 10 μl/min at 3000 psi, the time to refill and recompress methanol (compressibility 0.84% per 1000 psi) before flow begins out of the piston chamber can be as long as 22.7 s [15]. Different pumping systems have different methods for overcoming these compressibility effects.

The slight increase in system volume with increasing pressure, the system *compliance*, comes about by slight expansions of connecting tubing, Bourdon tubes in pressure gauges, pulse dampers, etc. The effect is that changes in pressure during a gradient,

with a constant-flow system over a short time period, may slightly
delay the gradients.

Minimum Flow Limits with High-Pressure Mixing

With high-pressure gradient generation, the flow accuracy and flow
reproducibility are limited by the minimum flow of the pump. With
high-pressure mixing, the solvents are delivered by two (or more)
high-pressure reciprocating piston pumps (each of which may have
one or two heads) to a low-volume dynamic mixer on the high-pressure side of the LC system. Systems using this approach include,
for example, Hewlett-Packard (1080), Hitachi, Shimadzu, and
Waters. With this type of gradient formation, dynamic and static
mixers have been used to achieve smooth, noise-free baselines.
The eluent may be viewed as two side-by-side parallel streams after
the mixing tee. If, instead of fast reciprocating pistons, a piston
displacement pump is used with microsteps, e.g., Applied Biosystems
(Brownlee) Microgradient system, only "radial mixing" across the tube
may be required. A column packed with nonporous particles can act
as an effective "static mixer." Recent work with sensitive refractive
index detection [16] showed that, with either gradient or isocratic
flow mixing, when two reciprocating piston pumps are used, the
parallel liquid streams, in fact, have superimposed a slight undulation
that corresponds to the cyclic pump pulsations. Even large-diameter
piston displacement pumps can show these undulations corresponding
to the stepping-motor movement of the pump, as shown by Powley
et al. [17].

The minimum flow limit with the high- and low-pressure mixing
was investigated by Sjodahl et al. [18]. With high-pressure mixing,
if the minimum flow per pump is 10 µl/min, at a flow of 300 µl/min
arrow in Fig. 2b), gradients could be generated from about 3 to
97% B. However, at a minimum flow of 100 µl/min, the actual
gradient was limited to only 25 to 75% B. With low-pressure mixing,
to be discussed in the next section, the full 0 to 100% gradient
range could be covered at a minimum flow of even 10 µl/min (Fig.
2a).

Minimum Flow Limit with Low-Pressure Mixing

With low-pressure gradient generation, the precise actuation time of
the proportioning valves that partition the eluents ultimately limits
the composition accuracy with low-pressure mixing [19]. In this
approach, two or more eluents are mixed on the low-pressure side
before the pump, the composition being determined in one of two
ways. First, an older approach, uses electrically actuated solenoid
valves to deliver segments of solvent A, then B, then A, etc.
Companies now using this approach include LKB, Nicolet, and
Varian. Another approach uses low-pressure feed pumps to propor-

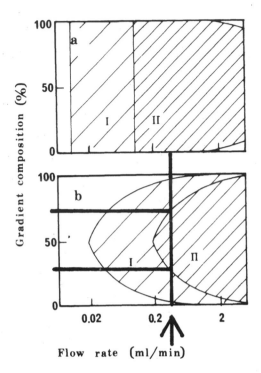

Fig. 2   Limitations of gradient systems using (a) a low-pressure
mixing proportioning valve system and (b) a high-pressure gradient
mixing two-pump system.   Zone I is for a pump with a minimum flow
of 10 μl/min, and zone II is for a pump with a minimum flow of 100
μl/min.   For the one-pump system, a minimum time of 25 ms in one
valve position and a switching cycle volume of 45 μl are assumed.
(Adapted and reprinted from Ref. 18.)

tion the eluent into a pre-chamber.   An example of this approach is
the Hewlett-Packard 1090 LC system.

Low-pressure mixing is gaining in popularity because often (a)
precision is better at the gradient extremes (%B) and (b) only a
single high-pressure pump is used, potentially reducing mechanical
complexity, reducing costs, and improving reliability.   However,
this single high-pressure pump may have one to three heads or low-
pressure pumps to deliver eluent into a preforming gradient section,
thus returning the mechanical complexity to the system.

This type of gradient formation can produce adjacent pulses of
solvent lined up in the tube.   Thus *longitudinal mixing*, along the
length of the column, is required.   Longitudinal mixing is also required for isocratic *flow mixing* of eluents.   Gradients made with

low-pressure mixing often use dynamic mixers to average out moment-to-moment compositional variations caused by the proportioning valves and the pump piston movement. With larger-volume mixers, generally steadier baselines are obtained. However, as pointed out above, gradient accuracy decreases with increasing mixer volume.

The minimum flow limit with high-pressure pumps results if the valve cycle time approaches the pumping speed. For gradients with a small total volume (at low flows), small valve cycle volumes are necessary to achieve accurate gradient profiles. In this case, the volume of the pump piston chambers, mixers, pressure transducers, and pulse dampers is important and may limit the composition accuracy. Another problem often encountered with low-pressure mixing systems is that they are generally more prone to air bubble blocking of the inlet check valve [2]. However, this problem may be eliminated by using a mechanically driven inlet check valve, as in the Varian 5500 or Hewlett-Packard 1050 pumps.

Expected Reproducibility of Peak Retention Times and Peak Areas

The reproducibility of retention time with gradients requires precise control with time of both (a) flow and (b) gradient composition. Both factors are a function of the instrument design. With gradient elution, conventional pumps yield retention time reproducibility of 0.1 to 0.5% relative standard deviation (RSD) [2,14] when both the eluent and columns are thermostated [19] or flow corrections are made [20].

Reproducibility of peak area and peak height depends on the design of the injector and the injection technique [21,22]. Reproducibility of peak area is only slightly affected by temperature and mobile phase composition but does depend on flow rate. However, peak height is dependent on mobile phase composition and temperature but is not affected much by flow rate variations. For most applications the reproducibility data mentioned above are adequate and may serve as a rough guideline for microbore gradient systems, namely: retention time, 0.1–0.5% RSD; area, 1.0–3.0% RSD.

ADAPTATION OF CONVENTIONAL INSTRUMENTATION FOR GRADIENT MICROBORE LC

Pumps and other components for LC were recently surveyed by McNair [23]. At present, many manufacturers consider an internal diameter of about 2 mm a practical size for microbore because fewer constraints are placed on injectors or detectors, which often can be used directly or after slight modification. These columns reduce solvent consumption by 80% compared to the conventional 4.6-mm columns.

The first work with microbore columns, by Ishii et al. in 1977 [24], involved the use of mechanically driven glass syringes to deliver flows of 1—10 µl/min (to 70 bars) through submillimeter i.d. "micro-HPLC" columns. However, in the United States in 1979, Scott and Kucera [25] slowed down the motor speed of conventional pumps (to 400 bars), thereby achieving flows of 100 µl through the 1- to 2-mm-i.d. columns. This work led to a generation of commercial pumps that could be operated at low flows suitable for 2-mm-i.d. columns. Besides slowed-down pumps and low-flow pumps, three other approaches with unmodified conventional pumps later permitted work with microbore columns; these approaches involved flow

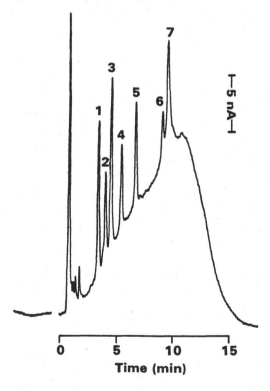

Fig. 3   Gradients with conventional pumps (Varian 5500) at low flows for peptides at 500 µl/min (gradient volume about 5000 µl) with a 150 × 2 mm, 4-µm C18 column, a linear gradient in about 10 min, and electrochemical detection of 3—9 pmol of (1) EAE, (2) (D-Ala 2) Met-enkephalin, (3) Met-enkephalin, (4) angiotensin II, (5) Leu-enkephalin, (6) bombesin, and (7) neurotensin. (From Ref. 26.)

splitting, gradient storage, and miniaturized pump heads. These five different approaches using conventional instrumentation are described in this section.

Conventional Pumps at Low Flows

Pumps from several manufacturers can generate reproducible gradients for flows of about 50 to 100 µl/min (see Table 1) suitable for the largest microbore columns of 1 to 2 mm i.d. Figure 3 shows a separation of peptides on a 150 × 2 mm column at flows of 500 µl/min using the Varian 5500 reciprocating piston, rapid-refill, low-pressure gradient generation pump [26].

R. Schuster of Hewlett-Packard showed that the model 1090 LC system could generate gradients for 2-mm-i.d. columns with good precision and accuracy [27–30]. In Fig. 4, a fast 18-min separation on a 100 × 2.1 mm column of 15 orthophthalaldehyde (OPA) amino acids is shown. Derivatization takes place on-line with a reproducibility for retention of 0.07–0.3% RSD and for area of 1.1–2.7% RSD [27]. A fast 12-min separation was achieved with a phenylthiohydantoin (PTH) derivative of 21 amino acids on a 100 × 2.1 mm column. A similar but longer separation (25 min) was made on a 200 ×

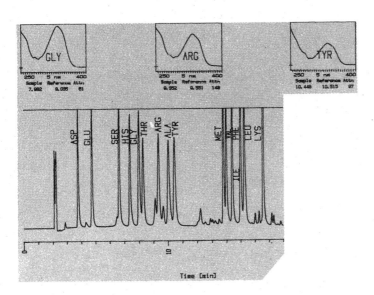

Fig. 4 Gradient separation with a conventional pumping system (Hewlett-Packard 1090) for OPA amino acids at 350 µl/min (gradient volume 6300 µl) with a 100 × 2.1 mm, 5-µm Hypersil ODS column. Nonlinear gradient in ca. 18 min from A (water with 100 mM sodium acetate) to B (methanol) at 40°C. Fluorescence detection of 1 µl with 1.25 nmol of each in 0.5 µl. (From Ref. 27.)

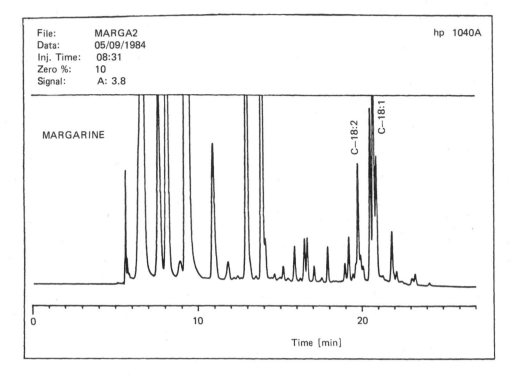

Fig. 5   Gradient separation with a conventional pumping system
(Hewlett-Packard 1090) at low flows for fatty acids in margarine
with automated derivatization of 450 μl/min (gradient volume 11,250
μl), a 100 × 2.1 mm column of 5-μm Hypersil ODS, a 30—100% nonlinear gradient in 25 min from A (water) to B (acetonitrile).   UV
detection (258 nm) of 4 μl of fatty acids from triglycerides in hot
MeOH/KOH, and on-line derivatization with bromophenacyl bromide
plus 18-crown-6-ether.   (From Ref. 29.)

2.1 mm column to produce a reproducibility for retention time of
0.11 to 1.0% RSD and for area of 0.7—2.1% RSD [28].   Detection to
lower picogram levels was possible with this longer separation.   At
higher flows, 300—500 μl/min, the Hewlett-Packard 1090 instrument
was used for fatty acid analysis using on-line derivatization on short
100 × 2.1 mm columns (Fig. 5).   In this technique, derivatizing
reagent, spacer solvents, and sample were drawn into the autoinjector syringe, the flow was decreased momentarily for reaction at
the head of the column, and then the flow was resumed for separation [29].

     The Hewlett-Packard system at lower flows (100 μl/min) was
used by Nice et al. [31] to separate several proteins by reversed

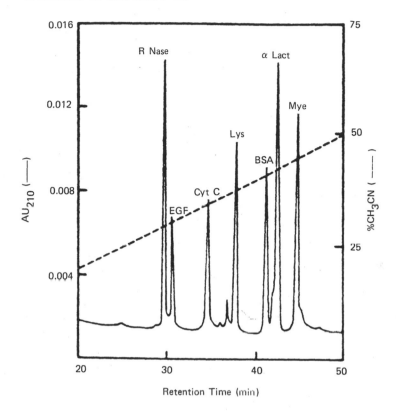

Fig. 6   Gradient with a conventional pumping system (Hewlett-Packard 1090) for proteins at 100 μl/min (gradient volume 5000 μl) with a 30 × 2.1 mm, 7-μm RP-300 C8 column.   20–50% linear gradient in 50 min from A (water with 0.9% NaCl, pH 2.1) to B (acetonitrile) at 45°C.   UV detection 210 nm of 80–134 ng/2 ml of (RNase) ribonuclease A, (EGF) murine epidermal growth factor, (Cyt C) cytochrome C, (Lys) lysozyme, (BSA) bovine serum albumin, (aLact) bovine alpha-lactalbumin, and (Myo) myoglobin. (From Ref. 31).

phase (Fig. 6) (gradient volume about 5000 μl) with a very short 30 × 2.1 mm column.

Two other examples of gradients with unmodified conventional pumps are presented in Figs. 7 and 8.   A separation of proteins at 400 μl/min using the Waters M6000 pump is shown in Fig. 7 [8], and Fig. 8 shows an essential oil separation [32].

For columns below 2.1 mm i.d., a useful compromise is to modify instrumentation originally designed for high flows of 1–4 ml/min with conventional pumps.   A number of approaches to this type of modification have been used, as described in the next section.

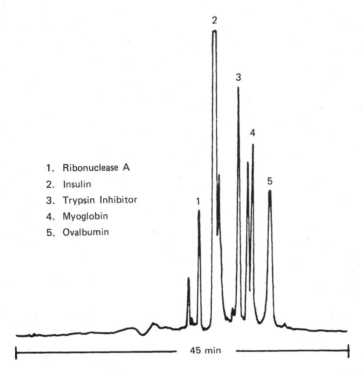

1. Ribonuclease A
2. Insulin
3. Trypsin Inhibitor
4. Myoglobin
5. Ovalbumin

45 min

Fig. 7   Gradient separation with a conventional pumping system
(Waters M6000A) of proteins at 400 µl) with a 300 × 2 mm Micro-
Bondapak C18 column.   0—100% linear gradient in 35 min at 45°C
from solvent A (water with 0.1% trifluoroacetic acid) to solvent B
(25% water, 75% acetonitrile, 0.1% trifluoroacetic acid).   UV detection (280 nm) at 0.2 AUFS.   (From Ref. 8.)

Slowed-Down Motor Speed of Conventional Pumps

In 1979, Scott and Kucera [25] showed that a high-pressure mixing
system, consisting of two dual-reciprocating piston pumps (Waters
M6000) with the standard heads (200 µl/stroke) and a simple mixing
tee, could generate gradient, with high-pressure mixing at flows
from 20 to 100 µl/min.   They used a frequency generator to drive
the pumps at low motor speed for 1-mm-i.d. columns.   Later,
Waters supplied system with lower flow capabilities, first the model
721 gradient controller in 1982 and then the model 680 gradient controller that could reduce the flow of the M6000 pumps to 10 µl/min.
The recently introduced Waters 590 pump has built in a low flow
capability to 1 µl/min, thus potentially permitting nearly full
gradients with high-pressure mixing from two pumps to about 50 or

0
min
120

Fig. 8   Gradient separation with a conventional pumping system
(Perkin-Elmer) of crude essential oil extract at 80 µl/min (gradient
volume 9600 µl) with a 250 × 1 mm reverse-phase C18 column,
20—90% gradient in 120 min from A (water) to B (acetonitrile), UV
detection of 220 nm.   (From Ref. 32.)

100 µl/min.   Scott and Kucera's system yielded reproducibility of
retention times of less than 1% and peak areas of 2—8% RSD using
50—100% methanol gradients [25].   However, under conditions of a
wider gradient range and using buffers and ion pair reagents, the
tee mixer may not be adequate.   Schwartz et al. [7] found large
baseline oscillations due to inadequate mixing.

These composition oscillations can be very problematic in some cases. For example, with LC of large protein molecules, in which retention falls much more rapidly with %B (compared to small molecules, the small compositional variations in poorly mixed gradient eluents were shown to split single protein peaks into many "sawtooth" peaks [33].

Problems with baseline oscillation are especially evident when

1. Detection is with low-wavelength UV or is electrochemical.
2. Mixer design is poor.
3. Eluents differ greatly in physical properties (e.g., density, viscosity, or polarity) which give rise to poor mixing.
4. Short (low-volume), efficiently packed analytical LC columns are used.

Schwartz et al. [7] showed that good composition reproducibility can be obtained with a system similar to that of Scott and Kucera except for the addition of a low-volume (38-µl) dynamic mixer. As expected, with a high-pressure mixing system, gradient accuracy decreased at the gradient ends. Repetitive runs over a range of 10 to 90% acetonitrile gave adequate reproducibility in retention times (0.2−0.5% RSD) and peak areas (2−3% RSD) for a separation of a mixture of phenols (Fig. 9). Similar reproducibility data were found with a 350 × 1 mm column for the separation of proteins at 40 µl/min (gradient volume, 1600 µl) and di- and tripeptides at 80 µl/min (gradient volume, 2880 µl) [7].

Conventional pumps in which the motor speed was slowed down were used by Hayes et al. [34] for gradient LC coupled to a mass spectrometer (LC/MS). The system was shown to be usable over a wide range of flows and gradients, including high water contents. The gradient permitted relatively large 5-µl samples to be loaded on a 1-mm-diameter microbore column. Spray deposition onto a moving belt was used for the analysis of phenols in a coal gasifier condensate sample. The separation in Fig. 10 shows the UV chromatogram (top) compared to the MS chromatogram created by the summation of selected ions characteristic of the sample component. This application demonstrates that gradient on-line liquid chromatography/MS can be a useful complement to the gas chromatography/MS analysis of complex environmental samples. Forster et al. [35] used a similar microbore LC/MS gradient system in the analysis of test well samples from a landfill site.

With conventional pump systems with a slowed-down motor speed such as those described above, it is often possible to obtain various convex and concave as well as linear gradient profiles by simply setting the profiles on the gradient controller. This is useful because many "real-world" samples require other than linear gradient shapes.

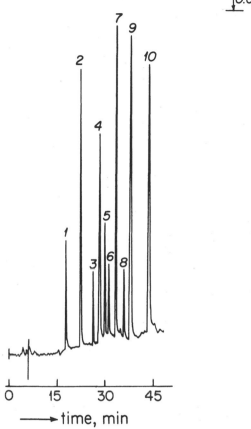

Fig. 9 Gradient separation with "slowed-down" conventional pumps (two Waters M6000's) of phenol standards at 40 μl/min (gradient volume 3200 μl) with a 350 × 1 mm 7.5-μm Zorbax-BP-ODS column, a 10—100% linear gradient in 80 min from A (95% water, 5% acetonitrile, 0.1% phosphoric acid) to B (50% water, 50% acetonitrile, 0.1% phosphoric acid), and 214 nm UV detection. Peaks represent 4—19 ng in 5 μl of (1) phenol, (2) 4-nitrophenol, (3) 2-chlorophenol, (4) 2,4-dinitrophenol, (5) 2-nitrophenol, (6) 2,4-dimethylphenol, (7) 4-chloro-3-methylphenol, (8) 2,4-dichlorophenol, (9) 4,6-dinitro-2-methylphenol, and (10) 2,4,6-trichlorophenol. (From Ref. 7.)

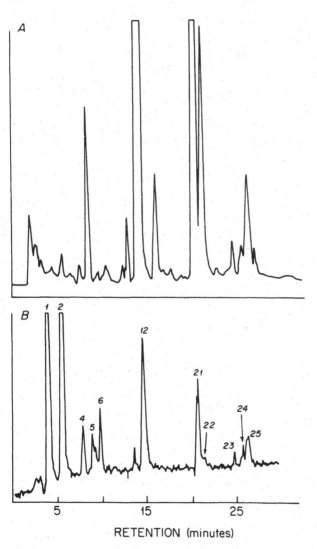

Fig. 10   Gradient separation with "slowed-down" conventional pumps
(two Waters M6000's) of a coal gasifier condensate sample at 100 µl/
min (gradient volume 4000 µl) with a 350 × 1 mm, 5-µm C18 column,
a 5–70% linear gradient in 40 min from A (95% water, 5% acetonitrile,
0.1% trifluoroacetic acid) to B (5% water, 95% acetonitrile, 0.1% trifluoroacetic acid).   (A) 280 nm UV detection at 0.05 AUFS and
(B) selected ion chromatogram using moving-belt LC/MS with positive
methane chemical ionization.   (From Ref. 34.)

In a different approach to slowing down conventional pumps, Powley et al. [17] constructed a microbore gradient system from two of the earlier Varian piston displacement pumps (model 8500), each with a syringe volume of 500 ml. Considerable baseline noise was found when no mixer was used at flows from 10 to 50 µl/min because of the low frequency (0—3 cycles/s) of the stepper motors driving the pistons. As did Schwartz et al. [7], the authors compared various static and dynamic mixers and examined parameters such as mixing chamber volume, gradient ramp rate, and flow. An 80-µl dynamic mixer gave the best results. A comparison with a standard gradient system under equivalent linear-velocity conditions showed that the "homemade" microbore system performed adequately; however, lower reproducibility in retention times was found with the microbore system.

Flow Splitting with Conventional Pumps for Submillimeter-I.D. Columns

Van der Wal and Yang [36,37] adapted a conventional low-pressure gradient system with a single-piston reciprocating pump (the Varian 5500) for flow splitting as shown in Fig. 11. By splitting the flow before the injector, the major part of the eluent is allowed to bypass the injector, column, and detector. This major flow passes a pressure transducer, a pulse damper, and a flow control device. The rest of the eluent flows through the injector, column, and detector. The authors found acceptable reproducibility for retention time (∿2% RSD) when using a 130 × 0.32 mm fused silica column packed with 3-µm particles. The precision with the instrument operating in the constant-pressure mode was found to be

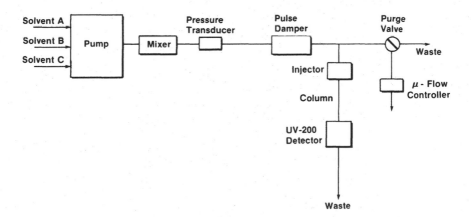

Fig. 11   Apparatus for flow splitting with the Varian 5000 LC for micro-HPLC. (From Ref. 37.)

strongly dependent on temperature effects. Varian offers an up-
grading kit for their conventional instrument (model 5000) to provide for flow splitting.

Flow-splitting with the Varian instrument was used for separations of polynuclear aromatics on a 660 × 0.25 mm column at 5 μl/
min with on-column fluorescence detection (gradient volume, 75 μl)
[36], underivatized amino acids by cation exchange on a 0.54-mm
column at 20 μl/min (gradient volume, 1200 μl), PTH amino acids on
a 0.54-mm column at 12 μl/min (gradient volume, 730 μl), and tri-
cyclic antidepressants on a 0.54-mm column (gradient volume, 360
μl) [37]. Einarsson et al. [38] applied a new primary and secondary
amino acid derivative (4-fluorenyl chloroformate) to the analysis of
up to 34 amino acids in urine in 90 min on a 500 × 0.226 mm fused
silica column (Fig. 12). Reproducibility of retention time was below
0.5% and of areas was below 7% for actual urine samples. They
used a new technique of "on-packing" detection in which the fluor-
esence excitation light was trained on the side of the packed fused
silica column (with the protective imide layer removed). Since
peaks are sharper before elution, this technique may eventually
provide better sensitivity than eluting the sample, as suggested by
Huber [39]. In addition, samples eluted in different levels of organic eluent may give a more constant fluorescence response in the
solid-phase environment of the packing, as suggested by a Varian
patent for a packed fluorescence detection cell.

Preformed Gradients from Conventional Pumps for
for Millimeter-I.D. Columns (Gradient Storage)

The large gradient delay volumes associated with conventional LC
systems have led to a number of workers to develop an approach in
which a preformed gradient is temporarily stored in a column, a
capillary, or the pump hydraulics prior to delivery to the column.

During work with fast LC, Katz and Scott [40] stored the
gradient generated by a Perkin-Elmer 3B pump in a 250 × 4.6 mm
column packed with 40-μm solid beads. The solid nonporous beads
make the gradient shape nearly independent of flow (see section on
Gradients from a Breakthrough Curve) [9]. Fast LC resembles
microbore LC in that both techniques are low-dispersion methods requiring minimal extracolumn dispersion. This gradient was displaced
at 5 ml/min with an isocratic flow in 20 s through a very short
analytical column (25 × 2.6 mm) packed with 3-μm C18 silica to separate 13 peaks. The 25−100% gradient was originally made and
stored over several minutes with the greater precision possible with
a lower flow.

Alfredson et al. [41] also described a gradient storage technique that employed a single-piston reciprocating Varian 5500 pump
A 1000-μl preformed gradient (20 min gradient at 50 μl/min) was

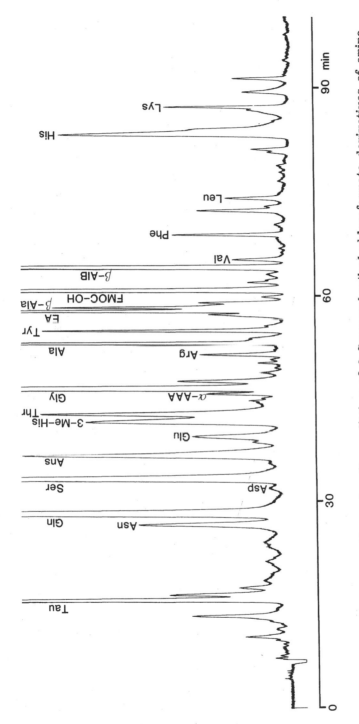

Fig. 12 Gradient separation using flow splitting of 9-fluorenylmethyl chloroformate derivatives of amino acids in urine at 2 μl/min with a 500 × 0.226 mm packed fused silica capillary column of 5-μm Spherisorb C8. 25–65% ternary gradient from 15 to 90 min (gradient volume 150 μl) from A (75% pH 4.08 acetate buffer, 25% acetonitrile) to B (35% pH 4.08 acetate buffer, 65% acetonitrile), and on-packing fluorescence detection (265 nm excitation, 315 nm emission). (From Ref. 38.)

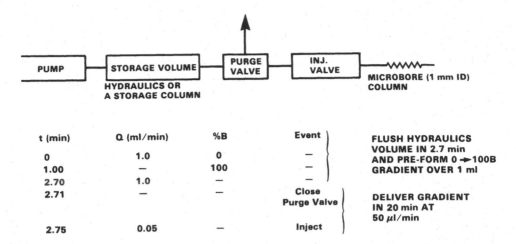

t (min)	Q (ml/min)	%B	Event	
0	1.0	0	—	FLUSH HYDRAULICS VOLUME IN 2.7 min AND PRE-FORM 0→100B GRADIENT OVER 1 ml
1.00	—	100	—	
2.70	1.0	—	—	
2.71	—	—	Close Purge Valve	DELIVER GRADIENT IN 20 min AT 50 µl/min
2.75	0.05	—	Inject	

Fig. 13   Diagram showing the apparatus for preformed gradient elution with the Varian 5000 liquid chromatograph.   Also shown are the time program for the flow (Q), the %B, and the purge valve operation for a gradient separation of the aromatic standards shown in Fig. 14. (From Ref. 40.)

stored in the pump hydraulics (2700 µl total volume) preceding the column (static mixer, pressure transducer, etc.).   Figure 13 shows a schematic diagram of the apparatus and the time/even program for the flow and %B necessary for separation of the aromatic standard shown in Fig. 14 [42].   Note that the gradient is nearly linear, although it is formed in only 11 strokes of the pump.

Using alkylphenone standards with the preformed gradient approach, Simpson and Schachterle [26] demonstrated that acceptable reproducibility of retention time could be achieved: 0.92—1.29% RSD over 22 runs (slightly worse than that found with the conventional system).

Examples of separations made with preformed gradients include PTH amino acids on a 150 × 1 mm column at 50 µl/min (gradient volume, 1900 µl) and a cough syrup formulation (quaiacol sulfonate, phenylephrine, dye, codeine phosphate, methyl paraben, and promethazine) on a 300 × 1 mm column at 50 µl/min (gradient volume, 1500 µl) [26].

Exponential Gradients Generated in Mixers

Gradient generation in mixing chambers was used as early as 1973 in work by Bombaugh [43].   Nice et al. [44] used short, narrow columns (75 × 2.1 mm) to generate the gradient within the 1.8-ml

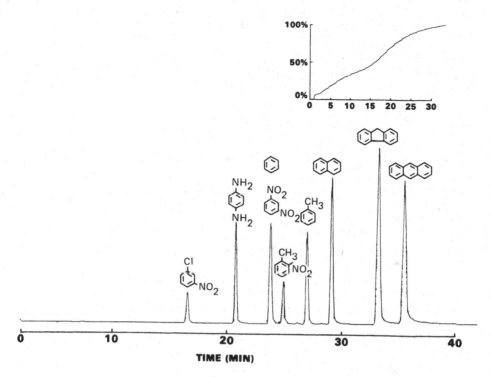

Fig. 14   Preformed gradient elution with a conventional pump
(Varian 5500) of aromatic standards at 50 µl/min on a C18 column.
0 to 100% gradient in 30 min from A (water) to B (acetonitrile) and
UV detection.   (From Ref. 26.)

dynamic mixer, using the principle of exponential dilution.   After
the mixer is flushed with initial eluent, the final eluent is pumped
through the mixer to exponentially dilute the initial eluent [as
described by Eq. (1)].   The slope of the gradient depends on
(a) the concentration (%B) of the initial eluent, (b) the concentration (%B) of the final eluent, (c) the volume of the mixer, and
(d) the flow rate.

Problems with dynamic mixers are that

1. Only gradients of one shape (convex) can be formed.
2. More linear gradients are obtained at the expense of the
    composition range of the gradient.
3. The onset of the gradient is abrupt.
4. The gradient steepness is not readily changed.

Despite these limitations, the method has been used for separations with the JASCO system for polynuclear aromatics (Fig. 15) [45] and with the Beckman-Altex system for proteins [44]. More versatile versions of this approach (i.e., the connected exponential dilution chamber) are discussed below.

Special Miniaturized Heads for Conventional Pumps

One solution to obtaining low flows is to miniaturize pumps. However, there are physical limitations to miniaturizing current designs. For example, check valves close by effects from the pressure differential, Bernoulli flow forces, flow friction, surface tension (wettability), and, in some cases, springs or mechanical actuators. Complex interactions of these effects make it difficult to make check valves smaller. Also, it is difficult to manufacture piston seals and other parts with acceptable tolerances when they are scaled down. This is a dilemma for designers. On the one hand, an "all-purpose" pump is desirable, but flow reproducibility and flow accuracy will be compromised at low flows. On the other hand, equipment redesigned for low flows can be used only for low flows.

Sjodahl et al. [18,46] used an LKB pump with a small displacement volume (36.5 µl) and miniaturized double check valves. Conventional pumps typically have much larger displacement volumes; e.g., the Beckman-Altex model 100 has 100 µl/stroke-head with two heads; the Beckman-Altex model 110A, 140 µl/stroke-head with a single rapid-refill head; and the Waters M6000, 200 µl/stroke-head with two heads. Examples of other microbore pumps with microbore heads having a low displacement volume are the Gilson model 302, 39.6 µl/stroke with a single rapid-refill head, and the Knauer model 64.00 with either 20 or 4 µl/stroke.

With the LKB low-pressure mixing system, Sjodahl et al. achieved reproducibility of retention times of 0.16—0.47% RSD and areas of 1.9—4.3% RSD for 50 to 100% gradients with phthalate standards at 50 µl/min (gradient volume, 1500 µl). With the same system, femtomole quantities of OPA amino acids were separated on a 1-mm-i.d., 3-µm columns [46]. The separation is shown in Fig. 16. The composition accuracy of this system was discussed in the section on Minimum Flow Limits with High-Pressure Mixing.

A number of companies (e.g., Beckman-Altex, Gilson, Hewlett-Packard, JASCO, Perkin-Elmer, Kontron, Kratos, Varian, and Waters/Millipore) upgraded previous conventional pumps to allow gradient elution with ca. 2-mm-i.d. columns and/or fast LC with short (<10 mm) columns and particles of 3 to 5 µm [48]. In some cases (Knauer, Gilson) different-size pump heads with different stroke volumes can be interchanged to allow LC with different-diameter columns (from microbore to preparative). An example of this approach is shown in Fig. 17. Here, a Gilson 302-5S pump equipped

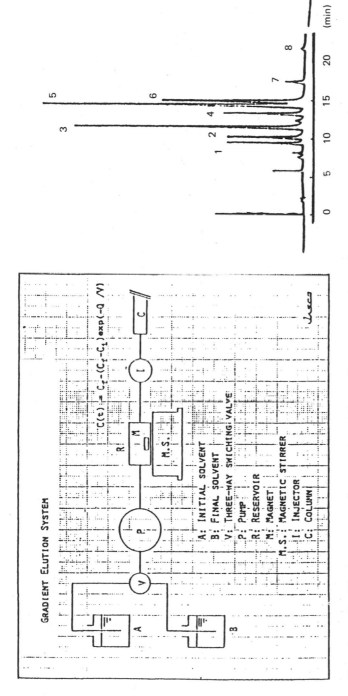

Fig. 15 (Left) Apparatus for exponential gradient generation with the JASCO liquid chromatograph. (Right) Exponential gradient generated in a 1700 μl mixing chamber with a conventional pump (JASCO) for polynuclear aromatics at 150 μl/min (gradient volume 3000 μl) with a 250 × 1.5 mm micro S-Finepak SIL C18 column. 50 to 100% exponential gradient in 20 min from A (water with 50% acetonitrile) to B (acetonitrile). UV detection of 254 nm. Peaks are (1) naphthalene, (2) 2-biphenyl, (3) fluorene, (4) anthracene, (5) pyrene, (6) triphenylene, (7) crysene, and (3) benzo[e]pyrene. (From Ref. 45.)

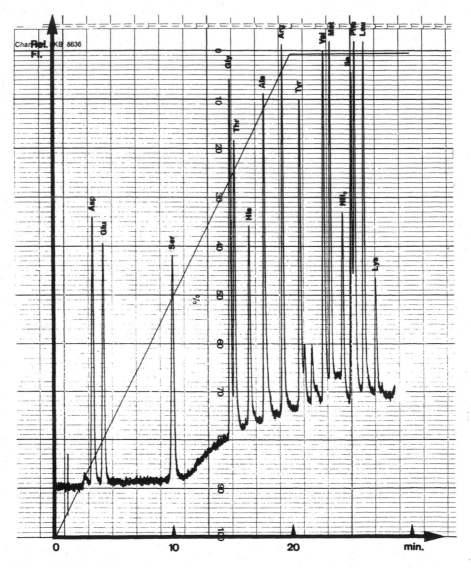

Fig. 16 Gradient separation with a miniaturized reciprocating piston pump based on low-pressure mixing (LKB model 2150 pump) of OPA amino acids at 80 µl/min (gradient volume 2000 µl) with a 250 × 1 mm, 3-µm Spherisorb ODS II column, a linear gradient in 25 min from A (water with 12.5 mM sodium phosphate, pH 7.2, 0.5% acetonitrile, 2.5% methanol) to B (water with 12.5 mM sodium phosphate, pH 7.2, 25% acetonitrile, 5% methanol) at 30°C, and fluorescence detection (330 nm excitation, 450 nm emission). Peaks represent 25 femtomole each of amino acids from a protein hydrolysate dissolved in 1 µl. (From Ref. 47.)

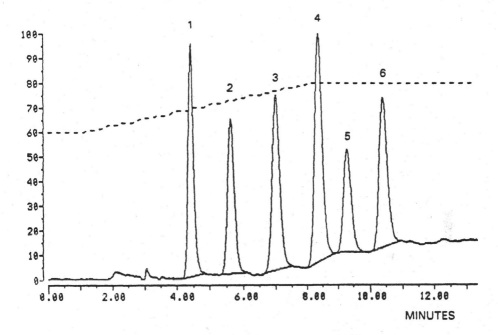

Fig. 17 Gradient separation using a conventional pump (Gilson 302-5S) with a microflow head at 50 µl/min (gradient volume 400 µl) with a 250 × 1 mm, 10-µm C18 column, a 60—80% linear gradient in about 8 min from A (water) to B (methanol), and 235 nm UV detection. Peaks represent 170 to 440 ng dissolved in 1 µl; (1) barbital, (2) phenobarbital, (3) butabarbital, (4) mephobarbital, (5) pentobarbital, and (6) secobarbital. (From Ref. 49.)

with a "microbore" head was used for the gradient separation (limited range) of barbiturate standards [49]. Although these systems perform well in most isocratic applications using microbore LC columns, insufficient precision and accuracy data have been published concerning performance under low-flow conditions with gradients.

INSTRUMENTATION SPECIFICALLY DESIGNED FOR GRADIENT MICROBORE LC

Some instrumentation has been designed specifically to perform microbore gradient work. This section deals mainly with some approaches to gradient elution with microbore columns which have been developed primarily in Japan. The last part covers syringe pumps for microbore LC which have been developed primarily in the United States.

Micro-HPLC Gradient Generators

Ishii and various co-workers have published five different approaches to generating gradients with 0.25—0.35 mm columns since first presenting their micro-HPLC work in 1977 [24].

Stored Step Gradients

This early micro-HPLC gradient method permitted step gradients to be delivered to the column. Ishii et al. [24] used a 1000 × 0.5 mm open tube for storing "steps" of increasing eluent strength. The steps were formed using a 50-µl glass syringe pump (usable to about 70 bars) to initially pull the steps of different solvents into a gradient storage tube. Subsequently, the solvents were displaced by the mechanically driven glass syringe pump. A similar approach was shown more recently by Novotny et al. [50] for a 3-hr high-resolution separation of benzoyl chloride-derivatized urinary steroids on a fused silica column (1000 mm × 0.25 mm) using a Shimadzu reciprocating piston pump (LC-5A) operated in the constant-pressure mode.

High-Back-Pressure LC Step Gradients

Developed by Takeuchi, Ishii, and co-workers [51], high-back-pressure LC used three glass syringe pumps to permit step gradient elution. Flow through the column was determined by a motor-driven constant-flow pump (a 100- or 250-µl glass syringe pump) "accepting" eluent after the detector. Flow to the column was supplied by two pumps, another constant-flow syringe pump (with water) and a constant-pressure syringe pump (with acetonitrile). The proportion of acetonitrile in the eluent was increased in steps by step increases in pressure (by adding weights) from the constant-pressure pump. Baseline stability was improved with a mixer made by filling the outlet arm of a tee connector with 15—30 µm porous particles. A unique application of this approach is in normal-phase separations using liquefied volatile gases (e.g., butane, isobutane, and propane) [51,52]. A step gradient separation of a cold medicine is shown in Fig. 18, and the very small column diameter, low flows, and small injection volumes used with micro-HPLC are indicated in the figure legend. The high-back-pressure approach potentially could be used with continuous gradients if the constant-pressure pump were programmed to change pressure continuously.

Gradients from a Single (Exponential) Mixer

A third micro-HPLC gradient approach from Takeuchi and Ishii used a miniaturized version (400 µl volume) of the previously discussed stirred (exponential) mixer (section on Exponential Gradients Generated in Mixers). Reproducibility of retention time was about 1%

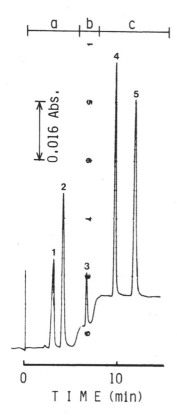

Fig. 18 High back-pressure step gradient separation of cold medicine at 4.2 µl/min (gradient volume 50 µl) with a 101 × 0.35 mm,
5-µm C18 column. 7.5% B (zone a), 15% B (zone b), and 30% B
(zone c) step gradient in about 12 min from A (water) to B
(acetonitrile) at 26°C, and 240 nm UV detection. Peaks represent
about 30 ng loaded in 0.02 ml: (1) barbital, (2) acetaminophen,
(3) caffeine, (4) phenacetin, and (5) *p*-chloroacetanilide. (From
Ref. 51.)

RSD. The low-flow capability (0.5 to 10 µl/min) of this approach
was shown by the separation of epoxy resin oligomers [53], coal
dust extracts, and dansyl-derivatized amino acids in Japanese sake
(Fig. 19) [54].

A similar gradient system designed for 1.5-mm-i.d. columns is
marketed in Japan by JASCO. In this case, a six-port valve is
used and the final eluent is delivered by a second pump (see Fig.
15) [55]. In an application with polynuclear aromatic hydrocarbons,

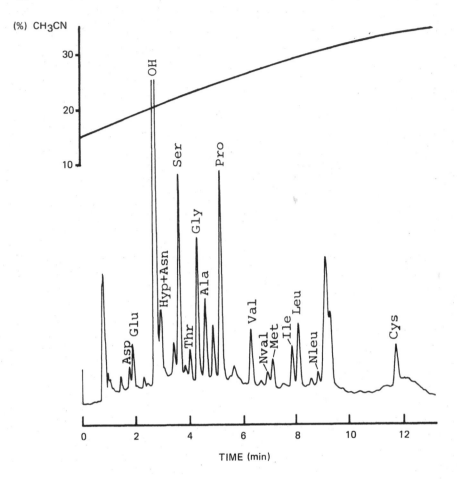

Fig. 19   Exponential dilution gradient separation of Japanese sake
at 10 µl/min with a 130 × 0.3 mm 3-µm Hypersil ODS column.
15—35% exponential gradient in 12 min from A (water with 130 mM
ammonium acetate) to B (acetonitrile), and 220 nm UV detection.
(From Ref. 54.)

this gradient system yielded retention time reproducibilities of better
than 0.9% RSD.

Karlsson and Novotny [56] showed a separation of polynuclear
aromatics using a similar small (84 µl) exponential mixer for very
low flows from 0.14 to 1.9 µl/min.   The large-volume (500 ml) older
piston displacement pump (Varian 8500) in the constant-pressure
mode was used for solvent delivery.   The typical nonlinearity of the
gradient was found to be partially offset by a continuous increase
in flow due to the increase in the less viscous strong eluent.   Typ-

ical reproducibility of retention times was from 0.97 to 1.52% RSD, with the error gradually increasing with retention time.

In the same paper [56], an interesting method for experimentally determining the volume, V, of this stirred mixer was shown. A version [Eq. (2)] of the mathematical expression describing %B versus time [Eq. (1)] was derived, in which F is flow, C_{in} is the concentration of the strong eluent, and C_t is the concentration at time t into the gradient. Plots of the left-hand side of this equation against time give the slope (F/V), which permits calculation of the mixer volume V.

$$\ln \frac{C_{in}}{C_{in} - C_t} = \left(\frac{F}{V}\right) t \tag{2}$$

Hirata and Nakata [57] developed a simple gradient system using an induction coil to stir the eluents in a 163-μl exponential mixer; valve switching was used to achieve the gradient. Reproducibility was <1.8% RSD for retention time and <2.6% RSD for areas at flows from 1 to 10 μl/min.

Gradients from Connected (Exponential) Mixing Chambers

A new method developed by Berry, Ishii, and Takeuchi [10,54,58] overcomes some of the problems mentioned earlier with the exponential mixer gradient generator. The single exponential mixer (described above) was replaced with a series of smaller mixers in series. The number of mixing chambers was varied from 1 to 12 without changing the total volume (598 μl) of this "connected exponential diluter." Since the entire device was made from a transparent high-pressure glass column (from Chrompack), a proper eluent mixing or the presence of interfering air bubbles could be observed to 200 bars. When a second chamber was added, the gradient slope changed dramatically from an exponential to a more linear profile. When using additional connected chambers, the linearity of the gradient improved only marginally (Fig. 20). This experimental observation was later theoretically confirmed by Ishii et al. [54]. Equation (3) expresses the gradient profile as a fraction of the gradient (fn) versus the run time (t), flow (F), and number of chambers (n) for a fixed total chamber volume (V) [54]. The summation is from m = 0 to m = n − 1. Ishii showed computer-generated gradient profiles using this equation for various numbers of chambers (n).

$$fn = 1 - \sum_{m=0}^{m=n-1} \frac{1}{m!} \left(\frac{nvt}{V}\right)^m e^{-nvt/V} \tag{3}$$

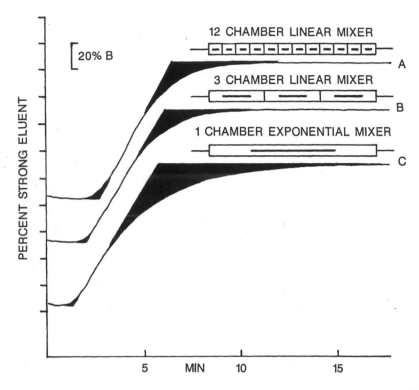

Fig. 20   Experimental relationship between the gradient profile and
the volume through the "linear generator" (run time, t, times flow,
v) for different numbers of chambers (n) for a fixed total chamber
volume (V, 598 μl).   Details of construction in Ref. 29.   (From
Ref. 10.)

In separations of epoxy resin oligomers with a two-chamber connected exponential mixer, the onset of the gradient was less steep,
giving improved initial resolution (Fig. 21).   The gradient achieved
the final compositions more rapidly, therefore shortening the analysis
time.   It can be seen that the central portion of the gradient is
almost linear.

In a modification of this linear generator, nine short (9-mm-long)
pins were used to separate a 100-mm-long mixing chamber into nine
mixing chambers [10].   The shafts of the pins act as the mixing
bar when the chamber is vibrated and the heads of the pins act to
segment the chamber.   Advantages of this design are that it can be
further miniaturized and that a near-linear gradient profile is ob-

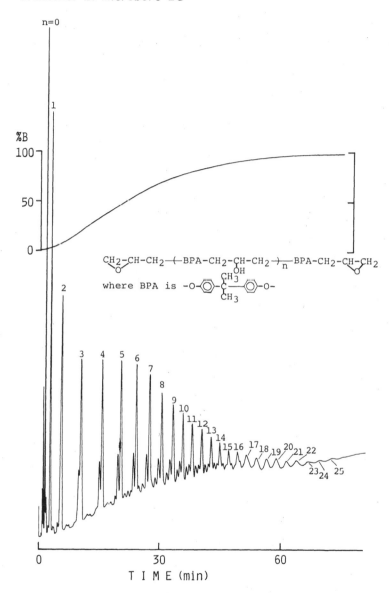

Fig. 21 Gradient separation with a "linear generator" of epoxy resin oligomers (Epicote 1004) at 10 µl/min on a 150 × 0.34 mm, 5-µm Develosil ODS-5 C18 column two chambers (130 µl each) were connected to form a gradient generator in 70 min gradient from A (water with 80% acetonitrile) to B (acetonitrile with 10% tetrahydrofuran); 225 nm UV detection. (From Ref. 54.)

tained when the mixer is used without vibration as a static mixer
with a large volume and high permeability. Hence, this design
could be an alternative to a bead-packed column.

Gradients from a Breakthrough Curve

Another method for micro-HPLC gradients developed in 1985 by
Berry, Takuichi, and Ishii [9] used the breakthrough curve produced when a valve is used to switch abruptly from weak to strong
eluent to form the gradient. The interface between eluents, when
passed through a small packed "gradient column" (30 × 0.21 mm),
spreads into a near-linear S-shaped gradient profile. The
automated system initially uses a timed high-pressure valve to vent
the gradient column, followed by low-pressure pump to quickly wash
the gradient column with water. Then a high-pressure pump rapid-

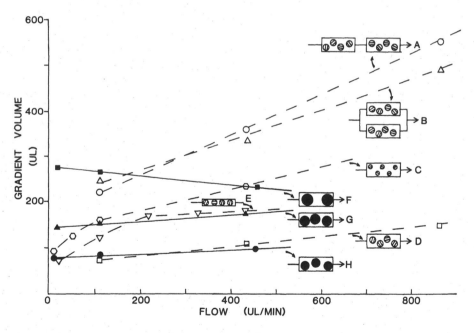

Fig. 22   Summary of gradient volumes versus (flush) flow rates
for gradient generator columns of different column dimensions, particle sizes, and particle porosities. Porous-particle generators
(dashed lines) were: A, two (30 × 4.6 mm) in series, 75—150 μm
dp; B, two (30 × 4.6 mm) in parallel, 75—150 μm; C, 30 × 4.6 mm,
15—30 μm dp; D, 30 × 4.6 mm, 75—150 μm dp; E, 30 × 3 mm, 75—150
μm dp. Solid-particle gradient generator columns (solid lines) were:
F, 30 × 4.6 mm, 1600 μm dp; G, 30 × 4.6 mm, 1200 μm dp; H, 30 ×
4.6 mm, 250—500 μm dp. (From Ref. 10.)

ly establishes a "flush flow" of acetonitrile until the beginning of the interface between the weak eluent (water) and strong eluent (acetonitrile), i.e., the breakthrough curve, just reaches the end of the gradient column. As the water/acetonitrile breakthrough curve starts to elute from the gradient column, it is connected by an automated valve to the micro-HPLC column; simultaneously the acetonitrile flow is reduced to rates compatible with micro-HPLC columns (5–40 µl/min). The ranges of gradient volumes possible with porous and nonporous particles in the gradient column are summarized in Fig. 22. If the gradient generator column is packed with nonporous particles (150–1600 µm), the volume of the gradient is nearly independent of the flush flow (Fig. 22, curves F–H). However, the gradient volume is increased by using larger solid particles in the gradient column (increased eddy diffusion).

If, instead of solid particles, the gradient column is packed with porous particles (75–150 µm), by varying the speed of the flush flow, the volume of the gradient can be varied dramatically from less than 220 µl to more than 550 µl (Fig. 23); volumes that are useful for micro-HPLC gradients (Fig. 22, curves A–E) [10]. This approach exploits the slow mass transfer of water out of the porous particles into the acetonitrile to change the slope of the gradient.

Advantages of the breakthrough curve method are that the system is automated, the slope of the gradient can be automatically changed by the flush flow rate, and only a single high-pressure pump is needed to generate gradients. In addition, the valve not

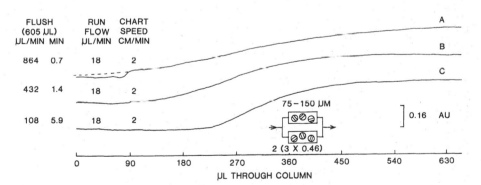

Fig. 23 Typical breakthrough gradients using two columns in series packed with large, 75–150 µm porous particles, showing the change in gradient volume with flush flows. For a flush volume of 1210 µl (to bring the gradient to the head of the column), achieved with flush flows ranging from 108–864 µl/min, the gradient volumes ranged from 220 to 554 µl. (From Ref. 10.)

only begins the gradient and eliminates the delay volume but also can be used to simultaneously inject large sample volumes. This approach might find its widest application with packed micro-HPLC columns (0.3–0.5 mm i.d.) or open-tubular columns, where very small submicroliter gradient volumes are required.

Syringe Pumps

Syringe pumps were among the first pumps developed for LC (e.g., by Nester-Faust, Perkin-Elmer, and Varian). Volumes of these pumps were very large, from 500 to 1000 ml. The current renewed interest in syringe pumps is illustrated by some recently introduced pumps for narrow-bore and micro-HPLC: LDC/Milton Roy (5 ml volume), ISCO (50 ml volume), and Brownlee Labs (two heads at 10 ml volume each). Only the last can be used directly in a gradient mode. For microbore LC, syringe pumps offer the advantages of

1. Freedom from pulsations (necessary with electrochemical, Fourier transform infrared, GC-type, high-sensitivity UV and mass spectroscopy detectors)
2. Capability of pumping at very low flows
3. Capability of pumping very viscous solutions such as glycerin [50]

The older generation, large-volume syringe pumps were difficult to use with gradient elution for several reasons. First, retention times were poorly reproducible as the syringes do not always begin the gradient with the same volume of eluent in each syringe. Eluent compressibility or the ability of the system to expand with pressure (compliance) results in the problem that the volume delivered by the pump per minute (i.e., a given linear piston movement) decreases as the volume in the pump increases or as back pressure increases [60,61]. Second, long periods of time are necessary to reach the final "steady-state" flow and pressure conditions if no special precautions are taken, such as prepressurizing the syringes. This steady-state time becomes larger with increases in (a) reservoir volume (and system compliance), (b) eluent compressibility, (c) final back pressure of the system, and (d) piston movement. Therefore, low-volume reservoirs are preferred.

The Brownlee Labs dual syringe pumping system is designed for gradient analysis with microbore columns. In this pump (and the LDC/Milton Roy isocratic pump) check valves are eliminated by using a microprocessor-controlled high-pressure valve to automatically refill each 10-ml pump cylinder at the end of every run. The final (steady-state) flow and pressure are quickly reached by having the piston speed up initially to prepressurize the cylinders to a target pressure, determined on previous runs, a principle used manually with early LC syringe pumps. The prepressurization and the automatic refill to precisely the same volume before each run greatly improve precision compared to that with earlier conventional syringe pumps. Schwartz and Brownlee [62] showed good reproducibility of retention times (0.2 to 0.5% RSD) for microbore gradient analyses.

As discussed above, most LC systems use dynamic or static mixers to eliminate reciprocating piston pump pulsations, which may cause baseline oscillation. The flow from syringe pumps is nearly pulse-free and the two eluents are delivered as parallel zones into the tubing; thus, only radial mixing (across the column) is required to homogeneously mix the eluent streams. With the Brownlee Labs pump, this is achieved with packed-bed mixers consisting of interchangeable cartridges filled with 100-μm nonporous glass spheres. The delay volume of these cartridges varies from 52 to 700 μl. Figure 24 demonstrates that with the static mixers smooth linear profiles at various flows and mixer sizes are obtained.

The utility of the Brownlee Labs microbore gradient pumps for

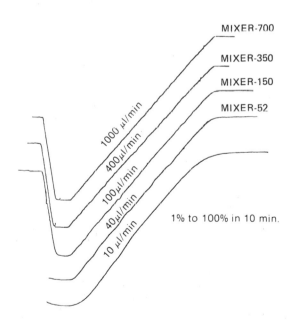

Fig. 24 Gradient profiles from 1 to 100% B with A (methanol) and B (methanol with the UV absorber acetone) for various flows with different static mixers using the Brownlee Labs Microgradient System. Slower flows give longer delay times, eliminated from the figure so that the gradients overlap for comparison. Details of the gradient static mixers are given in Ref. 62. (From Ref. 62.)

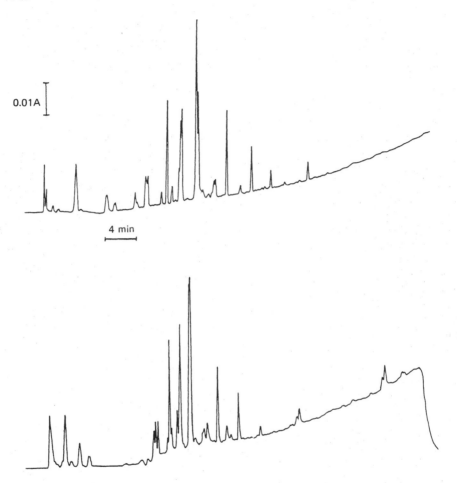

Fig. 25   Linear gradient separation with a dual-syringe pump
(Brownlee Labs model G) of peptides from a tryptic digest of the
protein apomyoglobin on a conventional 250 × 1 mm column at 1000 µl/
min (top) versus a micropore 250 × 1 mm column at 50 µl/ min (bottom).
In both cases the column material was Aquapore RP-300 and a linear
gradient was run from 10 to 60% B in about 45 min from A (water
with 0.1% trifluoroacetic acid) to B (40% water with 60% acetonitrile
and 0.1% trifluoroacetic acid).   220 nm UV detection (0.1 AUFS) of a
100 pmol injection of hydrolyzed protein.   (From Ref. 63.)

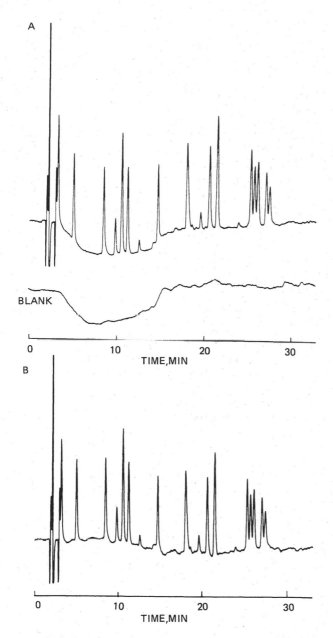

Fig. 26 Gradient separation with a dual-syringe pump (Brownlee Labs model G) of PTH aminoacids at 80 µl/min (gradient volume about 2400 µl) on a 250 × 1 mm 5-µm C18 column. Linear gradient in 30 min from A (95% 20 mM sodium acetate, 5% THF, pH 5.4) to B (acetonitrile). 254 nm UV detection (0.008 AUFS). Each peak represents 10 pmol of PTH amino acid mix. (From Ref. 64.)

protein and peptide isolation and structure determinations was demonstrated by Wilson et al. [63], who compared conventional (4.6 mm)
and smaller-bore columns (2.1 and 1.0 mm) in terms of detector
sensitivity. Figure 25 shows the chromatogram of a tryptic digest
of apomyoglobin with this system.

Schwartz and Brownled [64] compared the performance of static
and dynamic mixers under high-sensitivity conditions. A dynamic
mixer of 220 µl volume was required for the PTH amino acid separation shown in Fig. 26. The analysis of PTH amino acids represents
an important example of the utility of microbore chromatography
since in protein sequence analysis frequently only limited amounts
of sample are available. Advances in gas-phase sequence technology
have made it possible to obtain sequence information with amounts
of protein as low as 100 picomoles. The Brownlee Labs pump is incorporated in the Applied Biosystems model 120 PTH analyzer, which
can be used on-line with a gas-phase sequencer.

Several reports have described the advantages of micro-LC/MS
over conventional LC/MS [65]. With micro-LC/MS, the total effluent
can be introduced into the ionization chamber of the mass spectrometer. Lee and Henion [66] employed the Brownlee Labs pump for micro-
LC/MS with a direct liquid introduction (DLI) interface. Detection
limits in the low nanogram range routinely can be achieved, as demonstrated by the separation of pesticides (25 ng each) in Fig. 27.

Even with the new-generation syringe pumps, flow precision
and flow accuracy decrease at the extremely low flow rates (<5 µl/
min) required for micro- and open-tubular LC (see Table 1). However, by splitting the flow after injection, good precision and accuracy can be obtained in gradient analysis, as shown by McGuffin
and Zare [67]. An example of a gradient at 2 µl/min on a long
(1300 × 0.32 mm) packed column is shown in Fig. 28 [68]. Note
that in this case, on-column laser fluorescence was used for detection of 20 derivatized amino acids. The upper trace in Fig. 28 is
the chromatogram with UV detection. The system was also applied
to 4-bromomethyl-7-methoxycoumarin derivatives of saponified fatty
acids in sesame and peanut oils and of prostaglandins [69].

Fig. 27   Gradient separation with a dual syringe pump (Brownlee
Labs model G) of pesticide standards using micro-LC/MS at 40 µl/
min (gradient volume about 800 µl) on a 500 × 1 mm, Zorbax C18
column.   Linear gradient in about 20 min from A (45% water, 55%
acetonitrile) to B (acetonitrile).   Left trace: UV detection, 254 nm.
Right trace: total ion chromatogram.   Pesticides (from left to right)
are oxamyl, methomyl, monuron, carbofuran, carbaryl, prophan
prophan, and captan.   (From Ref. 65.)

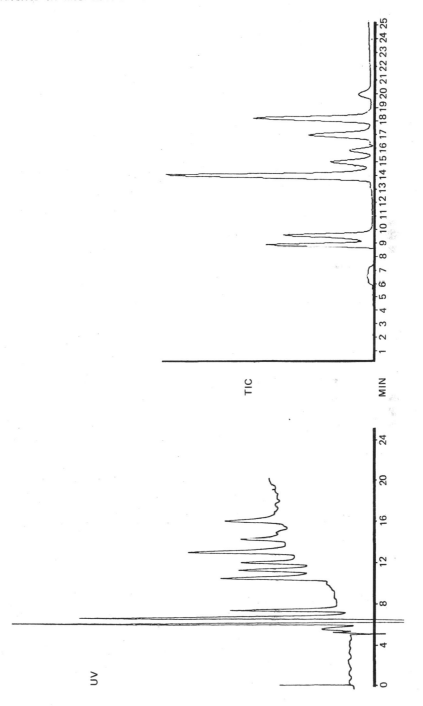

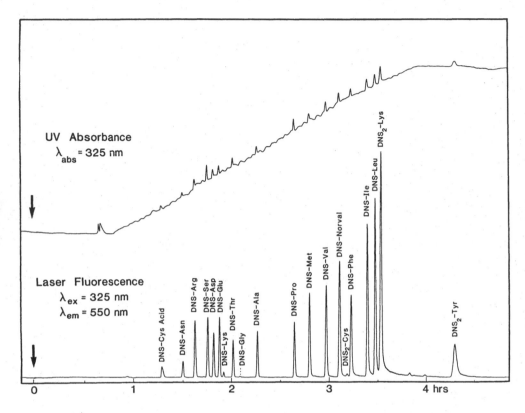

Fig. 28   Linear gradient separation with a dual-syringe pump
(Brownlee Labs model G) of dansyl amino acids with split flow at
2 µl/min (gradient volume 400 µl) on a 1300 × 0.32 mm fused silica
packed column (10-µm C18 Aquapore RP-300).   Linear gradient from
5 to 32% B in about 200 min from A (water with 50 mM formic acid,
60 mM acetic acid, pH 3.2) to B (2-propanol) at 55°C.   Top trace:
UV detection of 325 nm (0.02 AUFS).   Bottom trace: laser fluoro-
escence detection (325 nm excitation, 550 nm emission) of 50—100
pmol each in 0.130 µl of sample.   (From Ref. 69.)

Miniaturization of an on-line LC-electron capture (EC) detector
was achieved using the Brownlee Labs pump by Maris et al. [70].   As
in the above example of LC/MS, baseline stability was found to be high
and extracolumn peak spreading was comparable to that in conventional LC-EC.   The system was applied to the normal-phase gradient
elution of chloroanilines on an amino-bonded microbore column (Fig.
29) with detection limits of about 1 pg [70].

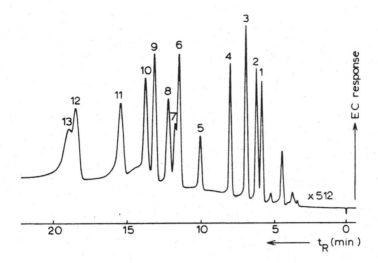

Fig. 29 Gradient separation with a dual-syringe pump (Brownlee Labs model G) of chloroanilines with electron-capture detection at 25 µl/min (gradient volume 250 µl) on a 200 × 0.7 mm, 5-µm Spherisorb-NH2 column. Linear gradient from 0 to 50% B in 10 min from A (hexane) to B (toluene). 0.2 to 100 ng in 60 nl of: (1) 2,6-dichloro-, (2) 2,4,6,-trichloro-, (3) 2,3,5,6-tetrachloro-, (4) pentachloro-, (5) 2,5-dichloro-, (6) 2,4,5-trichloro-, (7) 2,4-dichloro-, (8) 2,3-dichloro-, (9) 2,3,4,5-tetrachloro-, (10) 2,3,4-trichloro-, (11) 3,5-dichloro-, (12) 2,3,5-trichloro-, (12) 2,3,5-trichloro-, and (13) 3,4-dichloroaniline. (From Ref. 70.)

CONCLUSIONS

Although micro-HPLC is less than 10 years old, many approaches have been used to generate gradients since 1977, and even novel methods are still being developed (e.g., [9,10]). A number of commercially available instruments have been shown to generate reproducible gradients with 2- and 1-mm-diameter columns. They provide nonlinear gradient programming, from the control keyboard, as well as providing flows for conventional columns of 4–5 mm i.d. The split-flow gradient system has been applied successfully to submillimeter (0.2–0.5 mm) fused silica columns (as well as 1- to 2-mm columns). One instrument (Brownlee Labs) has been specifically designed for gradients with columns from about 0.5 to 2 mm i.d., and split flow with this instrument can be applied to even smaller-diameter columns. Other piston displacement isocratic microbore pumps (LDC and ISCO) might use the new methods of connected exponential mixers [10] or breakthrough curves [9] to provide gradients for submillimeter and even open-tubular systems (requiring gradient volumes of less than 1 µl).

Microbore gradients have been applied to the separation of many different materials, often as demonstration separations. However, recent applications capitalizing on the long, high-plate-count fused silica columns have been used with complex samples (coal tar derivatives) [34] or complex biochemical samples such as prostaglandins [70]. Interest in microbore LC with fused silica columns is likely to continue as new phenomena are discovered that are not found with conventional columns, e.g., on-column detection [69], reduced permeability for longer columns [71], and internal wall coating with polymers for improved plates [72]. The easy and safer use of microbore columns with supercritical fluids and gas chromatography-type detectors, as well as the advantages of new construction materials, will continue to stimulate this field. Gradients to simplify sample injection and to cover a wide sample polarity range, coupled with long (inexpensive) fused silica microbore columns to provide many plates, should show that gradient microbore LC can make an important contribution to separations of sample-limited and complex biological samples.

ACKNOWLEDGMENTS

The authors wish to thank Prof. M. Verzele and Dr. C. Dewaele of the University of Ghent, Belguim, for their support in developing this chapter and their helpful comments. V. Berry also wishes to thank Professors W. Mahaney, J. Engelke, S. Slater, and A. Stadthaus of Salem State College and Dr. Y. Fraticelli and Dr. R. Viavattini of Polaroid (Cambridge, Mass.) for their encouragement and support and Drs. D. Ishii and T. Takeuchi of the University of Nagoya, Japan, for their kind assistance.

NOTES ADDED IN PROOF

Since submitting this manuscript, Karel Slais has described two novel and very clever gradient generating devices. Their "tubular mixer" uses a strong eluent input feed into an outer tube [73]. Changing the location of several holes from the inner tube generates non-linear convex gradients of about 2000 microliters demonstrated with 0.7 mm i.d. columns at 60 µl/min [73]. Their "spiral-helical generator" gives a much more linear gradient, variable in volume from ca. 60—700 µl using 1 mm columns at 60 µl/min flows [74].

Gradient volume is varied by changing the fraction of the first solvent in a spiral, which produces various levels of column mixing to generate the gradient, depending on how far the weak eluent is pushed into the spiral during the loading step.

REFERENCES

1. H. E. Schwartz and V. V. Berry, *LC Magazine*, 3(2), 110 (1985).
2. L. R. Snyder, in *High Performance Liquid Chromatography: Advances and Perspectives*, Vol. 1 (Cs. Horvath, ed.), Academic Press, New York (1980).
3. P. Jandera and J. Chwacek, Gradient Elution in Column Liquid Chromatography, Theory and Practice, *J. Chromatogr. Lib.*, Vol. 31, Elsevier, New York (1985).
4. J. H. M. van den Berg, C. B. M. Didden, and R. S. Deelder, *Chromatographia 17*, 4 (1983).
5. J. H. M. van den Berg, H. W. M. Morsels, and R. J. M. Groenen, *Chromatographia 18*, 574 (1984).
6. M. A. Quarry, R. L. Grob, and L. R. Snyder, *J. Chromatogr. 285*, 1 (1984).
7. H. E. Schwartz, B. L. Karger, and P. Kucera, *Anal. Chem.* 55, 1752 (1983).
8. R. Cotter, *Am. Lab. 16*(1), 80 (1984).
9. V. V. Berry, T. Takeuchi, and D. Ishii, *J. High Resolut. Chromatogr. Chromatogr. Commun. 10*, 659 (1985).
10. V. V. Berry, T. Takeuchi, and D. Ishii, *J. Liq. Chromatogr.*, in press.
11. M. W. Dong, J. R. Gant, and P. A. Perrone, *LC Magazine 3*, 786 (1985).
12. J. W. Dolan and V. V. Berry, *LC Magazine 2*, 78 (1984).
13. S. M. McCown, B. E. Morrison, D. L. Southern, and D. Gartiez, Ninth International Symposium on Column Liquid Chromatography, Edinburgh, Paper PO 2.7, July 1 (1985).
14. P. Jandera and J. Churacek, *Advances in Chromatography*, Vol. 19 (J. C. Giddings et al., eds.), pp. 235–260, Dekker, New York (1981).
15. Brochure PB-18, Rainin Instrument Co., Woburn, Mass. (June 1984).
16. V. V. Berry, R. Reusch, M. Kaleja, and H. Knauer, Ninth International Symposium on Column Liquid Chromatography, Edinburgh, Paper PO 2.14, July 1 (1985).
17. C. R. Powley, W. A. Howard, and L. B. Rogers, *J. Chromatogr. 299*, 43 (1984).

18. J. Sjodahl, H. Lundin, R. Eriksson, and J. Ericson, *Chromatographia 16*, 325 (1982).
19. R. D. Conlon, L. S. Ettre, C. E. Schmid, and A. Schwartz, *Am. Lab. 14*(9), 104 (1982).
20. H. Joshua, *LC Magazine 3*, 442 (1985).
21. S. R. Bakalyar, Technical Note No. 5, Rheodyne, Inc., Cotati, Calif. (December 1983).
22. M. C. Harvey, S. D. Stearns, and J. P. Averette, *LC Magazine 3*, 434 (1985).
23. H. M. McNair, *J. Chromatogr. Sci. 20*, 537 (1984).
24. D. Ishii, K. Asai, K. Hibi, T. Jonokuchi, and M. Nagaya, *J. Chromatogr. 144*, 157 (1977).
25. R. P. W. Scott and P. Kucera, *J. Chromatogr. 185*, 27 (1979).
26. R. A. Simpson and S. D. Schachterle, Pittsburgh Conference and Exposition on Analytical Chemistry and Applied Spectroscopy, Abstract 650, Atlantic City, N.J. (1984).
27. R. Schuster, HPLC Application Note, No. 12-5954-0805, Hewlett-Packard, Palo Alto, Calif. (April 1984).
28. R. Schuster, HPLC Application Note, No. 12-5953-0087, Hewlett-Packard, Palo Alto, Calif. (December 1983).
29. R. Schuster, HPLC Application Note, No. 12-5954-0826, Hewlett-Packard, Palo Alto, Calif. (February 1985).
30. R. Schuster, HPLC Application Note, No. 12-5953-0085, Hewlett-Packard, Palo Alto, Calif. (July 1983).
31. E. C. Nice, B. Grego, and R. Simpson, *Biochem. Int. 11*(2), 187 (1985).
32. P. E. Deesen, *Essential Oil Analysis by Microbore Gradient LC*, Perkin-Elmer Liquid Chromatography Applications, No. FA-13, undated.
33. M. Rubinstein, D. Fischer, Z. Eshhar, and D. Novick, Eighth International Symposium on Column Liquid Chromatography, New York, Paper 4p-B2, May 20 (1984).
34. M. J. Hayes, H. E. Schwartz, P. Vouros, B. L. Karger, A. D. Thurston, and J. M. McGuire, *Anal. Chem. 56*, 1229 (1984).
35. M. G. Forster, O. Merejz, D. E. Games, M. S. Lant, and S. A. Westwood, *Biomed. Mass Spectrom. 10*, 338 (1983).
36. Sj. van der Wal and F. J. Yang, *J. Resolut. Chromatogr. Chromatogr. Commun. 6*, 216 (1983).
37. F. J. Yang, Sixth International Symposium on Capillary Chromatography, Riva del Garda, Italy, pp. 861–870 (1985).
38. S. Einarsson, S. Folestad, B. Josefsson, and S. Lagerkvist, Ninth International Symposium on Column Liquid Chromatography, Edinburgh, Paper PO 6.45, July 1 (1985).
39. J. F. K. Huber, Proceedings of Fifth Budapest Chromatography Symposium, Budapest, June 11 (1985).
40. E. Katz and R. P. W. Scott, *J. Chromatogr. 253*, 159 (1982).

41. T. Alfredson, S. Schachterle, B. Cunico, and L. Correia, 21st Eastern Analytical Symposium, New York, Abstract 211, November (1983).
42. R. Simpson, S. Schachterle, M. Tompkins, T. Schlabach, and S. Abbott, *Microbore Chromatography with the Varian Model 5000 and 5500 Liquid Chromatographs*, Varian Instruments, Walnut Creek, Calif., undated.
43. K. J. Bombaugh, *Am. Lab.* 5(5), 68 (1973).
44. E. G. Nice, C. J. Lloyd, and A. W. Burgess, *J. Chromatogr.* 296, 153 (1984).
45. *Semi Micro HPLC and Its Applications* JASCO International Co., Tokyo, Japan (February 1983).
46. J. Sjodahl, R. Eriksson, J. Ericson, and K. J. Dilley, *A New System for Microbore HPLC*, LKB Producter AB, Bromma, Sweden, undated.
47. H. Godel, T. Braser, P. Furst, and P. Foldi, *Amino Acid Analysis Through Microbore HPLC, 5—25 Femptamoles*, LKB Producter AB, Bromma, Sweden, undated.
48. *Chem. Eng. News 62*(12), 31 (1984).
49. F. Verillon, *Introduction to the Microbore Technique: Theoretical and Practical Aspects. The Use of a Microcomputer to Monitor Microbore HPLC*, Gibson Medical Electronics, Middleton, Wis. (January 1984).
50. M. Novotny, M. Alasandro, and M. Konishi, *Anal. Chem. 55*, 2375 (1983).
51. T. Takeuchi, Y. Watanabe, K. Matsuoka, and D. Ishii, *J. Chromatogr. 216*, 153 (1981).
52. T. Takeuchi and D. Ishii, *J. Chromatogr. 240*, 51 (1982).
53. T. Takeuchi and D. Ishii, *J. Chromatogr. 253*, 41 (1982).
54. D. Ishii, Y. Hashimoto, H. Asai, K. Watanabe, and T. Takeuchi, *J. High Resolut. Chromatogr. Chromatogr. Commun. 8*, 543 (1985).
55. M. Saito, A. Wada, K. Hibi, and M. Takahashi, *Ind. Res. Dev. 25*(4), 102 (1983).
56. K. E. Karlsson and M. Novotny, *J. High Resolut. Chromatogr. Chromatogr. Commun. 7*, 411 (1984).
57. Y. Hirata and F. Nakata, *J. Chromatogr. 294*, 357 (1984).
58. V. V. Berry, *Am. Lab. 17*(10), 33 (1985).
59. V. V. Berry and T. Waldron, Pittsburgh Conference and Exposition on Analytical Chemistry and Applied Spectroscopy, New Orleans (1985).
60. M. Martin and G. Guiochon, *J. Chromatogr. 151*, 267 (1978).
61. P. Achener, K. S. R. Abbott, and R. L. Stevenson, *J. Chromatogr. 130*, 29 (1977).
62. H. E. Schwartz and R. Brownlee, *Am. Lab. 16*(10), 43 (1984).

63. K. J. Wilson, D. R. Dupont, P. M. Yuan, M. W. Hunkapiller, and T. D. Schlabach, in *Modern Methods in Protein Chemistry* (J. L'Italien, ed.), Plenum, New York, in press.

64. H. E. Schwartz and R. G. Brownlee, *J. Chromatogr. Sci. 23*, 402 (1985).

65. J. D. Henion, in *Microcolumn High Performance Liquid Chromatography* (P. Kucera, ed.), pp. 260–300, Elsevier, New York (1984).

66. E. D. Lee and J. D. Henion, *J. Chromatogr. Sci. 23*, 253 (1985).

67. V. McGuffin, unpublished results.

68. V. McGuffin and R. Zare, in *Applications of Laser Fluorimetry to Microcolumn LC* (S. Ahuja, ed.), American Chemical Society, Washington, D.C. (1985).

69. V. L. McGuffin and R. N. Zare, *Appl. Spectrosc. 39*(5), 847 (1985).

70. F. A. Maris, A. Van der Vliet, R. B. Geerdink, and U. A. Th. Brinkman, *J. Chromatogr. 347*, 75 (1985).

71. V. V. Berry and H. G. Barth, *LC Magazine 3*(2), 178 (1985).

72. F. J. Yang, 24th Eastern Analytical Symposium, New York, Abstract 156, November 19 (1985).

73. K. Slais and V. Preussler, *J. High Resolut. Chromatogr. Chromatogr. Commun. 10*, 82 (1987).

74. K. Slais and R. W. Frei, *Anal. Chem., 59*, 376 (1987).

4

Advances in Optical Detectors for Micro-HPLC

EDWARD S. YEUNG / Iowa State University, Ames, Iowa

INTRODUCTION

The topic of detection in liquid chromatography (LC) has been extensively reviewed [1,2]. Indeed, quite a few other chapters in this monograph bear on this topic. Detection is a critical part of chromatography. In analytical applications, quantitation and identification are the ultimate goals and are conveniently achieved by putting appropriate detectors after the chromatographic column. Even in preparative applications, it is necessary to monitor the effluent to synchronize fraction collection.

In micro-HPLC and supercritical fluid chromatography (SFC) special requirements are imposed on the detectors. Frequently, standard detectors for LC (columns of 4.6 mm inside diameter) cannot easily be adapted for these situations. The most important factor is that of detector volume. It is generally accepted that the detector volume should be at most one-tenth of the peak volume in chromatography, to avoid introducing extracolumn broadening effects [3]. This means volumes below 1 µl for columns of 1 mm i.d. (commercial packed microbore) and below 1 µl for open-microtubular columns of 10 µm i.d. The volume must include all connections to the column, not simply the optical region. The shape of the

detector is also important. It turns out that the physical lengths
of the eluted chromatographic peaks are not very different from one
type of column to the next. They are all in the 10-cm range for
k' = 1, if the detector inside diameter is identical to that of the
corresponding column. This means that detector cell lengths up to
1 cm can be used. In fact, it is usually desirable to use the full
1-cm path length to maximize concentration detectability. Controlling
the shape of the detector is a challenge, since for ideal focusing
[4] one has

$$V = 16z^2\lambda \tag{1}$$

where V is the minimum volume, z is the path length, and λ is the
wavelength of light. For a 1-cm path at 300 nm excitation, V =
120 nl. To allow volumes of the order of 1 nl, the maximum path
length is 1 mm. Because of the small volume requirement, it is desirable to have multiple detection capabilities in the same optical
region if at all possible. This approach to increasing selectivity in
the measurement, by providing multidimensional data, is even superior to connecting multiple detectors in series, in which case
extracolumn band broadening can become important.

The operating conditions in micro-HPLC and SFC sometimes prevent simply scaling down the volume of conventional detectors for
use. The smaller columns mean that light intensity (per unit area)
can become much higher. Consequently, there is a larger heating
effect as a result of absorption by the analyte or by the solvent.
The thermal gradients and turbulence thus created may contribute
to inaccuracies in the measurement. In SFC, temperature and/or
pressure are used to program the density of the fluid during elution.
This means that density-sensitive detectors, such as refractive
index detectors, have limited utility in SFC. Measurements in the
fluid must be made under positive pressure even when the density
is kept constant. So, certain types of schemes that work well for
LC cannot be adapted for use in SFC. These include "cells" based
on a flowing droplet [5], free-falling jet [6], and sheath flow
streams [7].

Detection is not always degraded by the typical operating conditions of micro-HPLC and SFC. On the contrary, certain types of
detectors actually perform better under these conditions. The
smaller sizes of the columns allow smaller quantities of eluent to be
used. This then extends detector applications to those involving
high-purity or exotic solvents or exotic stationary phases. An interesting case is optically active solvents [8]. Other examples include mass spectroscopy [9], which is better coupled when the

mobile phase flow rates are reduced, infrared spectroscopy [10], where deuterated or fluorinated solvents can be substituted to reduce background absorption, and light scattering by particles [11], where the mobile phase can be more readily vaporized or separated. The large density change of a supercritical fluid as a function of temperature can potentially enhance thermo-optical measurements [12]. The small bore in the column and in the connecting tubing for micro-HPLC and for SFC further helps heat transfer. For a given volume of eluent, there is more surface area for contact with the outside. Temperature gradients are thus less likely to be a factor in, e.g., detector drift.

In what follows, we examine several optical detection methods in micro-HPLC and SFC. These are not intended to be comprehensive. Rather, a reasonable selection is discussed to illustrate the important features of detection. The reader is referred to the other chapters of this monograph to learn more about other detection schemes.

DETECTION SYSTEMS

Fourier Transform Infrared (FT-IR)

The ability to obtain structural information from IR spectra makes this method unique. Its importance as a detector in gas chromatography (GC) has already been well established [13]. Unlike the situation in GC, the eluent in LC and SFC in general absorbs strongly in the IR, obscuring spectra of the analyte.

One approach that avoids interference from solvent absorption is to remove the solvent [14]. The effluent from the column is deposited on KCl powder. The solvent can be evaporated, leaving the solute of interest. The KCl powder then passes into an FT-IR spectrometer in the diffuse reflectance mode; i.e., the backscattered radiation is monitored. The entire system can be automated [15], with 32 sample cups that collect the appropriate LC fraction as indicated by an ultraviolet (UV) detector in series with the collection system. This scheme is suitable for normal-phase LC, where the solvent is usually quite volatile, and for nonvolatile analytes at the submicrogram level.

Because there is less solvent to deal with in micro-HPLC, solvent removal is easier. A successful interface is the "buffer-memory" approach [16], where the column effluent is deposited on the KBr crystal continuously. The design of the transfer line is critical, so that band-broadening effects are minimized. The actual device is shown schematically in Fig. 1. The detectability is about 0.5 µg of material. A composite chromatogram with IR spectra is shown in Fig. 2, depicting the large spectral window that is available because

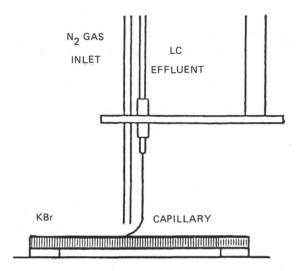

Fig. 1   Interfacing device for micro-LC/FT-IR via buffer-memory
technique.   (Reproduced with permission of K. Jinno.)

the solvent is removed.   Instead of using KBr, which is not compatible with common reverse-phase solvents, one can use a steel
wire screen [17] as the collecting medium.   Using this modification,
LC separations of almost any type are possible.   Problems still remain with micro-HPLC since the column capacities are not very different from the mass detectabilities of FT-IR, and only very limited
dynamic range is allowed.   To improve detection, one can combine
diffuse reflectance with micro-HPLC, since the former can be adapted
to small-area measurements [18].   For 4-chloronitrobenzene, one
expects detectabilities in the low nanogram range to be possible.

    If the solvent is not removed, an appropriate flow cell must be
used.   Typically, one must transmit a sufficient amount of light to
optimize the signal-to-noise ratio.   This means short path lengths
(e.g., 100 μm) even though sensitivity for the solute is also degraded.   Commercial cells of this type are available for conventional
LC, with detectabilities in the 0.5-μg range.   Significant gain is
possible if deuterated solvents are used [10], since they are com-
· pletely transparent in the C—H, O—H, and N—H stretch regions.
In practice, some hydrogen form is still present in commercial deuterated solvents.   Furthermore, exchange with water in the air
can gradually degrade these solvents.   Another group of solvents
that offer good windows for common IR vibration bands are the
halogenated hydrocarbons.   It is the small volumes of solvents
needed for micro-HPLC that make this a cost-effective approach.

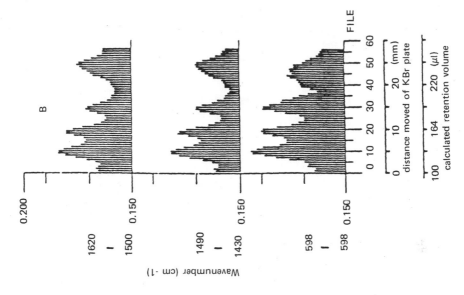

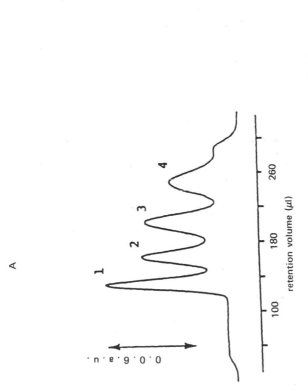

Fig. 2 (A) Size exclusion chromatogram of polystyrene standards measured with a UV detector at 254 nm. Micro-LC conditions: column, 1 mm i.d. × 22 cm PTFE tube packing, TSKGel 3000 H; mobile phase, THF 8 μl/min. Peaks: 1, MW 37,000, 2, MW 10,200, 3, MW 2800; 4, MW 500. (B) Fourier transform infrared chromatograms at various wave numbers. Detector, TGS; accumulation, ×64; resolution, 3 wave numbers. (From Ref. 15.)

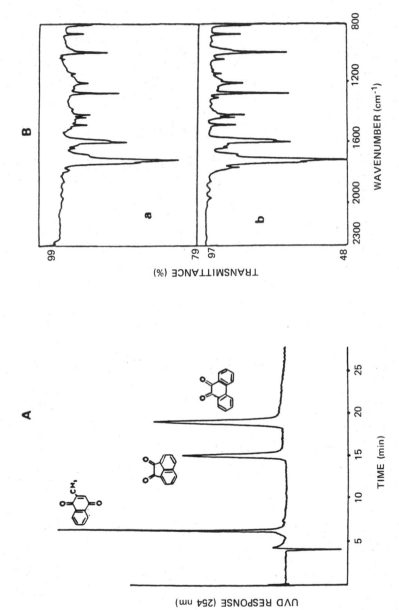

Fig. 3 (A) Chromatogram of quinone mixtures separated by SFC (5% methanol/CO$_2$, 200 atm, 50°C); 350 ng per component injected. (B) Infrared spectra of (a) second peak in chromatogram shown in (A) and (b) authentic acenaphthenequinone. (From Ref. 20.)

The previous techniques are readily adapted for detection in SFC. Similar flow cells can be used [19]. In fact, SFC fluids like CO_2 and Xe themselves provide large transport windows in the IR for detecting many analytes. Solvent elimination is also trivial once the fluid is brought back to atmospheric pressure [20]. Figure 3 shows impressive results with this last approach, which gives excellent spectra for 350-ng quantities injected.

Fluorescence

Since there are other chapters in this monograph that deal in detail with fluorescence detection, only a brief discussion is given here. Fluorescence intensity generally increases with excitation intensity. The first impression is that the higher the excitation intensity, the better the detectability. In practice, fluorescence detection is limited by the presence of background light, which includes various types of light scattering, luminescence from the flow cell walls, and emission from impurities in the solvent. All of these increase with excitation intensity to produce no net gain. When lasers are used, the situation is even worse since laser intensities are inherently much less stable than those of conventional light sources. It is the fluctuations in the background that limit detection. It is not surprising that the development of laser fluorometric detection is due to new designs of flow cells and optics that reduce stray light [21]. A simple approach is to use an optical fiber to collect fluorescence [22]. The limited collection angle of optical fibers allows discrimination against stray light originating from the cell walls. This system can readily be adapted for detection in micro-HPLC [4] or SFC. Figure 4 shows an arrangement that can be attached directly to the end of a microbore column. No loss of resolution due to extracolumn broadening is seen in Fig. 5. Subpicogram levels of molecules can be detected, all in a volume of about 10 nl. Even though fluorescence intensity increases with path length, high-power lasers can be used to provide sufficient signal levels that once again background (solvent fluorescence and solvent Raman) is the limiting factor. Thus, this is the easiest way to achieve small detector volumes. For SFC, Raman scattering is minimized when CO_2 or Xe is used, and fluorescence impurities are not expected to be present. An ideal situation exists for fluorescence detection.

With high light intensities, it is possible to observe fluorescence derived from two-photon excitation [23]. This process reaches a different set of electronic states, is polarization-dependent, is more restrictive, and can be reasonantly enhanced [21], so the information obtained is complementary to that from normal fluorescence. In the original demonstration [23], a continuous laser was used to provide detectabilities in the nanogram range. Since the two-photon

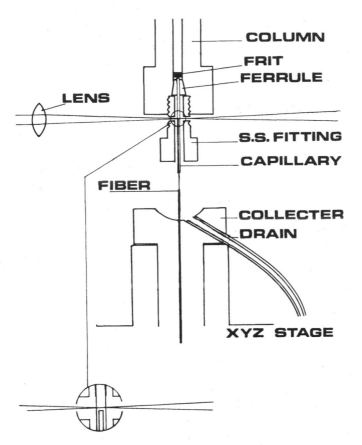

Fig. 4 Fiber-optic flow cell for microbore LC. An enlargement of
the optical region is shown at the lower left.

process is enhanced by incident intensity (peak power), we recently [24] redesigned the detector based on a copper vapor laser.
The optical region is shown in Fig. 6. The focused laser beam
(FLB) of a copper vapor laser passes through one Corning 3-71
sharp-cutoff filter and one Corning 4-96 wide-bandpass filter
mounted directly on the laser head. The purpose of the Corning
3-71 filter is to eliminate secondary laser radiation. The Corning
4-96 filter is used to block out the orange fluorescence created by
the 3-71 filter. No attempt was made to separate the 510/578 nm
lines from the laser. The radiation is then focused into the flow
cell by a lens of 8.5 cm focal length and 5 cm diameter. The flow
cell consists of a 1-mm-i.d. quartz tube (QT) directly attached to

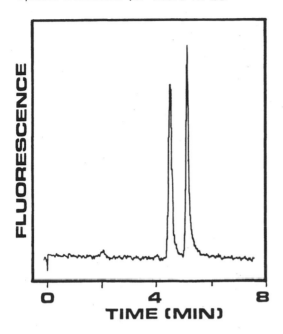

Fig. 5   Chromatogram from fluorescence detector shown in Fig. 4 for amino derivatives of NBD excited by 200 mW of 488 nm radiation. Column, 3-μm C_{18}, 2 mm i.d., 15 cm long; eluent, 35:65 H_2O: acetonitrile; flow, 0.17 ml/min.

the outlet (bottom) of the column (CS). A quartz rod (QR) is attached to the inside of the tube to limit the actual cell volume to approximately 0.8 μl. The purpose of the direct column attachment and the quartz rod is to lessen the effect of band spreading associated with HPLC detectors. The fluorescence leaving the flow cell passes through three UV bandpass filters: one Corning 7-54, one Corning 7-59, and one Corning 7-51. The fluorescence then enters a photomultiplier tube connected to a picoammeter. The instrument was set at 1×10^{-10} A to measure the fluorescence. The signal was adjusted with the damp control until a reasonably smooth baseline was obtained. This corresponds to a time constant of about 3 s.

A chromatogram of the test mixture is shown in Fig. 7. The injected quantities are 45 pg PPD and 3.0 pg PBD. The peaks are well separated and the manufacturer's specification for the column of 100,000 plates/m is achieved, showing negligible extracolumn band broadening. We find, however, that if the quartz rod in Fig. 6 is removed, substantial band broadening occurs, presumably

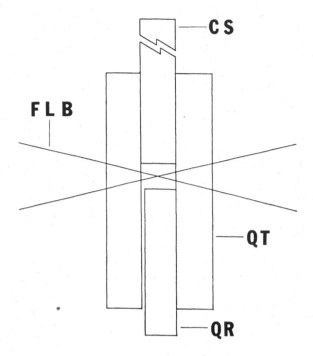

Fig. 6   A 0.8-µl flow cell for two-photon excited fluorescence.
CS, Microbore column outlet tube; FLB, focused region of copper
vapor laser; QT, quarts tube (1 mm i.d.); QR, quartz rod (0.8
mm o.d.).

because of turbulent mixing after the optical region. The quartz
rod can be replaced by an optical fiber [22] so that the visible
fluorescence can also be monitored to achieve two-dimensional detection in complex samples. Despite the filters, the background is
not at exact zero. So, laser intensity fluctuations (±1%) still contribute to baseline noise. The slow drift in the baseline also corresponds to a change in average laser power during observation.
Spikes in the baseline correspond to an occasional dust particle
crossing the laser beam and increasing the stray light. The PBD
peak is higher because of a better match in the absorption wavelength for this laser system. The detectability (signal/noise = 3)
of PBD is estimated from Fig. 7 to be 250 fg injected. Over the
range of 1×10^{-8} to 1×10^{-6} M PBD injected, the signal is found
to be linear after corrections for laser power variations.
     With the improved detectabilities here, two-photon excited
fluorescence is competitive with most other LC detectors in sensitivity. Normal fluorescence will still be slightly better in ideal

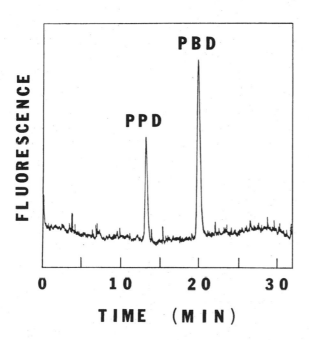

Fig. 7 Two-photon excited fluorescence chromatogram of test mixture. $[PPD] = 2 \times 10^{-7}$ M; $[PBD] = 1 \times 10^{-8}$ M; injection loop, 1 μl; flow rate, 30 μl/min; eluent, 90:10 acetonitrile/H_2O; column, 5-μm C_{18}, 1 mm i.d., 25 cm long.

cases where emission is far away from solvent Raman bands. The fact that there is so little background from the solvent here implies that solvent purity requirements can be relaxed. In fact, most biological matrices show very little two-photon fluorescence background, so the unique selectivity is valuable. The transverse excitation geometry and small optical volume make this detector suitable even for on-column measurements in capillary LC, supercritical fluid chromatography, or capillary zone electrophoresis. The 510 and 578 nm laser lines effectively excite bands around 255 and 289 nm, respectively, allowing coupling with a large group of chromophores. Future work in resonance enhancement and polarization selection can make this an even more powerful technique.

Optical Rotation

Optical activity is a particularly interesting molecular property, since it is generally associated with biological activity. Traditional polarimeters are not sensitive enough for LC detection. The instrument has recently been improved substantially with a new design based on a laser [25]. At present, rotations of the order of 1 microdegree (1 ng of a material injected with $[\alpha] = 100°$, a path length of 1 cm, and a peak volume of 10 μl) can be detected [26]. Cells as small as 1 μl have been used for microbore columns [8], the limit being 120 nl for a 1-cm cell. As biotechnology becomes more and more important, the need for this type of special sensitivity increases. In fact, it is exactly in this area, where sample sizes are limited, that micro-HPLC and SFC excel. It is possible that by moving toward shorter wavelengths in the measurement and/or by proper derivatization, another order of magnitude can be gained in sensitivity. Demonstrations have been successful for the quantitation of sugars in urine [27], cholesterol in blood [28], components in shale oil [29], extracts of coal [30], and enantiomers of amino acids [31].

The laser-based polarimeter can also be used in a nonselective mode very much like the refractive index (RI) detector, which is limited by its poor detectability and the relatively large volumes in commercial units. Recently, the concept of indirect polarimetry was demonstrated by us [8,32] as a universal LC detector. If an optically active solvent is used for LC, there will be a large constant background rotation in the absence of any analyte. This background can be compensated for by physically rotating the second polarizer (analyzer) so that once again a low light intensity reaches the phototube. The polarimeter still functions as before, but now a zero (baseline) signal is observed when the optically active solvent is present. When an optically inactive analyte elutes from the column, it displaces an equal amount of the solvent in the optical region. So fewer solvent molecules will be present in the optical region, and a decrease in optical rotation will result. This then registers as a negative signal in the polarimeter, i.e., opposite in sign to the rotation of the solvent. It can be seen that unless the analyte has exactly the same specific rotation as the solvent, a signal will be observed. Figure 8 shows a chromatogram obtained by indirect polarimetry for analytes that are not optically active. The detectability of this system is 4 ng of injected material. Below this concentration there is a problem with thermal noise, analogous to that of the RI detector.

The success of indirect polarimetry is a direct result of the low solvent consumption in microbore LC. Natural products like limonene are available in a variety of elution strengths. In fact, the cost of running a chromatogram such as that in Fig. 8 is about the same as the cost of using UV-grade acetonitrile in analytical scale (4.6 mm i.d.) LC. Purity of the solvent is not critical in indirect polarimetry, since one simply loses a bit of sensitivity. The small volume and the good detectability may eventually allow indirect polarimetry to replace RI detection for many LC applications.

Figure 8 shows clearly that the signal in the polarimeter depends on the difference in specific rotations of the analyte and the eluent and on the concentration of the analyte. This means that if we run the same sample through the same chromatographic system, once with the optically active solvent and once with the corresponding racemic mixture, two different sets of responses will be obtained. In fact, since the only two unknowns in the system are the concentration and the specific rotation of the analyte, these can be solved from the set of two simultaneous equations. This is the idea of quantitation without standards and thus without identification [33]; since the polarimeter has the largest dynamic reserve [34] of all common detectors, it gives the best results [8].

The concept of indirect polarimetry can be taken one step further. If an analyte absorbs light in the optical region, heat is generated. If an optically active solvent is used, the heating effect produces an *expansion* in the liquid and a *decrease* in optical rotation is detected because of the decrease in number density in the optical path. Reference [32] shows that one can use the same experimental setup for universal detection and for absorption detection simply by changing the laser power. Peaks that do not change with laser power correspond to indirect polarimetry, and peaks that vary with laser power are due to absorption processes. At 488 nm, N-methyl-o-nitroaniline has a molar absorptivity of 112 liter/mol•cm. The detectability in this scheme with 32 mW of radiation is 7 ng. At 458 nm, the molar absorptivity is 1040 liter/mol•cm, and [α] becomes larger. With 73 mW of radiation, the detectability is 36 pg of injected material. This corresponds to 1.8 ×

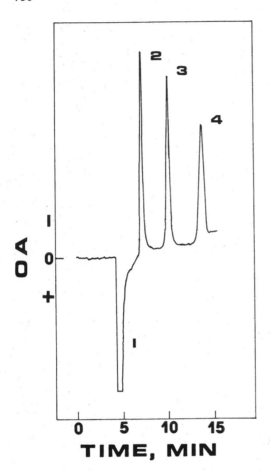

Fig. 8 Detection by indirect polarimetry; eluent, 95:5 (R)-(+)-limonene/isooctane; flow rate, 24 μl/min; column, 25 cm × 1 mm i.d. 10-μm silica. (1) Injection peak; (2) dioctyl phthalate; (3) dibutyl phthalate; (4) diethyl phthalate. (From Ref. 32.)

10^{-6} au (peak volume 28 µl, S/N = 3). Compared to other photo-thermal techniques, the absorption detectability is about the same. The limit seems to be that of solvent absorption. The advantage of polarimetry is that the bulk heating effect is probed, so beam quality and beam alignment create fewer problems.

It is interesting that the same instrumentation can be used in four different modes depending on the operation, as a chiral detector, as a universal detector, as an absorption detector, and as a means of quantitation without standards. The system still needs engineering to become a "black box" for the routine laboratory. But at least the cost of assembly is within the acceptable price range of other LC detectors.

Circular Dichroism (CD)

Circular dichroism also probes optical activity in molecules. However, this is the difference in absorption of left circularly polarized light (LCPL) and right circularly polarized light (RCPL) and is closely related to normal absorption. This then allows the "local" chirality to be interrogated—i.e., the presence of both asymmetry and an absorbing chromophore. Commercial CD spectrometers are not sensitive enough for the low concentration levels in LC. Since the signal corresponds to the difference between two large intensities, stability of the light source is the key. This is emphasized by the fact that CD is a very small (10^{-3} to 10^{-4}) effect on top of normal absorption. It turns out that one needs a moderate intensity ($\sim$10 mW) to stay above the shot noise limit, so a laser is required. To overcome the stability problem in laser sources, one can incorporate modulation at high frequencies. Noise often varies as the reciprocal of the frequency. For example, at 1 kHz the stability of ion lasers is about 10^{-3}. This improves to 10^{-5} at 100 kHz and 10^{-6} at 1 MHz. The last figure corresponds to an absorbance detectability of 10^{-6} au.

The detection system [35] can be seen in Fig. 9. The 488 nm light from the argon ion laser is sent through the center of a lens of 33 cm focal length. The light from the lens is directed to an electro-optic modulator and through a Fresnel rhomb prism, comes to a focus in the detection cell, and finally diverges to a larger diameter at the detector, where the laser power measured was typically near 20 mW. The signal from the detector is sent to a high-frequency lock-in amplifier and the output is sent to a chart recorder. The electro-optic modulator functions via a modulation driver operated at 500 kHz, which in turn is synchronized with the lock-in amplifier via a wave generator. By modulating the Pockels cell appropriately, on the first half-cycle of the modulation frequency RCPL is produced, and on the second half-cycle LCPL is

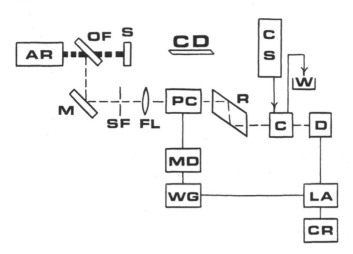

Fig. 9 Circular dichroism experimental configuration for conventional HPLC: AR, argon ion (488 nm) continuous-wave laser; OF,
optical flat; S, beam stop; M, mirror; SF, pinhole spatial filter;
FL, 33-cm focal length lens; PC, Pockels cell; MD, modulation
driver; WG, waveform generator; R, rhomb prism; C, detection
cell; D, photodetector; LA, lock-in amplifer; CR, chart recorder;
CS, chromatography system; W, waste.

produced. An optimum noise/signal ratio of 1.75×10^{-6} (500 kHz,
1-s time constant) was obtained with the CD system. The differential absorbance detectability is $\Delta A = 1.5 \times 10^{-6}$ AU. A chromatogram obtained in this system for an optically active inorganic complex, $(+)\text{-Co(en)}_3^{3+}$, eluted from a microbore column is shown in
Fig. 10. The cell used is 1 cm long and 2.6 µl in volume, which is
adequate for this separation. Using a lens of shorter focal length
and more careful alignment, a 1-µl detector volume [8] is certainly
feasible. The detectability here is 5.6 ng of injected material,
corresponding to an absorbance LOD of 5×10^{-6}. This is not as
good as that demonstrated for conventional columns [35] because of
the poorer optical quality of the smaller cell used.
    Circular dichroism is a difficult measurement to make. When the
intensities of RCPL and LCPL are not identical, a false signal is
produced by the background absorption [35]. So, the system must
be checked with highly absorbing but optically inactive species.
Useful detectability is obtained in Fig. 10. To improve detection,
one can use fluorescence-detected CD, i.e., monitoring the difference in fluorescence intensity when the LC effluent is excited by
RCPL and LCPL alternately. Particularly for microcolumns, the

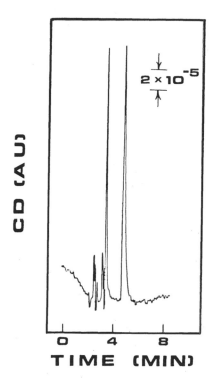

2×10^{-5}

Fig. 10   Microbore HPLC-CD detection of (+)-Co(en)$_3^{3+}$ at 2.5 ×
10^{-3} M.   Conditions: 5-µm C$_{18}$, 250 × 1 mm column, 20:80 aceton-
itrile/water, 0.5 µl injected, 40 µl/min, 1-s time constant, 500 kHz
modulation frequency, 1-cm path length cell.   Retention time at peak
is 4.8 min.   Solvent disturbance is before 4.0 min.   (From Ref. 35.)

same gain in detectability for fluorescence versus absorption can be
realized.   Also, the small-volume advantage can be maximized.   However, the false signal must be carefully isolated by proper adjustment or by using optically inactive standards.

Thermal Lens Spectrometry

Thermal lens is one of several thermal optical methods for absorption
measurements [36].   When light is absorbed, heat is eventually produced in the system.   The temperature of the eluent increases,
leading to a refractive index change.   This is analogous to the
density change for absorption detection using polarimetry.   When
the excitation is due to a laser beam with a Gaussian cross-sectional
intensity distribution, the center part of the optical region will be

heated up more than the sides. This typically results in an RI distribution that resembles a diverging lens. A related effect is photothermal deflection [37], which relies on the RI gradient to deflect a second laser beam. A signal is then generated in a position sensor some distance away. It is also possible to cross two laser beams at the detector to create an interference fringe pattern. Absorption then leads to lines of high and low RI alternately, following the interference pattern. A third laser beam can then probe this region by diffraction. This is known as photothermal diffraction [38]. All of these are particularly suitable for micro-HPLC because of the small optical volumes. In fact, the smaller volume (area) allows concentrating the laser beams to produce a larger temperature change at the same input energy. A detectability of 7×10^{-4} au was demonstrated in a volume of 0.17 μl [38] using photothermal diffraction, and a detectability of 1×10^{-6} was reported for photothermal deflection [37].

There have been many publications on thermal lens detection in LC, because the effect is very easy to observe with moderate laser intensities and moderate absorptions. The experimental arrangement calls for a flow cell placed beyond the focal point of the laser beam and an aperture at far field in front of a photodiode. Without the aperture, we have basically a normal transmission type of absorption measurement. With the aperture, we limit the spatial region through which the beam can pass. As the beam diverges because of the thermal lens in the cell, a *smaller* fraction of the total beam is able to pass the aperture, so *more* light is lost to the detector compared to normal absorption. This enhancement essentially gives a larger signal at the detector than simple transmission. In fact, for a given absorbance, the lens increases in strength with laser power. Enhancements of several hundred have been observed [12], and absorption detectability is improved. However, because an intensity is being measured, some of the gain is negated by the intensity instability of the laser. We note that the better commercial UV absorption detectors for LC can achieve limits of detection of 1×10^{-4} au at S/N = 3 [39]. Continuous lasers with sufficient power for thermal lens spectrometry have inherent stabilities of 1%. This is about two orders of magnitude worse than well-regulated conventional light sources. Power stabilization circuits can reduce fluctuations to 0.1%. It is possible to use a reference photodiode to correct for intensity changes [40], but electronic normalization circuits can produce compensation only to the 0.1% level. The time dependence of a thermal lens [41] can be used to reduce certain types of noise, with some sacrifice in the time response because of the computation procedure. Alternatively, a second laser with better intensity stability (e.g., an HeNe laser) can be used to probe the thermal lens generated by a more powerful, but less stable, pump laser [42,43]. The main difficulty is the alignment and the matching of the beams in the probe region, especially when small sample volumes are of interest. Even then, HeNe lasers are stable only to the 0.1% level at typical operating frequencies. It is fair to conclude that thermal lens spectrometry has not achieved its full potential for detecting small absorptions because of the above problems.

The "ideal" thermal lens spectrometer, especially one suitable for LC, seems to be one that has the following features. First, it should be based on a single laser [40,41,44—48] with sufficient output power to cause a definite increase in signal. A second laser only complicates the alignment without necessarily bringing real reductions in noise. Second, since intensity is measured, one must be able to achieve an intensity stability of 1×10^{-4} or better; i.e., the baseline must be stable to that extent. Otherwise, the enhancement factor [36] in the thermal lens effect will not be fully utilized. Third, since the thermal lens takes time to develop to its full strength [49] and time to relax [36], the maximum signal is obtained when continuous excitation (vs. chopped excitation) is employed. Furthermore, since a high chopping frequency degrades the thermal lens signal [47] and a very low chopping frequency is not compatible with LC, the steady-state thermal lens seems to be the best approach. Fourth, only one beam shall pass through the sample cell. This avoids any critical alignment procedures or instabilities in alignment, especially in small-volume applications. Fifth, it is preferable, although not essential, to base detection on a lock-in amplifier, which conveniently provides a differential measurement with a narrow bandwidth.

The optical arrangement that satisfies most of these requirements [50] is shown in Fig. 11. An argon ion laser operated between 50 and 500 mW at 514.5 nm is sent into an acousto-optical modulator driven by a square wave from a signal generator. About 60% of the beam is modulated into the first-order Bragg component at 150 kHz. A lens of 15 cm focal length is placed immediately after the Bragg cell. This serves to focus the laser beam into the microcell and to further separate the deflected beams from the Bragg cell by forming a virtual image. The flow cell for the LC effluent is derived from an earlier design [8], except that a pie-shaped wedge is cut on one side to allow passage of the positive orders of the deflections by the Bragg cell. The cell inside diameter is 0.61 mm, giving a volume of 2.9 μl. A home-made aperture 1.2 mm in diameter is located 31 cm from the lens to analyze the thermal lens signal. To reduce the intensity of light on the photodiode, a glass optical flat is used at 45° to send about 8% of the laser light to the photodiode. The active area of the photodiode is large enough (1 cm^2) to receive both the zeroth-order beam and the first-order beam. Output from the photodiode is sent directly to a lock-in amplifier.

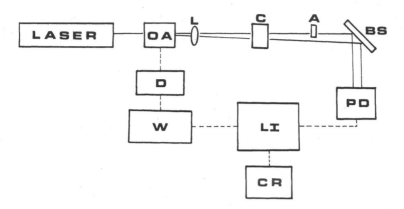

Fig. 11   Experimental arrangement for high-frequency modulation in
thermal lens measurements: LASER, Ar ion laser; OA, Bragg cell;
L, lens; C, chromatographic flow cell; A, aperture; BS, beam
splitter; PD, photodiode; W, square-wave generator; LI, lock-in
amplifier. (From Ref. 50.)

A typical thermal lens chromatogram is shown in Fig. 12. At a
power level of 90 mW, we injected a solution containing 1.5×10^{-7} M
benzopurpurin (BPP) and 3.9×10^{-7} M methyl red dye. The detectabilities were found to be 3.0 and 14 pg, respectively. The absorbance detectability, as judged from the BPP peak, is 4×10^{-6} au.
The system is presently limited by the pointing stability of the laser
and by solvent absorption, which limit all thermal lens experiments.

Three-Dimensional Detection

We stated earlier the advantages of obtaining multiple information
from the same optical volume in micro-HPLC and SFC. The three
most common detection methods in LC are absorption, refractive
index, and fluorescence. It is interesting to combine the three in
one detector, particularly if the path length can be of the order of
1 cm to preserve concentration detectability. Recently [51], we
demonstrated an optical cell in which the light enters the side of
the flow stream and yet propagates nearly parallel to the flow stream
before reaching the opposite window and exiting. This behavior is
accomplished by controlling the incidence angle of the light so that
it strikes the window-liquid interface at an angle just smaller than
the critical angle, thus giving a large angle of refraction. Light
incident on an optical interface at an angle just smaller than the
critical angle is very sensitive to changes in the refractive index
of the media making up the interface, and this has been recognized

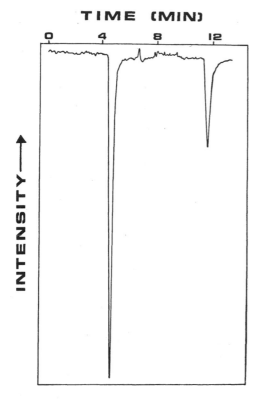

Fig. 12   Thermal lens chromatogram for 7.5×10^{-7} M BPP and 1.95×10^{-6} M methyl red. Injection loop, 1 µl; column, 25 cm × 1 mm i.d. 5-µm C_{18}; eluent, 90:10 acetonitrile:H_2O; intensity, 69 mW. (From Ref. 50.)

as a sensitive means of measuring changes in RI. Since it is the change in the refractive indices at the interface which causes the change in light intensity, this mode of detection has the potential for making a very small volume refractometer. A cell for RI detection in LC based on transmission near the critical angle is, in fact, commercially available. However, it has a volume too large for microbore LC. Furthermore, the transmitted light is detected by backscattering off a second surface. The resulting collection efficiency is low and the discrimination against the reflection from the first surface is poor. We used a design based on coupling out the transmitted beam by a second prism, symmetrically placed. A laser source ensures a small volume and a high beam intensity. Laser flicker noise is minimized by modulation and a reference flow cell.

The path length of the optical region is 1 cm with a 1-μl volume. This cell further allows the measurement of absorption and flourescence in the same volume, with path lengths optimized to maintain good concentration detectability. This combination allows one to obtain information in three dimensions with the minimum cell volume.

The behavior of light at a dielectric interface is described by both Snell's law of refraction and Fresnel's laws of reflection. Fresnel's laws describe the reflection and transmittance of light at an interface for the two types of linearly polarized light. The polarizations are p when the electric vector of light is parallel to the incident plane made by the normal and the incident light ray, and s when the electric vector is perpendicular to the incident plane. So,

$$R_s = \left| \frac{\cos\Theta - \sqrt{n^2 - \sin\Theta^2}}{n\cos\Theta + \sqrt{n^2 - \sin^2\Theta}} \right|^2 \tag{2}$$

$$R_p = \left| \frac{-n^2\cos\Theta + \sqrt{n^2 - \sin^2\Theta}}{n^2\cos\Theta + \sqrt{n^2 - \sin^2\Theta}} \right|^2 \tag{3}$$

where R_s and R_p are the reflectivities for the s and p polarized light, respectively, n is the ratio n_2/n_1, and Θ is the incidence angle.

The experimental arrangement for detecting both RI and absorbance change is shown in Fig. 13. An Ar ion laser is modulated between the two flow cells by a Bragg cell, which is driven by a signal generator at 100 kHz. A lens of 50 cm focal length focuses the beam into the optical cell. An optical flat is used to balance the intensities of the two beams reaching the detector, based on the variation in natural reflection with angle and with polarization. After passing through the cell, light is detected by a photodiode. The output from the diode is sent into a lock-in amplifier. The lock-in amplifier converts the modulated signal to a DC signal with a 3-s time constant, which is in turn sent to a strip chart recorder. The 100-kHz high-frequency modulation allowed us to improve the intensity stability of the laser from a noise-to-signal ratio of 5 × 10^{-3} to 2 × 10^{-5}. With the cell adjusted near the critical angle (allowing 10% of the light to be transmitted) we see an increase in noise, making the ratio 2 × 10^{-4}. This increase in noise is caused by the beam being partially clipped by the optical beam aperture of the glass-liquid interface. To detect fluorescence, a photomultiplier tube operated at 500 V by a power supply was placed above the optical path of the transmitted beams and as close to the cell as

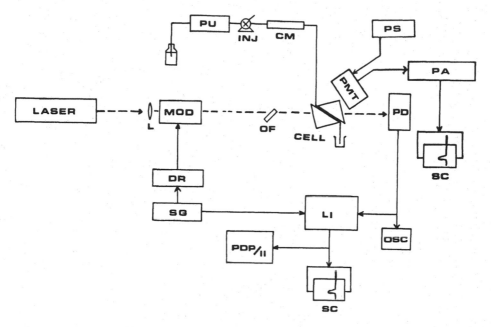

Fig. 13 Three dimensional detector and chromatographic system:
PU, pump; INJ, injecter; CM, microbore column; L, lens; MOD,
Bragg cell modulator; OF, optical flat; CELL, optical cell; PD,
photodiode; DR, driver; SG, signal generator; LI, lock-in amplifier;
PDP/11, minicomputer; SC, strip chart recorder; OSC, oscilloscope;
PS, power supply; LASER, argon ion laser; PMT, photomultiplier
tube; PA, picoammeter.

possible. Three 540 nm line filters and one colored-glass filter
were used to block the excitation light and reduce the background
emission from the prisms. The output from the phototube was sent
into a picoammeter with current suppression and with a 3-s time
constant. The output from the picoammeter was sent to a strip
chart recorder for data collection. The cell consists of a Teflon
gasket squeezed in between two high-quality right-angle prisms.
The Teflon tape defines the flow channels for both the reference
and sample chambers. The inlet and outlet tubes are made of stainless steel tubing of 1/16 in. outside diameter and are filed down to
a wedge at the cell end to minimize dead volume. Epoxy is used to
seal the tubes to the prisms, and braces hold the tubes rigidly and
minimize the danger of breaking the seal by accidental bumping.
The prisms are pressed together by additional braces placed at the

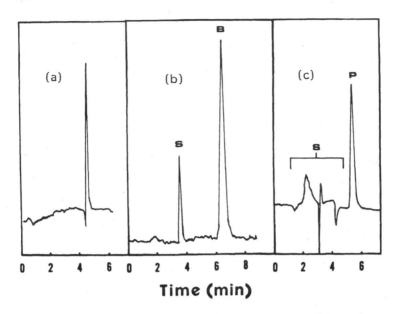

Time (min)

Fig. 14   (a) Refractive index chromatogram of benzene; peak represents 370 ng of benzene.   (b) Absorbance chromatogram of bromocresol green: S, solvent peak; B, 3.5 ng of bromocresol green. (c) Fluorescence chromatogram of derivatized product of propylamine and NBD-Cl: S, reproducible injection disturbance, P, 125 pg of fluorescent propylamine derivative.

apex of each.   This homemade assembly is mounted in a rotation stage with a resolution of 10^{-3} degree.

Figure 14a shows a chromatogram of benzene being eluted by the acetonitrile eluent at a flow rate of 30 µl/min.   To obtain this peak, 0.5 µl of 5×10^{-4} (v/v) benzene in acetonitrile was injected. Taking into account the dilution to a peak volume of 3 µl, this peak shows a detectability of 2.0×10^{-7} RI unit (S/N = 3).   This is comparable to the detectability of commercial RI detectors.   The detectable mass of injected benzene, however, is only 6 ng, which shows the advantage of using microbore columns for increased mass detectability.   The chromatogram also shows the high number of plates possible with a microbore column, 78,000 plate/m in this case with k' = 1.2.

Figure 14b shows a chromatogram of bromocresol green being eluted by an eluent of 3:1 methanol in an aqueous buffer of 1 mM citric acid at pH 7.4 at a flow rate of 30 µl/min.   In the chromatogram the first peak is a solvent peak detected as an RI change. The second peak represents a 0.5-µl injection of 9.9×10^{-6} M bromocresol green, showing an absorbance detectability (633 nm) of 2.1×10^{-4} AU. This compares well with the 2.2×10^{-4} AU detectability value one can calculate from the flicker noise of $N/S = 1.7 \times 10^{-4}$. The mass detectability corresponds to 48 pg of injected bromocresol green. It should be pointed out that a peak detected by absorbance will decrease the intensity of the probe beam, whereas a peak due to ΔRI will increase the intensity of the probe beam for an increase in RI. Since the LC eluent can be chosen to have a small RI, one can ensure that ΔRI is always positive. Thus, one can distinguish an RI peak from an absorbance peak.

The test molecule for fluorescence was a compound formed by derivatization of propylamine by 7-chloro-4-nitrobenzo-2-oxa-1,3-diazole (NBD-Cl). Figure 14c is a chromatogram of propylamine-NBD being eluted by 3:1 methanol in water at a flow rate of 33 μl/min. The peaks before the fluorescence peak are reproducible injection disturbances, which are also seen when the methanol solvent is injected. The injected solution was a 1.2×10^{-6} M solution of propylamine-NBD with a ninefold excess of NBD-Cl. The unreacted NBD-Cl has a very small fluorescence quantum yield and elutes after the propylamine-NBD peak. The analyte peak shows a detectability of 0.8 pg.

SUMMARY

It is clear that optical detectors, especially those that are based on lasers, have made substantial progress in applications involving microbore LC. Since the techniques should be readily transferable to other micro-HPLC and SFC systems, future expansion into those areas is almost a certainty. The ability of many of the examples of detectors given here to function in small volumes will make them indispensable tools, despite the nonroutine nature of their operation. Naturally, with further engineering and developments in those areas, they can become standard detection methods to support the growth of micro-HPLC and SFC.

ACKNOWLEDGMENTS

The author thanks the many co-workers in his group who have contributed to various parts of this work, particularly M. J. Sepaniak, W. D. Pfeffer, S. D. Woodruff, J. C. Kuo, D. R. Bobbitt, B. H. Reitsma, R. E. Synovec, K. J. Skogerboe, and S. A. Wilson. He also thanks K. Jinno for preprints of his work. The Ames Laboratory is operated for the U.S. Department of Energy by Iowa State

University under Contract No. W-7405-eng-82. This work was supported by the Office of Basic Energy Sciences.

REFERENCES

1. R. P. W. Scott, *Liquid Chromatography Detectors*, Elsevier, Amsterdam (1977).
2. T. M. Vickrey, ed., *Liquid Chromatography Detectors*, Marcel Dekker, New York (1983).
3. J. H. Knox and M. T. Gilbert, *J. Chromatogr. 186*, 405 (1979).
4. E. S. Yeung, in *Microcolumn Separations* (M. V. Novotny and D. Ishii, eds.), p. 135, Elsevier, Amsterdam (1985).
5. G. J. Diebold and R. N. Zare, *Science 196*, 1439 (1977).
6. S. Folestad, L. Johnson, B. Josefsson, and B. Galle, *Anal. Chem. 54*, 925 (1982).
7. L. W. Hershberger, J. B. Callis, and G. D. Christian, *Anal. Chem. 51*, 1444 (1979).
8. D. R. Bobbitt and E. S. Yeung, *Anal. Chem. 56*, 1577 (1984).
9. J. Henion, in *Microcolumn Separations* (M. V. Novotny and D. Ishii, eds.), p. 243, Elsevier, Amsterdam (1985).
10. K. Jinno and C. Fujimoto, *J. Liq. Chromatogr. 7*, 2059 (1984).
11. A. Stolyhwo, H. Colin, and G. Guiochon, *J. Chromatogr. 265*, 1 (1983).
12. J. M. Harris and N. J. Dovichi, *Anal. Chem. 52*, 695A (1980).
13. P. R. Griffiths, J. A. deHaseth, and L. V. Azarraga, *Anal. Chem. 55*, 1361A (1983).
14. D. Kuehl and P. R. Griffiths, *J. Chromatogr. Sci. 17*, 471 (1979).
15. D. Kuehl and P. R. Griffiths, *Anal. Chem. 52*, 1394 (1980).
16. K. Jinno, C. Fujimoto, and D. Ishii, *J. Chromatogr. 239*, 625 (1982).
17. C. Fujimoto, T. Oosuka, and K. Jinno, *Anal. Chim. Acta 178*, 159 (1985).
18. C. C. Conroy, P. R. Griffiths, and K. Jinno, *Appl. Anal. Chem. 57*, 822 (1985).
19. K. Shafer and P. R. Griffiths, *Anal. Chem. 55*, 1939 (1983).
20. K. H. Shafer, S. L. Pentoney, and P. R. Griffiths, *J. High Resolut. Chromatogr. Chromatogr. Commun. 7*, 707 (1984).
21. E. S. Yeung and M. J. Sepaniak, *Anal. Chem. 52*, 1465A (1980).
22. M. J. Sepaniak and E. S. Yeung, *J. Chromatogr. 190*, 377 (1980).
23. M. J. Sepaniak and E. S. Yeung, *Anal. Chem. 49*, 1554 (1977).

24. W. D. Pfeffer and E. S. Yeung, *Anal. Chem. 58*, 2103 (1986).
25. E. S. Yeung, L. E. Steenhoek, S. D. Woodruff, and J. C. Kuo, *Anal. Chem. 52*, 1399 (1980).
26. D. R. Bobbitt and E. S. Yeung, *Appl. Spectrosc. 40*, 407 (1986).
27. J. C. Kuo and E. S. Yeung, *J. Chromatogr. 223*, 321 (1981).
28. J. C. Kuo and E. S. Yeung, *J. Chromatogr. 229*, 293 (1982).
29. J. C. Kuo and E. S. Yeung, *J. Chromatogr. 253*, 199 (1982).
30. D. R. Bobbitt, B. H. Reitsma, A. Rougvie, E. S. Yeung, T. Aida, Y. Chen, B. F. Smith, T. G. Squires, and C. G. Venier, *Fuel 64*, 114 (1985).
31. B. H. Reitsma and E. S. Yeung, *J. Chromatogr. 362*, 353 (1986).
32. D. R. Bobbitt and E. S. Yeung, *Anal. Chem. 57*, 271 (1985).
33. R. E. Synovec and E. S. Yeung, *Anal. Chem. 54*, 1599 (1983).
34. S. I. Mho and E. S. Yeung, *Anal. Chem. 57*, 2253 (1985).
35. R. E. Synovec and E. S. Yeung, *Anal. Chem. 57*, 2606 (1985).
36. D. A. Cremers and R. A. Keller, *Appl. Opt. 21*, 1654 (1982).
37. A. C. Boccara, D. Fournier, W. Jackson, and N. M. Amer, *Optics Lett. 5*, 377 (1980).
38. M. J. Pelletier, H. R. Thorsheim, and J. M. Harris, *Anal. Chem. 54*, 239 (1982).
39. S. A. Wilson and E. S. Yeung, *Anal. Chim. Acta 157*, 53 (1984).
40. N. J. Dovichi and J. M. Harris, *Anal. Chem. 51*, 728 (1979).
41. N. J. Dovichi and J. M. Harris, *Anal. Chem. 53*, 106 (1981).
42. R. L. Swofford and J. A. Morrell, *J. Appl. Phys. 49*, 3667 (1978).
43. J. P. Haushalter and M. D. Morris, *Appl. Spectrosc. 34*, 445 (1980).
44. Y. Yang, *Anal. Chem. 56*, 2336 (1984).
45. Y. Yang and R. E. Hairrell, *Anal. Chem. 56*, 3002 (1984).
46. T.-K. J. Pang and M. D. Morris, *Anal. Chem. 56*, 1467 (1984).
47. T.-K. J. Pang and M. D. Morris, *Appl. Spectrosc. 39*, 90 (1985).
48. J. P. Gordon, R. C. C. Leite, R. S. Moore, S. P. S. Porto, and J. R. Whinney, *J. Appl. Phys. 36*, 3 (1965).
49. R. O. Carman and P. L. Kelley, *Appl. Phys. Lett. 12*, 241 (1968).
50. K. J. Skogerboe and E. S. Yeung, *Anal. Chem. 58*, 1014 (1986).
51. S. A. Wilson and E. S. Yeung, *Anal. Chem. 57*, 2611 (1985).

5

Principles and Applications of Photodiode Array Fluorescence Detection in Microcolumn LC

JENNIFER C. GLUCKMAN* and MILOS V. NOVOTNY / Indiana
University, Bloomington, Indiana

INTRODUCTION

Capillary column liquid chromatography can provide the resolution
needed to separate the complex mixtures of large, polar, and thermally labile compounds which are frequently found in technologically
and biologically important samples [1–3]. However, it is often desirable to know the identity of these substances, especially since
increasing chromatographic efficiencies and detection sensitivities
may reveal a multitude of previously unsuspected substituents in
many of these samples. A detection technique providing detailed
on-line structural information for the eluting compounds would be
valuable for the identification of these species.

    Among the several detector types which have been miniaturized,
fluorescence emission appears the most promising candidate for such
a detector. Not only is it an inherently sensitive and potentially
selective technique, but also fluorescence spectra can frequently
yield valuable information on the structure of an unknown compound.

*Current affiliation: Pfizer, Inc., Groton, Connecticut.

Indeed, many complex mixtures, such as those derived from fossil fuel materials or physiological fluids, contain numerous isomeric or structurally similar compounds, which may often coelute within a single chromatographic peak [1,4,5] and for which suitable standards are rarely available. Thus, optimal imaging detection, which readily permits spectral subtraction, is particularly attractive as an initial step toward the identification of these substances. The complexity of this problem will, no doubt, require an integrated analytical approach including, for example, retention time data, ultraviolet (UV), Fourier transform infrared, nuclear magnetic resonance and mass spectral information, data on elemental composition, etc.

As described in this chapter, a miniaturized fluorescence imaging detector has been constructed and applied to the analysis of a variety of complex mixtures [6,7]. This detector, which can utilize either broadband or single-wavelength laser excitation, incorporates a rapid-scanning intensified photodiode array (IPDA) and a miniaturized fiber-optic flow cell to achieve high detection sensitivities and good optical resolution without sacrificing chromatographic peak shape. In addition, design flexibilities allow both its future use with capillary supercritical fluid chromatography (SFC) and the potential for acquiring excitation, as well as emission, spectra.

Background Information

Optical multichannel imaging spectroscopy [8–10] has recently undergone many technological improvements [11] that have allowed both the silicon vidicon tube [12–15] and the linear photodiode array (PDA) [11,16–20] to be successfully employed as detectors in conventional-scale HPLC. Historically, vidicon tubes have traditionally been used for fluorometric detection [15,21], while the more recent linear photodiode arrays, or their intensified counterparts, have been utilized primarily as UV-visible absorbance detectors [11,18–20,22]. Indeed, many such detectors are now commercially available.

It is interesting, however, that although fluorescence-based detection is currently the most sensitive method available to modern HPLC, linear photodiode arrays have not yet been tested in this capacity, despite their reduced lag times and increased response linearity compared to vidicon tubes. Sensitive static fluorescence measurements, on the other hand, have been achieved with the PDA [23,24].

To date, there has been very little effort to utilize miniaturized optical multichannel imaging detectors for high-resolution chromatography. Thus far, the only reported work is that of Takeuchi and Ishii [20], who detected constituents from a commercial cold medicine using a PDA absorbance detector with a 50 nl detection volume. Lee and co-workers [25] have taken an alternative approach in the construction of an "on-the-fly" fluorescence scanning detector, which has been utilized in capillary column SFC. Two important limitations of photodiode array devices that may account for their scarcity in microcolumn separations are their lower sensitivity and higher noise compared to photomultiplier tubes. These characteristics often necessitate both long integration times and spectral averaging, leading to large detection time constants which are incompatible with the high efficiencies of capillary column chromatography. Recent developments in array technology have, however, led to rapid-scanning intensified PDA devices, which are more thermally stable and have higher detection efficiencies than did previous versions. These improvements, in conjunction with both the increased mass sensitivity of the microcolumns in concentration-sensitive detectors [26] and the increased sensitivity of fluorometric detection, permit sufficiently high sensitivities and low noise levels to be achieved when multichannel fluorescence imaging detection is coupled with high-resolution capillary column separations.

In the application of a miniaturized PDA fluorescence detector, samples which contain a complex mixture of structurally similar aromatic compounds derive the most benefit from its unique capabilities. Although a variety of such mixtures exist, technologically important fossil fuel materials typically include many large polyaromatic substances, whose structural similarity and many isomeric forms make optical imaging detection particularly attractive.

INSTRUMENTATION

Sample Excitation

Both the wavelength and the intensity of the sample excitation light are fundamental considerations in the design of an optical detector. As it is often necessary to find a suitable compromise between the two, the IPDA detector was constructed with the options of either broadband excitation or the more intense excitation provided by a laser beam.

As illustrated in Fig. 1, broadband light is provided by an air-cooled arc lamp mounted in a multipurpose housing (model LH 150, Kratos Analytical Instruments, Ramsey, N.J.) and powered by a regulated, low-ripple power supply (model LPS 251HR, Kratos Instruments) to yield an intense and stable excitation signal. An adjustable reflector mounted behind the light source and a UV-transmitting fused silica condensing lens (f/1.5) maximize the total output power, while forced-air cooling and feedback regulation within the power supply minimize power fluctuations due to arc wander. Fluctuations which do occur are monitored and compensated through the ratio circuitry described below. The focused output beam passes

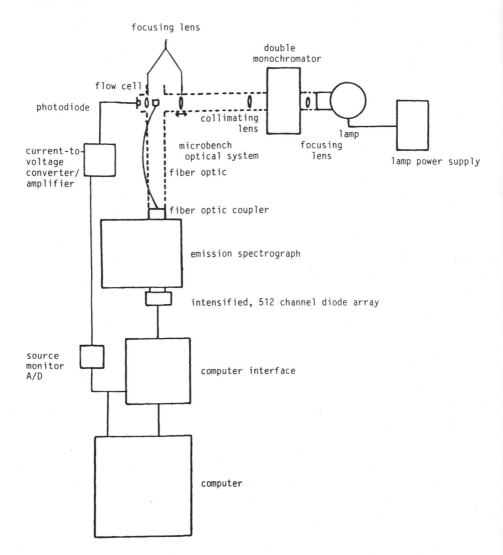

Fig. 1   Schematic diagram of the photodiode array fluorescence detector.   (Modified from Ref. 7.)

through a 0.2-m double-grating monochromator optimized for the UV region (240 nm blaze) and with an f/4 aperture and adjustable slits to provide bandwidths from 1 to 20 nm (model GM-200-3, Kratos Instruments). In addition, the entire lamp housing/monochromator unit is rigidly secured on a kinematic mount to allow precise alignment of the arc image with the excitation optical train.

Because of the necessarily small detection volumes, it is important both to minimize the size of the arc image (which measures 1.4 × 3.0 mm at the exit of the monochromator when a 200 W HgXe lamp and a 20 nm bandwidth are employed) and to maintain careful control of its alignment with respect to the flow cell. The image reduction is accomplished by means of an f/4 fused silica collimating lens (Oriel Corp., Stamford, Conn.) followed by an f/1 fused silica condensing/focusing lens (Physitec Corp., Norfolk, Mass.), reducing the arc size to 0.7 × 1.5 mm at the detection cell. Both the excitation and emission optics are rigidly mounted on an optical microbench (Physitec Corp.) to facilitate precise and reproducibile alignment. In addition, the excitation focusing lens is mounted on a precision micrometer (Physitec Corp.) to allow for changes in focal length with changing excitation wavelength. A second lens (f/1, Oriel Corp.), placed beyond the flow cell, focuses excitation light passing through and around the cell onto a UV-enhanced photodiode (type 5D/PSB, United Detector Technology, Culver City, Calif.). The photodiode monitors fluctuations in the lamp intensity and responds with a proportional current, which is converted to a voltage and then digitized. This digital value is subsequently ratioed with the fluorescence signal from the diode array detector to compensate for changes in excitation power.

Alternatively, the beam from a laser source can be directed into the excitation optical train and onto the condensing/focusing lens via two UV-enhanced beam-steering mirrors. This configuration offers a great potential gain in detection sensitivity because of the increased power density of the laser source.

Signal Detection

The fluorescence signal can be transferred to the emission spectrograph via two alternative pathways. The first was designed for use without a fiber-optic probe and is primarily useful in preliminary alignment procedures. It consists of an f/4 fused silica collimating lens (Oriel Corp.) followed by an f/1.5 focusing lens (Oriel Corp.) to image the emission signal onto the entrance slit of the spectrograph. In chromatographic applications, interference from scattered or stray light is minimized by transmitting the emitted fluorescence via a 20-cm segment of single fiber-optic probe (P/N WF00200, Maxlight Fiber Optic Division, Raychem Corp. of Arizona, Phoenix,

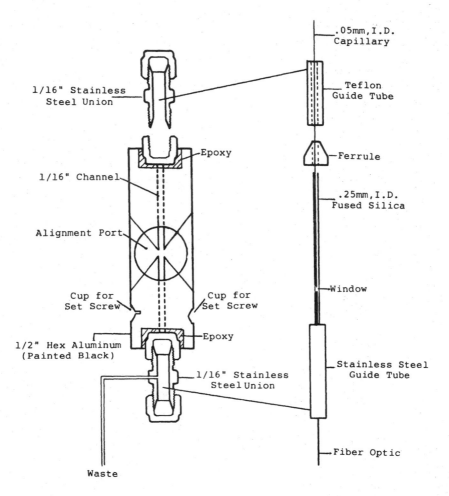

Fig. 2   Cross-sectional view of the flow cell body, showing expanded detail of the cell interior.   (From Ref. 6.)

Ariz.), inserted into the base of the flow cell as illustrated in Fig. 2. This fiber terminates at the entrance slit of the spectrograph, where it is secured by a fiber-optic coupler, configured as shown in Fig. 3. It should be noted that, since the core diameter of the fiber optic is 200 μm, this serves as the effective entrance slit width of the spectrograph when larger entrance slits are employed.

The remaining signal detection components are the emission spectrograph, the intensified diode array, and the computer system. The emission spectrograph (f/3.85, model TN 1149-9, Tracor Northern, Middleton, Wis.) is also kinematically mounted and contains a 150 groove/mm holographic grating. The fluorescence signal is dispersed onto the face of a 512-element photodiode array coupled to a proximity-focused microchannel plate intensifier (model TN6123, Tracor Northern). The array is Peltier-cooled to −3°C to promote baseline stability, and a continuous nitrogen purge flows across its surface to prevent moisture condensation.

Data Acquisition and Processing

The analog output from the array enters the detector controller/interface (model TN-6200, Tracor Northern), where it is digitized. This module, is, in turn, capable of transmitting the digitized information to a data acquisition computer at rates of up to 100 kHz. To ensure that the speed of data acquisition and transfer does not limit the minimum overall time constant of the detector, the Tracor Northern interface is hardwired to the personal computer used for data collection (IBM-PC, IBM Corp., Boca Raton, Fla.). This direct memory acquisition (DMA) configuration does not require software mediation to write the data in unstructured form onto floppy disks (formatted in the operating system of the computer), and hence data are stored at the 100 kHz transfer rate. When double-sided, double-density disks are used, approximately 375 spectra can be recorded on a single disk. The minimum scan time of the array (i.e., the time required to read the signal level of all 512 diodes) is 5.13 ms, making this the limiting rate factor of the detector.

A bidirectional assembly language program allows the user to control the acquisition parameters (i.e., the length of time the signal is integrated on each diode and the number of scans averaged per spectrum). It also computes the ratio of the array signal with the digitized output of the lamp-monitor photodiode. In addition, the fluorescence spectrum, two single-wavelength chromatograms, and the total fluorescence chromatogram are displayed in real time. It is the display portion of this program which restricts the number of continuous spectra that may be acquired according to the data buffer capacity of the program. Currently, 4094 spectra can

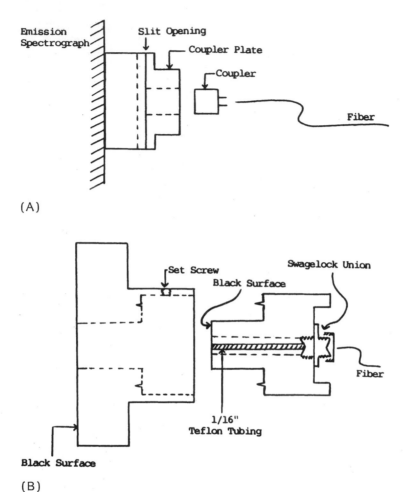

Fig. 3   Schematic diagram of the fiber-optic coupler.   (A) Fiber coupling with emission spectrograph; (B) expanded detail showing interior of coupler and coupler plate.   (From Ref. 6.)

be collected, but this number could readily be increased. Since only averaged spectra are stored, this does not limit the number of spectra that can be averaged. In general, signal integration times of 200 to 300 ms/scan provide adequate sensitivity, while averaging 8 scans/spectrum yields low background noise levels. This results in total detection time constants of 1.6 to 2.5 s. Since slurry-packed capillary columns generally exhibit chromatographic peak widths of 30 s or more, little band broadening is introduced by the detector. At the end of a chromatographic run, the initial data disk is reinserted into the computer, and the file information stored in the core memory is recorded.

For postrun processing, the final information stored on the first disk is reentered into the computer. This provides both information on the acquisition parameters and access to the various chromatograms. Individual spectra can then be viewed in detail by placing a cursor at the desired location in the chromatogram. In addition, the postrun software enables background subtraction, wavelength calibration, and nine-point Savitsky-Golay smoothing of the acquired spectra. It also allows smoothing of the chromatograms, although some peak distortion does result. In addition, single-wavelength chromatograms can be reconstructed and saved or not as desired. Plotting routines are also included for both spectra and chromatograms. Again, assembly language programs speed data manipulation. The specific programming details are lengthy and have not been included in this chapter.

Detection Cell

The detection cell is based on the fiber-optic design described by Sepaniak and Yeung [27,28] and Fjeldsted and Lee [29] and is similar in configuration to that reported by Gluckman et al. [30]. A short (5 to 10 cm) segment of 50-μm-i.d. (210-μm-o.d.) fused silica tubing is connected to the outlet of the column and then inserted into a 5-cm piece of 250-μm-i.d. fused silica capillary from which 6 mm of the polyimide coating has been removed, as shown in Fig. 2. The column effluent, thus, enters the flow cell 1 mm above the arc image. The fiber-ootic probe is similarly inserted through the bottom of the cell to a distance 3 mm below the column outlet, yielding a detection volume of 150 nl. The cell is mounted on a precision x—y translator to facilitate centering at the focal point of the arc image. In this way, a maximum amount of fluorescence emission can be collected, while the interference from scattered or stray light is kept to a minimum.

DETECTOR CHARACTERIZATION

Since the diode array system was designed as a detector for microcolumn chromatography, it was evaluated and characterized in this
role, using model compounds typifying those whose detection was
ultimately sought (i.e., polynuclear aromatic substances). Extrapo-
lation from these data to sensitivity values for acidic quinine sulfate,
the standard traditionally employed to evaluate spectrometers, can
be accomplished using the fluorescence quantum efficiencies of the
various compounds at an excitation wavelength of 365 nm. The conversion is not presented in this chapter since the system was considered to be a chromatographic detector, rather than a classic
spectrometer.

Detection Limits

The minimum detectable quantity, determined using Student's t-test
and the procedure described by Hieftje [31], was found to be an
injected amount of 15.5 pg of anthracene in 9:1 (v/v) acetonitrile:
water or 11.6 pg of the more highly fluorescent 1,3,6,8-tetraphenyl-
pyrene in 100% acetonitrile. These detection limits correspond to a
concentration of 76.2 µg/liter (ppb) anthracene and 57.0 µg/liter
(ppb) tetraphenylpyrene in the detection cell at the peak maximum
and were obtained at a signal-to-noise (rms) ratio of 5 (99% confidence level).

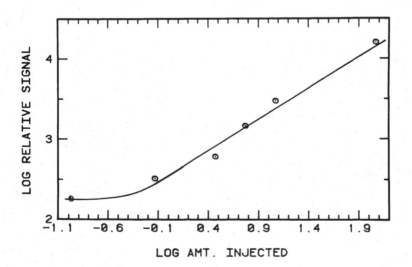

Fig. 4   Detection linearity for anthracene in 9:1 (v/v) acetonitrile:
water. (From Ref. 6.)

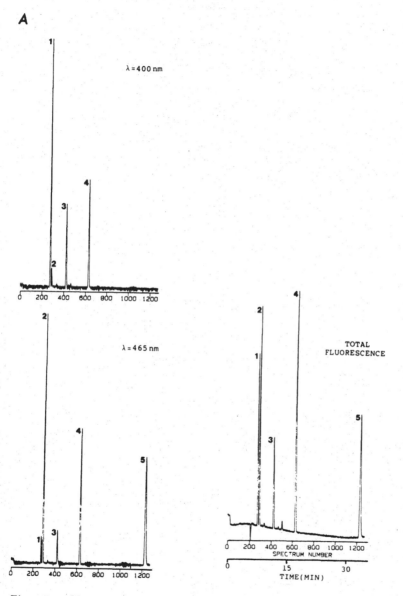

Fig. 5 Chromatogram of a standard mixture of polycyclic aromatic hydrocarbons, showing on-line fluorescence spectra obtained for each. Solutes: (1) anthracene (10 ng), (2) fluoranthene (8 ng), (3) rubrene (16 ng), (4) 1,3,6,8-tetraphenylpyrene (1 ng), (5) decacyclene (7 ng). Array exposure time = 0.2 s/scan; 8 scans averaged per spectrum; excitation at 365 nm. (From Ref. 6.)

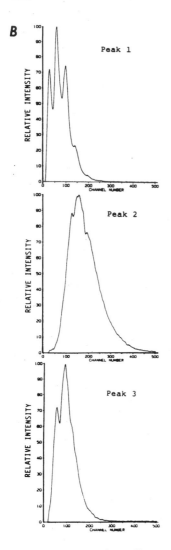

Fig. 5   (Continued)

Linearity

The detector was found to be linear (correlation coefficient of 0.99) over nearly three orders of magnitude when anthracene in 9:1 (v/v) acetonitrile:water was used as a model compound (Fig. 4). For these measurements, the lamp power was monitored during data acquisition, and minor variations were corrected for by normalizing all peak height measurements to correspond to a constant excitation power.

Chromatographic and Spectral Resolution

In the separation of complex mixtures, it is important to maintain the full resolving power of the slurry-packed capillary column, and therefore the detector itself can add little extracolumn broadening to the chromatographic system. This has traditionally been a problem for diode array-based detectors, because their lower sensitivities and relatively low scan speeds have necessitated long detection time constants. The miniaturized, rapid-scan IPDA system described here has, however, been able to maintain excellent chromatographic efficiency without sacrificing detection sensitivity. As is evident in Fig. 5A, which presents the separation of a standard mixture of polycyclic aromatic compounds monitored at two selected wavelengths and through the total fluorescence signal, high chromatographic efficiencies (on the order of 150,000 theoretical plates for anthracene) are well preserved.

In addition to chromatographic considerations, an evaluation of the detector's spectral resolution is important to its use for compound characterization and identification. Since many structural differences between similar compounds produce only subtle changes in their respective fluorescence emission spectra, the higher the resolution available, the more useful the detector is as an ancillary identification technique. Using a mercury pen ray lamp to provide a well-defined line source, peak widths on the order of 6 nm (full width at half-maximum) were achieved for lines between 360 and 650 nm. The high spectral quality of the detector is clearly evident in Fig. 5B, which presents the emission spectra obtained for each peak in the standard chromatogram. In particular, the high resolution is apparent in the structure of the anthracene spectrum and in the subtle spectral features which are clearly visible in the other spectra.

ANALYSIS OF COMPLEX MIXTURES

Detection Selectivity

Many factors play an important role in the successful separation and analysis of complex samples. In addition to high chromatographic

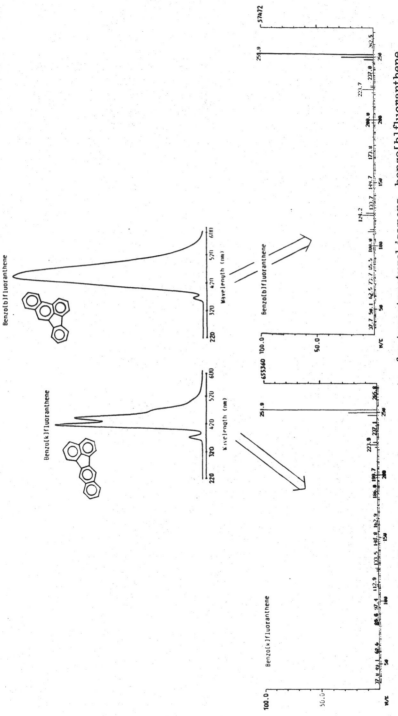

Fig. 6 Comparison of fluorescence and mass spectra for two-structural isomers, benzo[b]fluoranthene and benzo[k]fluoranthene (λ_{ex} = 365 nm).

and spectral resolution, the ability to detect selectively the analyte(s) of interest within a single chromatographic run and without the use of multiple detectors is frequently desirable. The IPDA provides flexible detection selectivity as well as detailed spectral information. This combination can be important in elucidating the composition of complex samples. For example, the fluorescence spectra of the standard compounds (Fig. 5) illustrate the relationship between molecular structure and spectral detail which forms the basis of this inherent selectivity and which can allow even closely eluting compounds, such as anthracene and fluoranthene, to be distinguished through a judicious choice of emission wavelength. Indeed, the selection or exclusion of a given compound type can often be predicted on the basis of known spectral trends, such as the tendency for larger aromatics like rubrene and decacyclene to fluoresce at longer wavelengths than smaller molecules.

Structural Isomers

Fluorescence emission spectra can also provide information complementary to that available from mass spectrometric data. For instance, many of the isomeric polycyclic arenes appear identical solely on a mass spectral basis [1,32] and yet may be readily distinguished by their fluorescence spectra. Benzo[b]- and benzo[k]-fluoranthene are two such compounds. As is evident in Fig. 6, the two species differ only in the location of one aromatic ring and show very similar mass spectral properties. Their fluorescence emission profiles, on the other hand, are distinctly different and hence helpful for compound identification.

Applications to Fossil Fuel-Derived Samples

Fossil fuel-derived fluids are known to contain many polycyclic aromatic compounds (PACs), ranging in size up to nine-ring systems [1,32,33], whose identification is of great technological and environmental importance. The large number of possible structural isomers for the higher-molecular-weight substances and the complex sample matrix in which they are found place a great demand on chromatographic resolution or spectroscopic specificity. In addition, the low volatility of these molecules frequently prevents their analysis by high-resolution capillary gas chromatography. Although it is also true that standard reference compounds are typically lacking for these species, the molecular fluorescence of polycyclic aromatic compounds has been extensively discussed in the literature [34–40] and is known to exhibit a relatively high degree of detail, which is frequently indicative of molecular structure. For example, increasing ring conjugation, alkyl and phenyl substitution, and the inclusion of heteroatoms in the ring all produce distinct spectral changes [34,36,37]. Indeed, preliminary studies have been conducted on the high-molecular-weight PACs using both their chromatographic properties and their fluorescence excitation and emission spectra [33,36,41]. The IPDA offers the unique ability to obtain on-line fluorescence emission spectra in conjunction with high-resolution microcolumn liquid chromatography (LC) separations so that detailed spectral information may be obtained for each chromatographic peak without the need for large amounts of sample and without the inconvenience of collecting fractions for subsequent spectral analysis. Thus, by correlating the information obtained from several sources, such as chromatographic, mass spectrometric, and UV-visible spectrophotometric data, with that obtained from fluoresence spectra, the identification of very large and complex polycyclic aromatic molecules may become feasible.

Carbon Black

Figure 7 illustrates the separation of a carbon black extract as monitored by the PDA detector [7]. It should be noted that, in addition to the selected emission wavelengths, alternative wavelengths may be examined after a run to allow a complete study of the data obtained.

Figure 8 shows the fluorescence spectra obtained for three of the separated peaks. Although mass spectral data indicated that peak A contained six- and seven-ring compounds with rather condensed ring structures and molecular weights ranging from 276 to 326 [32], the highly structured fluorescence emission spectrum suggests that the compounds were probably catacondensed species having a great deal of bond conjugation. Similarly, the mass spectra of peaks B and C were indicative of fairly condensed seven-ring compounds having molecular weights of 326 [32]. The fluorescence spectra, on the other hand, show distinctly different compounds to be eluting in the two peaks. The fluorescence emission of peak B is highly reminiscent of that obtained for rubrene (Fig. 5), although the higher degree of spectral structure may indicate greater ring conjugation in the unknown substance. Peak C, however, has a fluorescence spectrum very similar in both structure and location to that obtained for decacyclene [7].

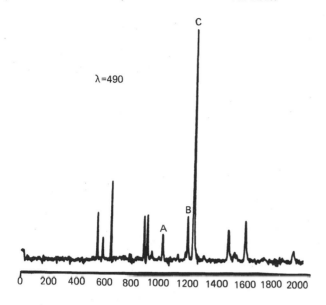

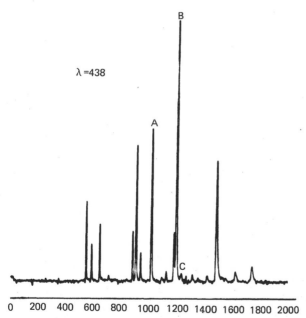

Fig. 7 Chromatogram of large polyaromatic compounds extracted from carbon black as monitored at 438 and 490 nm and through the total fluorescence signal. (From Ref. 7.)

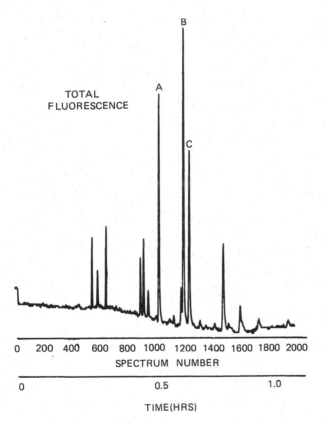

TOTAL
FLUORESCENCE

Fig. 7 (Continued)

Three Fractions of a Fuel Oil Blend

Coal-derived fluids contain a complex assortment of polyaromatic
species, frequently in many isomeric forms and containing a variety
of heteroatoms. The fuel oil blend whose analysis appears in Figs.
9—12 is just such a sample [6]. Following a preliminary fractionation into its neutral, phenolic, and nitrogen-containing components,
each fraction was subjected to high-resolution separation by microcolumn LC and detection with the IPDA system. Figure 9 presents
the total fluorescence chromatogram of the high-molecular-weight
neutral species from this mixture, including both the fluorescence
emission and the mass spectra for two representative peaks. The
utility of fluorescence information as a complement to mass spectral
data is clearly evident in this example. The chromatograms obtained
for the phenolic and nitrogen-containing fractions of the fuel oil

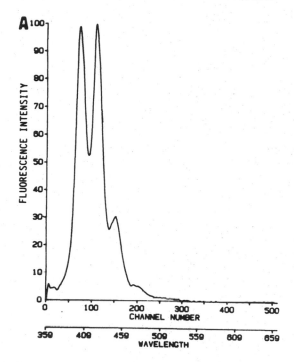

Fig. 8   On-line fluorescence spectra obtained for peaks A, B, and C in Fig. 7.   (From Ref. 7.)

blend appear in Figs. 10 and 11, respectively, with the inserts illustrating the on-line fluorescence spectra obtained for two of the peaks in each chromatogram.

Spectral Isolation

Coelution of many substances within one chromatographic peak is a frequent occurrence in complex chromatograms, such as that of the fuel oil blend. When the migration pattern is such that the concentration of one component falls to nearly zero over the course of the peak, spectral subtraction provides a convenient means of isolating and identifying this compound.   An example of the spectral resolution of a chromatographically unresolved peak appears in Fig. 12.   In this case, the first component of a single 30-s-wide chromatographic peak, which appears in expanded view of Fig. 12A, fell in concentration as the elution band passed through the detection cell, leaving the remaining components to form a "shoulder" region.   Therefore, by subtracting the spectrum obtained at the height of this

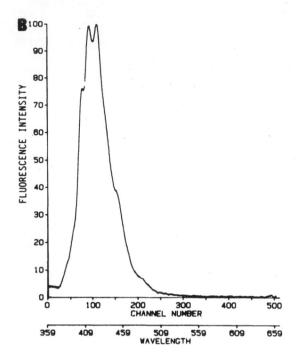

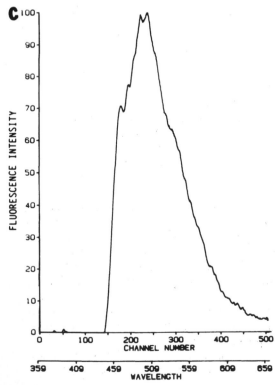

Fig. 8 (Continued)

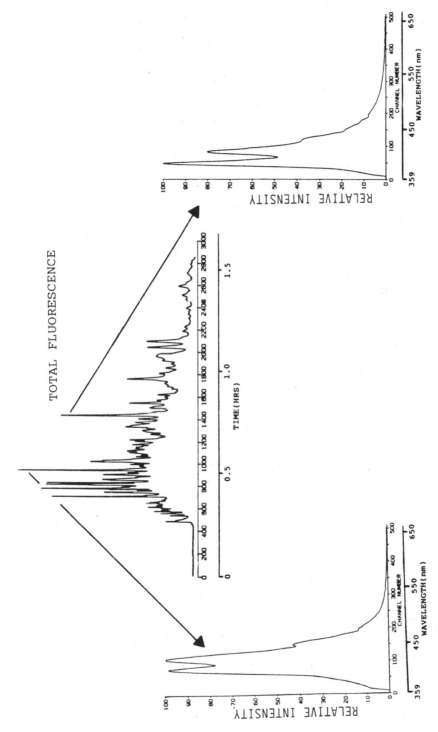

Fig. 9 Total fluorescence chromatogram of the high-molecular-weight neutral compounds extracted from a fuel oil blend. Insert shows the very similar mass spectra and the well-differentiated fluorescence spectra for two representative peaks. (From Ref. 6.)

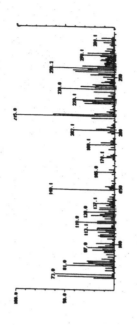

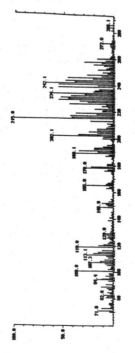

Fig. 9 (Continued)

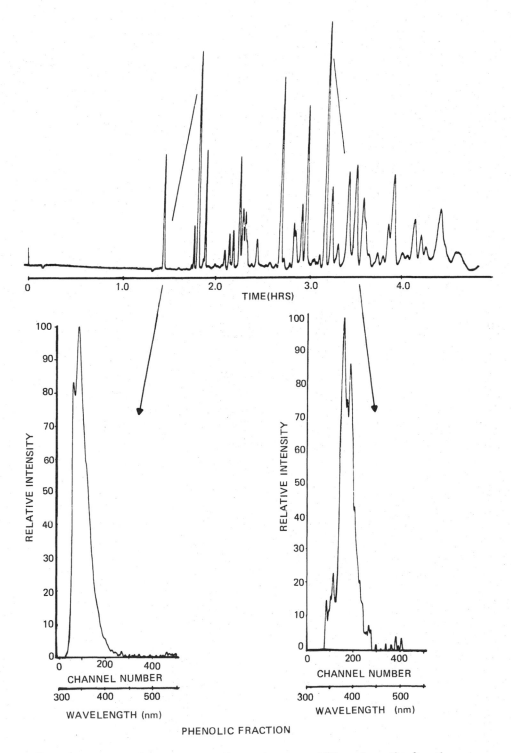

Fig. 10 Total fluorescence chromatogram of the phenolic fraction of a fuel oil blend, showing the on-line fluorescence spectra obtained for two representative peaks.

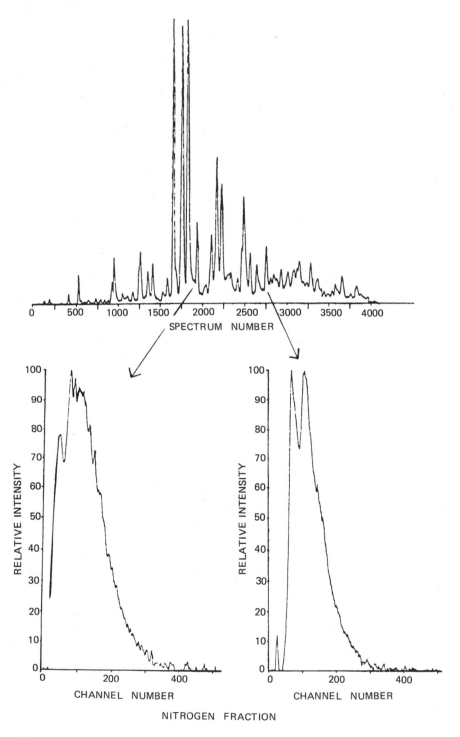

Fig. 11   Total fluorescence chromatogram of the nitrogen-containing fraction of a fuel oil blend, showing the on-line fluorescence spectra obtained for two representative peaks.

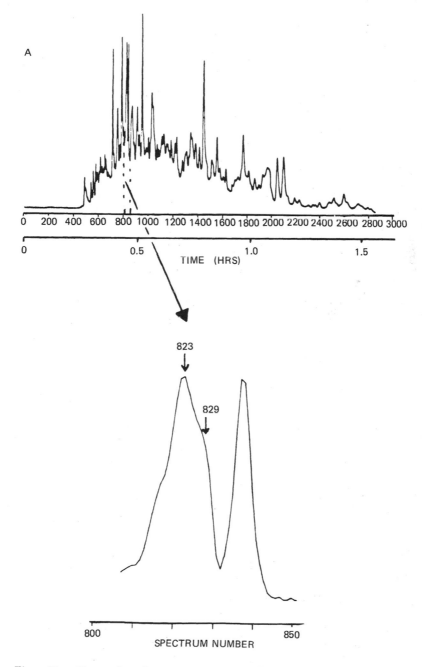

A

Fig. 12   Example of spectral subtraction used to resolve coeluting compounds from the fuel oil blend. (A) Expanded view of chromatographic region of interest; (B) component and residual spectra obtained. (From Ref. 6.)

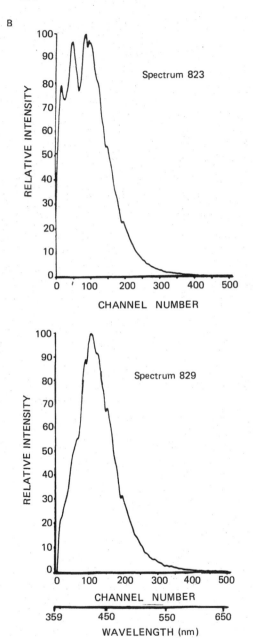

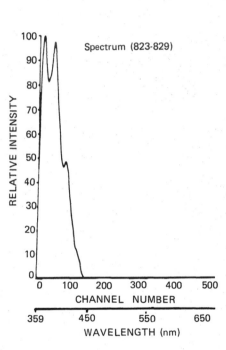

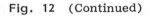

Fig. 12 (Continued)

shoulder (No. 829) from that obtained at the maximum concentration of the first component (No. 823), a residual spectrum providing qualitative information on the first component is obtained. (Note that the same approach can equally well be applied to the analysis of the shoulder on the leading peak edge.) This feature of fluor-esence data, which allows spectral deconvolution through subtraction, can be very important in the identification of compounds contained in complex mixtures, where chromatographic resolution is frequently incomplete.

Conclusion

Identification of compounds contained in highly complex mixtures, such as those of the fossil fuel-derived fluids, presents a serious analytical problem, whose resolution will, no doubt, come in the form of a multiparameter analysis scheme. The high resolving power of capillary column LC coupled with the on-line imaging capabilities of the IPDA can provide valuable information in the identification process. In particular, the inherent selectivity of fluorescence measurements combined with multiwavelength detection can greatly simplify the chromatograms obtained. In addition, spectral decon-volution may resolve many of the remaining coelution problems, and spectral data can be used to augment the information obtained from other sources. Indeed, the initial steps shown here for the analysis of fossil fuel samples illustrate the potential utility of the system.

SUMMARY

The construction of a novel rapid-scan photodiode array fluorescence detector for capillary liquid chromatographic separations represents the beginning of an integrated multidimensional approach to the iden-tification of large involatile substances contained in complex matrices. When the detector is interfaced to high-resolution slurry-packed capillary columns, on-line spectral information can be obtained for a variety of compounds not readily studied by other means. In the future, compilation of a complete spectral library of known com-pounds will facilitate the assignment of definitive compound struc-tures. In addition, by modifying the system to permit the acquisi-tion of excitation as well as emission spectra, the amount of informa-tion available from the photodiode array detector can be significantly increased.

ACKNOWLEDGMENT

Support for this work was provided by the U.S. Department of Energy under contract DE-FG2-Er 84 60215.

REFERENCES

1. M. Novotny, A. Hirose, and D. Wiesler, *Anal. Chem. 56*, 1243 (1984).
2. M. Novotny, K.-E. Karlsson, M. Konishi, and M. Alasandro, *J. Chromatogr. 292*, 159 (1984).
3. K.-E. Karlsson, D. Wiesler, M. Alasandro, and M. Novotny, *Anal. Chem. 57*, 229 (1985).
4. I. Molnar and C. Horvath, *J. Chromatogr. 143*, 391 (1977).
5. T. Hanai and J. Hubert, *J. Liq. Chromatogr. 7*, 1627 (1984).
6. J. C. Gluckman, D. Shelly, and M. Novotny, *Anal. Chem. 57*, 1546 (1985).
7. J. C. Gluckman and M. Novotny, *J. High Resolut. Chromatogr. Chromatogr. Commun. 8*, 672 (1985).
8. G. Horlick, *Appl. Spectrosc. 30*, 113 (1976).
9. Y. Talmi, D. Baker, J. Jadamec, and W. Saner, *Anal. Chem. 50*, 936A (1978).
10. Y. Talmi, *Multichannel Image Detectors*, Symp. Ser. No. 102, American Chemical Society, Washington, D.C. (1979).
11. A. F. Fell, B. J. Clark, and H. P. Scott, *J. Pharm. Biomed. Anal. 1*, 557 (1983).
12. Y. Talmi, *Anal. Chem. 47*, 658A (1975).
13. Y. Talmi, *Anal. Chem. 47*, 699A (1975).
14. A. McDowell and H. Pardue, *Anal. Chem. 49*, 1171 (1977).
15. M. Fogarty, D. Shelly, and I. Warner, *J. High Resolut. Chromatogr. Chromatogr. Commun. 4*, 561 (1981).
16. M. Milano, S. Lam, and E. Grushka, *J. Chromatogr. 125*, 315 (1976).
17. Y. Talmi and R. Simpson, *Appl. Opt. 19*, 1401 (1980).
18. K. W. Jost, Th. Crispin, and I. Halasz, *Erdoel Kohle Erdgas Petrochem. 37*, 178 (1984).
19. D. J. Desilets, P. T. Kissinger, F. E. Lytle, M. A. Horne, M. S. Ludwiczak, and R. B. Jacko, *Environ. Sci. Technol. 18*, (1984).
20. T. Takeuchi and D. Ishii, *J. High Resolut. Chromatogr. Chromatogr. Commun. 7*, 151 (1984).
21. M. Fogarty, D. Shelly, and I. Warner, *J. High Resolut. Chromatogr. Chromatogr. Commun. 4*, 616 (1984).
22. A. Fell, H. Scott, R. Gill, and A. Moffat, *J. Chromatogr. 282*, 123 (1983).

23. M. A. Ryan, R. J. Mller, and J. D. Ingle, Jr., *Anal. Chem.* *50*, 1772 (1978).
24. Y. Talmi, *Appl. Spectrosc.* *36*, 1 (1982).
25. J. C. Fjeldsted, B. E. Richter, W. P. Jackson, and M. L. Lee, *J. Chromatogr.* *279*, 423 (1983).
26. M. Novotny, in *Microcolumn High Performance Liquid Chromatography* (P. Kucera, ed.), pp. 194–259, Elsevier, Amsterdam (1984).
27. M. Sepaniak and E. Yeung, *J. Chromatogr.* *190*, 377 (1980).
28. M. Sepaniak and E. Yeung, *J. Chromatogr.* *211*, 95 (1981).
29. J. Fjeldsted and M. Lee, *Anal. Chem.* *56*, 619A (1984).
30. J. Gluckman, D. Shelly, and M. Novotny, *J. Chromatogr.* *317*, 443 (1984).
31. G. Hieftje, *Anal. Chem.* *44*, 81A (1972).
32. A. Hirose, D. Wiesler, and M. Novotny, *Chromatographia 18*, 239 (1984).
33. P. Peaden, M. L. Lee, Y. Hirata, and M. Novotny, *Anal. Chom.* *52*, 2268 (1980).
34. I. Berlman, *Handbook of Fluorescence Spectra of Aromatic Molecules*, Academic Press, New York (1971).
35. J. Kropp and C. Stanley, *Chem. Phys. Lett.* *9*, 534 (1971).
36. J. McKay and D. Lantham, *Anal. Chem.* *44*, 2132 (1972).
37. J. McKay and D. Lantham, *Anal. Chem.* *45*, 1050 (1973).
38. E. Clar and W. Schmidt, *Tetrahedron 33*, 2093 (1977).
39. E. Clar and W. Schmidt, *Tetrahedron 34*, 3219 (1978).
40. E. Clar and W. Schmidt, *Tetrahedron 35*, 1027 (1979).
41. R. C. Pierce and M. Katz, *Anal. Chem.* *47*, 1743 (1975).

6

Practice and Application of Microcolumn LC with Inductively Coupled Plasma Atomic Emission Spectrometric Detection

KIYOKATSU JINNO / Toyohashi University of Technology, Toyohashi, Japan

INTRODUCTION

As the importance of knowing the chemical form of metal elements in medicinal, environmental, biological, and pharmaceutical applications has increased, it has become necessary to know not only the concentrations but also the chemical states of metal analytes. A great deal of research has been generated to meet this demand. One area that has received major emphasis is the use of atomic spectrometry as a metal-specific detector. When combined with high-performance liquid chromatography (HPLC) techniques, atomic spectrometry as a metal-specific detector offers identification capability for metal elements. Therefore, the combination of HPLC with atomic spectrometry has been studied by a number of workers. A wide range of chromatographic columns and mobile phases can be coupled with atomic spectrometric instruments.

   Conventional flame atomic absorption spectrometry (AAS) has been used as an element-specific detector in the separation and determination of copper chelates in wastewater samples [1], chromium-(III) and chromium (VI) in natural waters [2], alkyl and aryl zinc compounds in lubricating oils [3], and heavy metal complexes in urine from patients undergoing chelating agent treatment for heavy metal poisoning [4]. Furnace atomic absorption (FAA) was used as an HPLC detector in the identification of arsenic metabolites in marine mollusks [5]. At the current level of technology, however, atomic absorption detection is restricted to single-element monitoring.

Inductively coupled plasma atomic emission spectrometry (ICP-AES, ICP) is being developed as a tool for the analysis of a number of elements and is particularly attractive for its simultaneous multi-element monitoring capability [6]. Furthermore, the analyte atoms in the hot argon plasma tend to behave as an optically thin emitting source [7]. This phenomenon fortunately provides a linear emission signal for each analyzable element over several orders of magnitude in concentration, a virtually necessary characteristic for chromatographic detectors. Because of its simultaneous multielement detection capability, element specificity, and low detection limits, ICP is a very promising mode of detection for elemental species separated by HPLC.

Farely et al. [8] and Morita et al. [9] have attempted to maximize analytical performance in designing interfaces. Investigations have been performed to characterize the effects of mobile phase composition and flow rates on HPLC-ICP methods [8,10]. The effects of pneumatic nebulizer-spray chamber configurations on the response of the ICP detector when interfaced with HPLC have been evaluated by Whaley et al. [11].

Size exclusion chromatography and ICP were combined by Gradner et al. [12] to fractionate and detect dissolved Ca and Mg in natural waters. Several distinct forms of each metal element were identified in natural water samples and selected species were found to coelute with UV-absorbing dissolved organic material. Hausler and Taylor [13] used size exclusion chromatography with ICP detection in a nonaqueous mobile phase system to separate and determine different organometallic species. Using toluene as the mobile phase, the detection limits were generally higher than those obtained under direct ICP aqueous operating conditions.

Other HPLC-ICP studies that have been reported include mixtures of nucleotides [14], arsenic compounds in biological samples [15], mixtures of phosphates [16], mixed amino acid solutions [17], and rare earth elements in U.S. Geological Survey (USGS) rock standard samples, rare earth ores, and high-purity lanthanide reagents [18].

To date, the coupling of HPLC to ICP has been accomplished only with conventional cross-flow, concentration, or Babington-type pneumatic nebulizers. Detection limits under these conditions have generally been poor when compared to conventional continuous sample flow conditions. These limitations have been due to sample losses associated with conventional nebulizers and band broadening of eluents from HPLC columns prior to entering the nebulizer unit, which have reduced detection limits to only 5 to 10% of the theoretical values. Efforts to improve the system for sample introduction into the ICP are continuing. Lawrence et al. [19] developed a microconcentric nebulizer which interfaces a 1-mm-i.d. microcolumn directly into the tip of the conventional sample introduction tube of an ICP torch.

In HPLC, over the past several years, much effort has been directed toward increasing speed and efficiency. Recent development in HPLC has been focused on two approaches: (a) reduction of the column length and use of small particle sizes for high-speed HPLC [20,21] and (b) reduction of the column diameter in microcolumn LC [22–26]. The former approach is along the lines of conventional HPLC and the latter, because of its low flow rate, offers the potential for more efficient HPLC-ICP interfacing. The performance of various combinations of microcolumn LC with ICP detection systems is discussed in the following.

MICROCOLUMN LC-ICP INTERFACINGS

Simple Connection of Microcolumn and ICP

The simplest method of interfacing HPLC and ICP is to connect the outlet of the columns to the inlet of the nebulizing system of the ICP. This approach in microcolumn LC, where a simple tee connection was used between a microcolumn and a pneumatic nebulizer for ICP, was first reported in 1982 [27] and subsequently in 1983 [28, 29]. A schematic diagram of this interface is shown in Fig. 1. The eluent of microcolumn LC can be carried into the nebulizer through the tee interfacing device with water carrier for the reversed-phase mode and methyl isobutyl ketone (MIBK) carrier for the normal-phase mode. Typical operating conditions for this system are given in Table 1.

Basic performance of this system has been tested with the reversed-phase separation mode. In order to retain the resolution povided by the microcolumn, the peak broadening caused by the interfacing device, connecting tube, nebulizer, spray chamber, and plasma torch must be kept to a minimum. To compare the peak broadening in this system to that obtained with UV detection, ratios of the peak width in ICP detection, W_I, to that of UV detection, W_U, were calculated and compared in Table 2. The ratios are almost constant for various mobile phase flow rates, with slightly

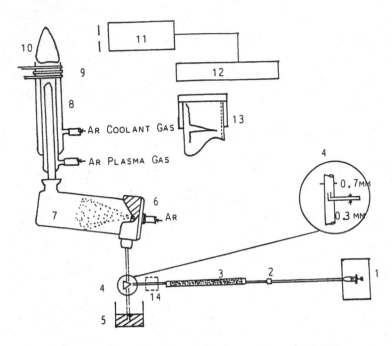

Fig. 1 Schematic diagram of microcolumn LC-ICP system. 1, Microcolumn LC pump; 2, sample injection point; 3, microcolumn; 4, interfacing device; 5, reservoir; 6, cross-flow pneumatic nebulizer; 7, spray chamber; 8, plasma torch; 9, coil; 10, plasma flame; 11, Ebert mounting monochromator; 12, R-456 photomultiplier; 13, chart recorder; 14, UV detector when used.

TABLE 1 ICP-AES Conditions for the Micro-HPLC-ICP Combination

	Reversed-phase mode	Normal-phase mode
Operating conditions		
Frequency (MHz)	27.12	27.12
Power (kW)	1.2	2.0
Ar gas flow rate (liters/min)		
Coolent	16	25
Plasma	1	1.5
Sample	0.50	0.45
Carrier		
Solution	Water	Methyl isobutyl ketone
Flow rate (ml/min)	0.50	0.45

Source: From Ref. 28.

TABLE 2   Ratio of Bandwidth at Half-Maximum in ICP Detection to That in UV Detection at Various Mobile Phase Flow Rates: $H = W_i/W_U$*

Mobile phase flow rate (μl/min)	W_i (μl)	W_U (μl)	H
25	11.0	9.5	1.16
20	10.4	8.8	1.18
15	9.6	8.4	1.14
10	7.6	6.6	1.15
8	6.7	6.1	1.10

*Experimental conditions: column, SC-01; mobile phase, methanol; sample and sample size, Cu-ACAC, 1.5 μl; detection wavelength, UV 254 nm, ICP 324.7 nm; water carrier flow rate, 0.50 ml/min.

Source: From Ref. 28.

broader peaks for ICP detection. It is apparent that the interface of microcolumn LC and ICP works well.

Sensitivity and dilution factors were also examined. The results indicate that sensitivity is essentially independent of flow rates above 15 μl/min, and the dilution factor increases as the mobile phase flow rate increases. The dilution effect caused by the interfacing system is, however, very small at a flow rate above 15 μl/min. The results also indicate that in a higher flow rate range, the ICP is a concentration-sensitive detector, whereas in a lower flow rate range it becomes a mass-sensitive detector. This basic investigation indicates that a simple interface between microcolumn and ICP using carrier solvents for better nebulization is useful for flow rates above 15 μl/min, i.e., suitable for microcolumns larger than 0.5 mm i.d.

Two examples are given below to demonstrate the capability of the system, one in the reversed-phase mode and the other in the normal-phase mode.

A mixture of cupric acetylacetonate (Cu-ACAC) and cupric diethyldithiocarbamate (Cu-DDTC) was injected into a microcolumn. The flow rate of the mobile phase used (methanol-water, 4:1) was

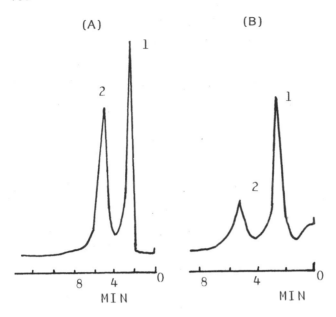

Fig. 2   Chromatograms detected by UV and ICP. (A) UV detection at 254 nm; (B) ICP detection at 324.75 nm. Mobile phase, methanol; flow rate, 25 µl/min; carrier solvent, water, 0.50 ml/min. Peaks: (1) Cu-ACAC, (2) Cu-DDTC. (Copyright Society for Applied Spectroscopy, 1983.)

25 µl/min. The UV chromatogram and the ICP chromatogram obtained are shown in Figs. 2A and 2B, respectively. It is apparent that the ICP chromatogram monitored at 324.75 nm is very similar to the UV chromatogram and that this technique has great potential in microcolumn LC-ICP applications.

   A mixture of Cu-DDTC, Cr-DDTC, and Co-DDTC was injected to test the system in the normal-phase mode. The mobile phase used was toluene and its flow rate was 16 µl/min. The results are shown in Fig. 3. Each element was detected at its specific emission line and the ICP and UV chromatograms are consistent.

   No serious restriction was found on the mobile phase compositions in normal- and reverse-phase spearations, although it is generally considered that ICP cannot be used with organic solvents because of instability of the Ar plasma. Addition of water or MIBK seems to stablize the Ar plasma. Because eluent from a microcolumn was diluted with the sample carrier solvent at the interfacing tee

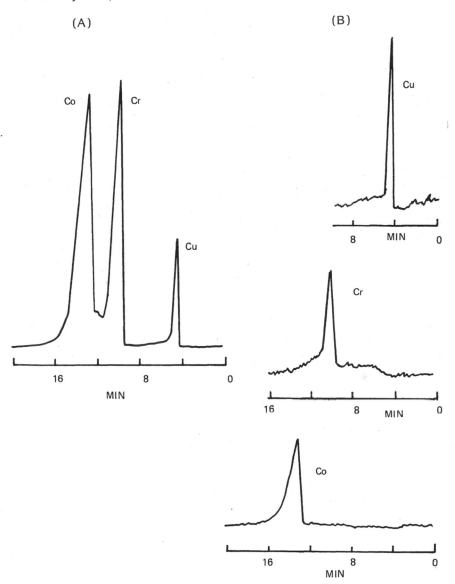

Fig. 3 Chromatograms detected by UV and ICP. (A) UV detector at 300 nm; (B) ICP detection at 324.75 nm for Cu, 267.72 nm for Cr, and 228.62 nm for Co. Mobile phase, toluene; flow rate, 16 μl/min. (Copyright Society for Applied Spectroscopy, 1983.)

device, the sensitivity was lower than that obtained with a conventional HPLC-ICP system.

Direct Connection Between Microcolumn and
ICP Cross-Flow Nebulizer

To improve the low sensitivity of the technique previously described,
a new direct-connecting system between a microcolumn and an ICP
cross-flow nebulizer has been attempted [30,31].

A cross-sectional view of the nebulizer directly connected to a
microcolumn is shown in Fig. 4. In this figure, the stainless steel
capillary tubing (0.13 mm i.d. × 40 mm) is directly connected to
the outlet of the microcolumn by shrinkable Teflon tubing. The
eluent from the column is introduced into the nebulizing point (P
in the figure), then nebulized and carried into the plasma torch

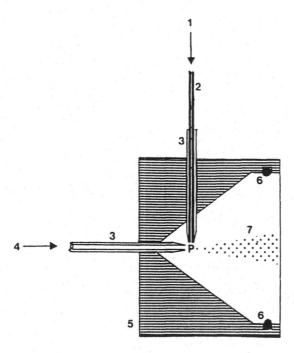

Fig. 4   Cross-sectional view of cross-flow nebulizer. 1, Eluent
inlet from a microcolumn; 2, stainless steel capillary tubing, 130 μm
i.d. (310 μm o.d.) ×40 mm; 3, glass capillary, 320 μm i.d. × 30 mm;
4, sample Ar gas inlet; 5, Teflon nebulizer body; 6, O-ring; 7,
nebulized eluent; P, nebulizing point. (Copyright American Chemical Society, 1984.)

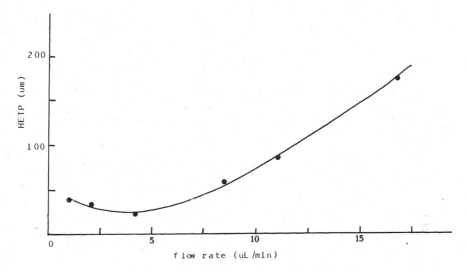

Fig. 5　HETP versus flow rate curve.　Column, 0.5 mm i.d. × 30 cm long packed with KC-811; sample, raffinose; mobile phase, water; detection, ICP at 247.86 nm carbon emission line.

through a spray chamber.　The sample Ar gas carries the eluent as a fog into the plasma torch.　It was recognized for preliminary experiments that the flow rate of the sample Ar gas should be optimized to the range of 0.44 to 0.46 liter/min (these subjects will be discussed in the next section in more detail) to maximize the signal intensity of carbon emission lines.

The performance of the system was evaluated using a Van Deemter plot.　Methanol was used as the model solute in the micro-column LC-ICP system examined, and Fig. 5 shows that the shape of the curve is similar to those well known in HPLC.　In this example, gel permeation chromatography with a water mobile phase was used because all the organic compounds present can be measured by monitoring the carbon emission line at 247.86 nm.　The minimum HETP was obtained at a flow rate of 4 μl/min, which is considered to be the best combination of eluent flow and sample Ar gas flow for columns of less than 0.5 mm i.d.

Figure 6 compares the chromatograms of two three-saccharide mixtures obtained with the previous simple interfacing system and the present cross-flow nebulization system.　A significant improvement in resolution and sensitivity is noted with the cross-flow nebulization system that excludes the dilution at the connecting tube.　In the previous system, it was very difficult to use the mobile phase at a flow rate lower than 10 μl/min, but the direct

PREVIOUS SYSTEM

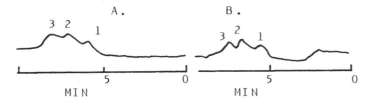

PRESENT SYSTEM

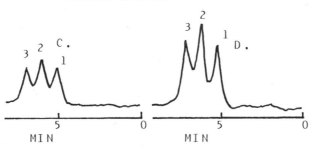

Fig. 6   Carbon selective chromatograms for saccharides with a direct microcolumn LC/pneumatic nebulization ICP system and a microcolumn LC/cross-flow nebulization ICP system.   Column, 1 mm i.d. × 20 cm long packed with SC-220 gel; temperature, 80°C; mobile phase, water at 16 µl/min; 80 µg of each saccharide for A and C, 160 µg of each saccharide for B and D; peaks (1) raffinose, (2) glucose, (3) arabinose.   (Copyright American Chemical Society, 1984.)

connection system suggested here can be used even at 2 µl/min with appreciable sensitivity and good peak shape.   It is apparent that the performance of the microcolumn LC-ICP system is improved by direct interfacing of the microcolumn to the cross-flow nebulizer.

Figure 7 illustrates the separation and ICP detection of saccharides.   Good detection can be obtained even at a 4 µl/min flow rate.   The response depends on the amount of carbon in the sample. These chromatograms show that a few micrograms of saccharides could easily be detected by monitoring carbon with ICP emission spectrometry.   The plots of carbon content and the peak heights showed linear relationships in the range of 5 to 200 µg of various saccharides examined.   The detection limit of this system was

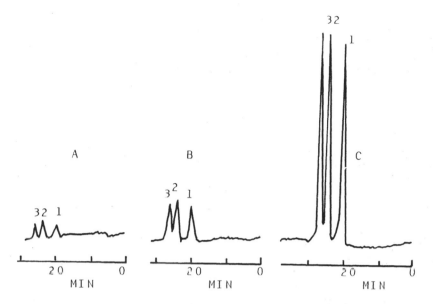

Fig. 7 Carbon selective chromatograms for saccharides with the microcolumn LC-ICP system. Column, temperature, and mobile phase as in Fig. 6; flow rate of mobile phase, 4 μl/min; sample, 12 μg of each saccharide for A, 24 μg for B, and 160 μg for C. Peaks: (1) raffinose, (2) glucose, (3) arabinose. (Copyright American Chemical Society, 1984.)

obtained as the amount of raffinose that provided a peak height corresponding to twice the baseline noise level; this value was 800 ng of carbon, which is superior to the value obtained by the conventional HPLC-ICP system (a few micrograms of carbon) described by Yoshida and Haraguchi [18].

In Fig. 8, chromatograms of PEG 200 at various column temperatures are shown. Increasing the column temperature increases resolution. At a column temperature of 80°C, the oligomers of PEG 200 are resolved as individual peaks.

Direct Sample Introduction for Microcolumn LC-ICP [32,33]

In an effort to achieve matched performance of microcolumn LC, a sample introduction method which has a "no-spray" chamber has been developed for microcolumn LC-ICP. In this section, three sample introduction systems are compared. The potential of the systems is also demonstrated by an example.

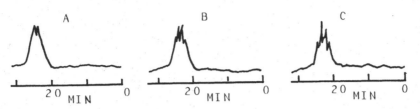

Fig. 8   Carbon selective chromatograms of poly(ethylene glycol) (PEG 200) with the microcolumn LC-ICP system.   Column, mobile phase, and flow rate as in Fig. 7; temperature, 40°C for A, 60°C for B, and 80°C for C.   (Copyright American Chemical Society, 1984.)

The three sample introduction systems compared are (a) the nebulizing system (shown in Fig. 10) with a modified cross-flow nebulizer and a no-spray chamber, (b) a conventional ICP nebulizing system using a modified cross-flow nebulizer and a spray chamber whose size is 31 mm i.d. × 82 mm long with a volume of 62 ml, and (c) a conventional ICP nebulizing system using a modified cross-flow nebulizer and a spray chamber whose size is 31 mm i.d. × 164 mm long with a volume of 124 ml (the system described in the preceding section).

Interfacing Between Microcolumn and ICP [33,34]

In microcolumn LC, extracolumn band broadening has a severe influence on column efficiency.   Therefore, to minimize extracolumn band broadening, the microcolumns were connected with the cross-flow nebulizer as shown in Fig. 9 in which the no-spray chamber was used.   The nebulizer was described in the preceding section.   For direct connection of fused silica capillary columns to the nebulizer, a stainless steel capillary (0.31 mm o.d., 0.13 mm i.d. × 25 mm long) was used as the interfacing tube to a glass capillary (0.32 mm i.d. × 20 mm long) which was the sample uptake tube for normal ICP operation.   The same stainless steel capillary extended close to the tip of the glass capillary, where nebulization was accomplished, as shown in Fig. 10.   The total column effluent from the fused silica capillary column was nebulized by the sample Ar gas flow over the capillary orifice (point 9 in Fig. 10).   A 0.35 mm i.d. × 400 mm long fused silica capillary column packed with KC-811 cation exchange resin (Showa Denko, Tokyo) was used with distilled water as the mobile phase to investigate the effect of sample Ar gas flow rate on emission intensity observed by the ICP system.   The sample Ar gas flow rate was varied from 0.3 to 0.6 liter/min.   The flow rate of the mobile phase was also changed from

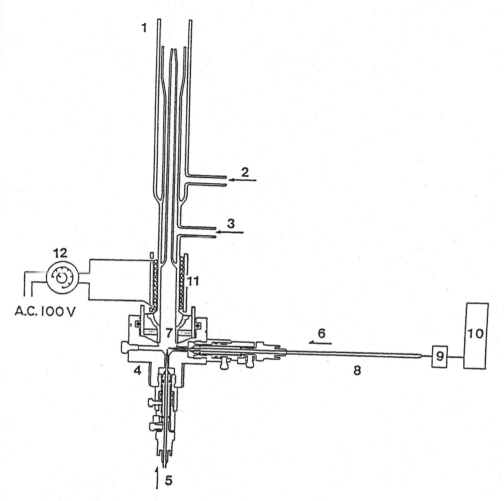

Fig. 9   Schematic diagram of no-spray chamber microcolumn LC-ICP system.   1, ICP torch; 2, coolant Ar gas flow; 3, plasma Ar gas flow; 4, cross-flow nebulizer; 5, sample Ar gas flow; 6, microcolumn LC effluent flow; 7, torch-nebulizer connecting parts; 8, microcolumn; 9, microsample loop injector; 10, microfeeder; 11, nichrome heater; 12, slide rheostat.

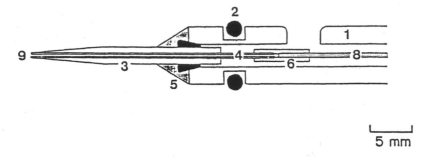

5 mm

Fig. 10   Schematic diagram of direct connecting device for sample introduction into nebulizer. 1, Device body; 2, O-ring; 3, fused silica capillary column; 4, stationary phase; 5, connecting tube made of Teflon; 6, quartz wool; 7, stainless steel capillary; 8, glass capillary; 9, nebulizing point. (Copyright American Chemical Society, 1985.)

16 to 1 µl/min. The methanol solution, which contained 0.8 µg methanol in water, was injected as the test probe into the column and the emission from the carbon atomic line at 247.86 nm was monitored.

The effect of various sample Ar gas flow rates on the emission intensity has been studied. The emission signals were recorded with changing viewing heights above the turns of the load coil between 5 and 25 mm. Representative results obtained with the three systems at mobile phase flow rates from 16 to 1 µl/min are shown in Figs. 11 to 13. The spatial emission profiles of the 247.86 nm carbon emission line were similar to the results described by Browner and Smith [35]. It appears that the carbon emission intensities decreased with increased viewing height, but further research is necessary to explain the excitation mechanism more clearly.

Figure 14 shows the results for a viewing height of 5 mm above the load coil. The maximum emission intensity of carbon was observed at this 5-mm height. It appears that the maximum emission intensity was measured at a sample Ar gas flow rate of about 0.45 liter/min, regardless of the flow rate of the mobile phase and the volume of the spray chamber. Similar results have been obtained in that reducing the diameter of the sample introduction tube to the nebulizer from 2.3 to 0.8 mm i.d. caused a shift in the maximum of the emission intensity observed at an Ar gas flow rate from 0.72 to 0.47 liter/min [34].

The effect of the spray chamber volume on the signal intensity is also shown in Fig. 14, which compares results obtained with a chamber volume of 124 ml (Fig. 14A), with a chamber volume of

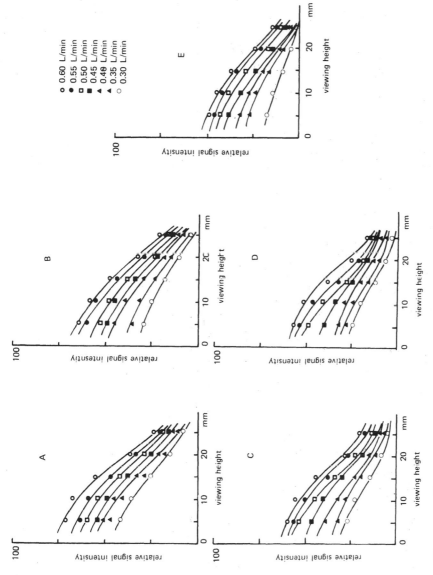

Fig. 11 Effect of viewing height on emission signal intensity of carbon atomic line at 247.86 nm with various sample Ar gas flow rates when spray chamber volume of 124 ml was used. Water flow rate: (A) 16 µl/min; (B) 8 µl/min; (C) 4 µl/min; (D) 2 µl/min; (E) 1 µl/min.

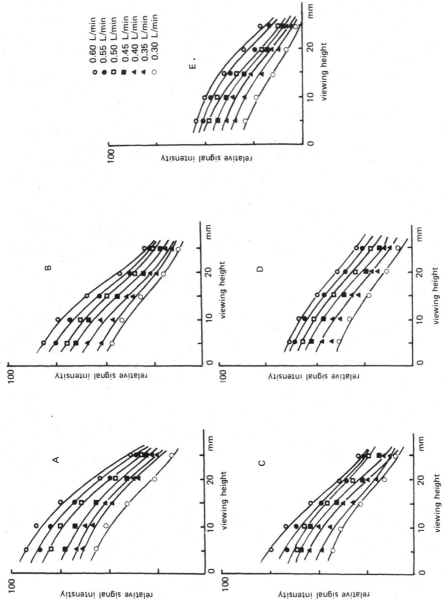

Fig. 12 Effect of viewing height on emission signal intensity of carbon atomic line at 247.86 nm with various sample Ar gas flow rates when spray chamber volume of 62 ml was used. (A to E) As in Fig. 11.

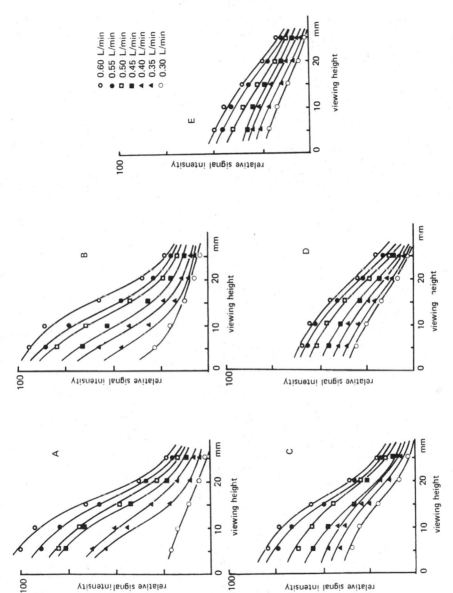

Fig. 13 Effect of viewing height on emission signal intensity of carbon atomic line at 247.86 nm with various sample Ar gas flow rates when no-spray chamber was used. (A to E) As in Fig. 11.

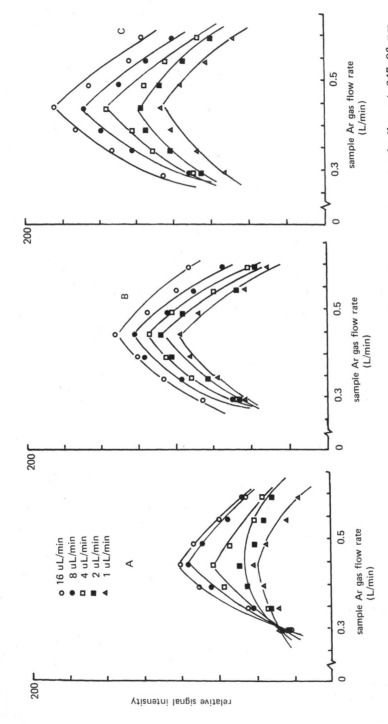

Fig. 14 Effect of spray chamber volume on emission signal intensity of carbon atomic line at 247.86 nm. Spray chamber volume: (A) 124 ml; (B) 62 ml; (C) 0 ml.

64 ml (Fig. 14B), and with a no-spray chamber (Fig. 14C). Figure 15 compares the results obtained on injecting a test solution containing 0.8 μg methanol (300 ng of carbon) into three different sample-aerosol introduction systems. The spray chamber has a basic function in removing aerosol droplets that are not suitable in diameter for introduction into the plasma. Unfortunately, it produces band broadening and mixing effects on microcolumn LC separations in the case of a direct interface to a nebulizer. In addition, as a result of adherence of sample aerosol to the inner wall of the spray chamber and the sample introduction tube of the torch, loss of ICP sensitivity is expected. Therefore, it is expected that more enhanced emission intensities will be observed with smaller-volume spray chambers. However, Hausler and Taylor [13] reported that a spray chamber having a "very small" volume did not eliminate waste solvent effectively. They reached this conclusion based on the fact that solvent eventually rose above the nebulizer outlet and caused spurious results. Because of the very low mobile phase flow rates used in microcolumn LC (e.g., 1 10 μl/min), severe problems due to nebulization of conventional solvent flow rates (e.g., 1 ml/min) should not be occurring. It has been observed that use of a small spray chamber volume or no-spray chamber nebulization is very effective in microcolumn LC-ICP interfacing.

The effect of spray chamber volume on the separation of a sample mixture of seven alkyl alcohols was evaluated and is shown in Fig. 15. Chromatograms A, B, and C were obtained with microcolumn LC-ICP systems having spray chamber volumes of 124, 62, and 0 ml, respectively. With decreased spray chamber volume, the chromatographic separation was improved and the carbon emission signal intensity was also increased. The intensity ratios are 1.4 and 1.7 for 62 and 0 ml, respectively, relative to the intensity observed with the 124-ml chamber, which is assumed to be 1.0. The result of increased signal intensity is also supported by the data of Fig. 14. In addition, it has been shown that remixing and loss of sample components in the spray chamber will severely affect the performance and sensitivity of the microcolumn LC-ICP system.

Reproducibility in Direct Sample Introduction System

When aqueous mobile phase was used in the no-spray chamber system, nonreproducible signal intensity was measured. This non-reproducibility was investigated by injecting 0.3 μl of methanol-water solution (0.8 μg methanol) directly into the nebulizer without passing through an LC column so that quick and efficient evaluation of the performance of the nebulizing systems could be achieved without additional complications imposed by retention on the LC column. In addition, to have a clear understanding of this phenomenon,

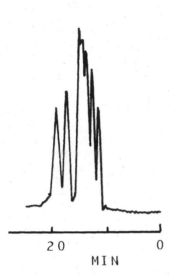

(A)

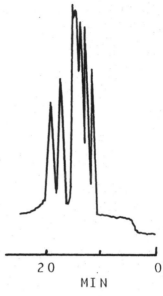

(B)

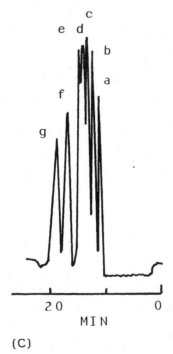

(C)

experiments were attempted in which the lower part of the sample introduction tube of the plasma torch was heated by a nichrome heater.

The effect of the heating on the intensity of the carbon emission line at 247.86 nm is shown in Fig. 16, in which the transient signals to 0.8 µg of methanol increase as the temperature for multiple injections is controlled in the range of 60 to 80°C. As shown in Fig. 16, the reproducibility becomes poor with either an increase or a decrease in the temperature. At higher temperatures, complete vaporization of methanol is achieved, but some quenching effects reduce emission intensity. At lower temperatures, the nebulization efficiency is relatively low and aerosols stick to the inner wall of the sample introduction tube of the plasma torch.

The transport efficiency for the present nebulizing system without a spray chamber was determined by conventional silica gel tube method. The amount of aerosols sprayed by the cross-flow nebulizer was determined gravimetrically by the difference in weight of the silica gel tube (10 mm l.d. × 18.5 cm long) before and after adsorbing aerosols. The transport efficiency is defined as the ratio of the amount of aerosols adsorbed onto the silica gel to the amount of liquid nebulized by the cross-flow nebulizer. Data on the transport efficiency expressed as a percentage are listed in Table 3.

Because silica gel collection procedures always give a positively biased transport efficiency [35], the values listed in Table 3 do not give the true transport efficiency, but they serve to indicate the relative transport efficiency for various nebulizing methods. At 16 µl/min water flow, the transport efficiency increased with reduction of the spray chamber volume. In the no-spray chamber, since aerosols stick to the inner wall at the lower part of the sample introduction tube of the plasma torch as described earlier, the efficiency was reduced. The efficiency increased with increasing temperature. At 4 µl/min water flow, the efficiency increased with decreasing spray chamber volume to 0 ml. Furthermore, the transport efficiency at 4 µl/min was superior to that obtained at 16 µl/min.

Fig. 15   Effect of spray chamber volume on emission signal intensity and chromatographic resolution: (A) 124 ml; (B) 62 ml; (C) 0 ml.   Column, 0.5 mm i.d. × 300 mm long packed with KC-811; mobile phase, water; flow rate, 4 µl/min; temperature, 60°C. Sample: (a) methanol; (b) ethanol; (c) 2-propanol; (d) 2-methylpropanol; (e) 1-propanol; (f) 2-butanol; (g) 1-butanol.   (Copyright Dr. Alfred Huethig Verlag, 1985.)

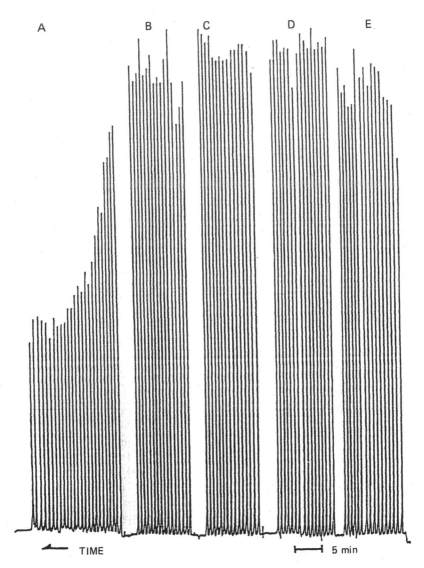

Fig. 16   Reproducibility of emission signal intensity with use of
water as mobile phase.   Sample: 0.8 μg methanol in water.   Heater
temperature: (A) room temperature;   (B) 40°C;   (C) 60°C;   (D) 80°C;
(E) 100°C.   (Copyright American Chemical Society, 1985).

TABLE 3 Transport Efficiency Measured by the Silica Gel Method[a]

Mobile phase flow rate for Micro-LC (μl/min)	Transport efficiency (%)			
	124 ml RT[b]	62 ml RT	0 ml RT	0 ml 60°C
16	78.2	81.8	69.7	87.7
4	83.9	87.2	90.9	97.6

[a]Sample Ar gas flow rate, 0.45 liter/min. Conventional ICP nebulizing with uptake rate of 0.48 liter/min was measured as 6.0% of the transport efficiency with this method.

[b]Room temperature.

Source: From Ref. 31.

Detector Characterization

The response of the ICP detector was characterized under optimum conditions using ethanol and ethylene glycol as model solutes and water as the mobile phase. The response was linear from the detection limit (100−200 ng) to at least 300 μg of carbon. The linear dynamic range was greater than 10^3. Typical signals observed are shown in Figs. 17 and 18, where the emission intensities of carbon at 247.86 nm are compared for ethanol and ethylene glycol at water flow rates of 16 and 4 μl/min, respectively. The results in Figs. 17 and 18 indicate that the ICP detection system is a mass-sensitive detector, because the signal intensity is linearly dependent on the flow rate of the mobile phase, distilled water.

The effect of solute structure and physical properties such as volatility on the response of the ICP was also expected, as shown in the intensity difference between ethanol and ethylene glycol in Fig. 17. To ascertain whether the chemical structure or physical properties of solute molecules influenced the detector response, the signal intensities for some solutes were measured under the same experimental conditions, and the results are summarized in Table 4. It seems that solutes with higher volatility give higher signal intensities. McGuffin and Novotny [36] also stated that some discrimination occurs on the basis of volatility in their report on a thermionic detector for microcolumn LC.

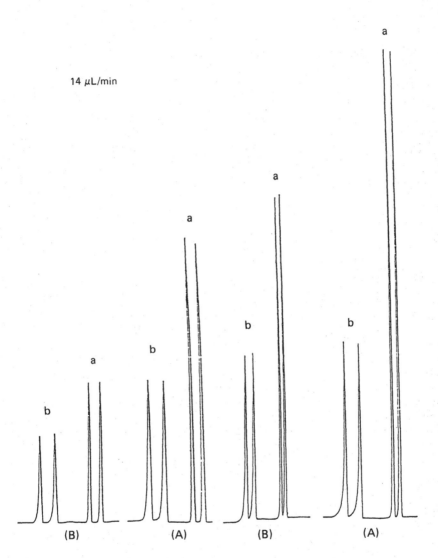

Fig. 17   Effect of mobile phase flow rate on emission signal intensity
of carbon atomic line at 247.86 nm.   Sample Ar gas flow rate: (A)
0.45 liter/min; (B) 0.60 liter/min.   Sample: (a) ethanol (120 µg of
carbon); (b) ethylene glycol (120 µg of carbon).

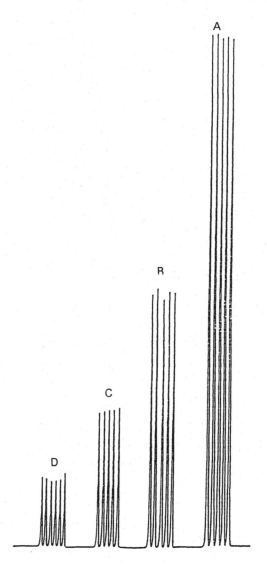

Fig. 18   Relationship between emission signal intensity and amount of carbon.   Sample, ethanol; (A) 120 µg of carbon; (B) 60 µg of carbon; (C) 30 µg of carbon; (D) 15 µg of carbon.   (Copyright American Chemical Society, 1985.)

TABLE 4   Emission Signal Intensities of Various Organic Compounds Under the Optimized ICP Conditions[a]

Compound	Intensity / μg of carbon (mm/μg)	bp (°C)	Viscosity (cP)	Vapor pressure (mm Hg)
Chloroform	1.45	61.3	–	160
Methanol	1.37	64.7	0.611	99
Ethanol	1.16	78.3	1.19	120
1-Propanol	1.00	97.5	2.20	15
2-Propanol	1.07	82.4	2.39	33
1-Butanol	0.55	117.5	2.95	4
Acetone	1.13	56.3	0.322	175
Benzene	0.36	80.1	–	76
Toluene	0.38	110.6	–	21
Acedic acid	1.02	117.8	1.22	12
n-Hexane	0.61	68.8	0.32	120
Cyclohexane	0.54	80.8	0.97	82
Ethyl acetate	0.48	76.8	–	74
Ethylene glycol	0.33	197.3	21	–

[a]Sample, 0.3 μl injection; flow rate, water, 16 μl/min; sample Ar gas flow rate, 0.45 liter/min.

Source: From Ref. 31.

Performance in Use of Organic Solvents as Mobile Phase

The no-spray chamber nebulizing system can be used in normal-phase liquid chromatography with pure organic mobile solvents such as toluene and chloroform.   The difficult problems encountered, for example, in conventional HPLC-ICP when organic solvents are used as mobile phases are overcome by the use of microcolumn separations and no-spray chamber systems.

To investigate the effect of sample Ar gas flow rate and viewing height on emission signal intensity, toluene was used as the mobile phase and the range of the sample Ar gas flow rate was varied from 0.3 to 0.6 liter/min. The Cu-DDTC solution, which contained 150 ng of copper, was injected as the test probe into the toluene stream (as organic solvent mobile phase) and the Cu emission line at 324.75 nm was detected. The results are shown in Figs. 19 and 20. The maximum Cu emission intensity was obtained at 15 mm above the load coil regardless of the sample Ar gas flow rate and the flow rate of the mobile phase.

Figure 20 shows the results for a viewing height of 15 mm above the load coil with the no-spray chamber system. As can be seen, the maximum emission signal intensity was measured at a sample Ar gas flow rate of about 0.45 liter/min when using organic solvents as the mobile phase.

The reproducibility of emission signals obtained when a no-spray chamber nebulizing system was used with an organic solvent as the mobile phase was also evaluated by repetitive injections of 0.3 µl of toluene solution containing 150 ng Cu-DDTC into a fused silica capillary column packed with 5-µm silica particles at 8 µl/min toluene mobile solvent flow rate. The signal intensity of the Cu atomic emission line at 324.75 nm was monitored. The results in Fig. 21 show that very high reproducibility of the emission signals can be obtained with the no-spray chamber nebulizing system and microcapillary LC-ICP combination using an organic solvent mobile phase. In this case, it is not necessary to heat up the bottom of the plasma torch because of the high volatility of organic solvents.

The response of the ICP for separations with organic solvents as mobile phases was measured, and typical results are illustrated in Figs. 22 and 23. The emission intensities of Cu at 342.75 nm are compared at toluene flow rates of 16 and 4 µl/min with two different sample Ar gas flow rates, 0.45 and 0.60 liter/min, respectively. Figure 23 also shows that the ICP detector with toluene mobile solvent is the same as that in the water mobile phase system. The detection limit of Cu is less than 1 ng and the dynamic range is greater than 10^3. It indicates that no restructions are found on the mobile phase utilized, whether aqueous or organic.

Application

One of the most interesting fields of analytical chemistry is the isolation and identification or organometallic components in samples such as petroleum oils and biological fluids. The study has been focused on separation, identification, and quantitation techniques. Fish and Komlenic [37,38] have reported on the molecular characterization and profile identification of vanadyl compounds in heavy crude petroleums by LC-AAS.

Microcolumn LC-ICP was applied in the separation of a shale oil (NBS SRM-1580). An SEC column (0.50 mm i.d. × 600 mm long

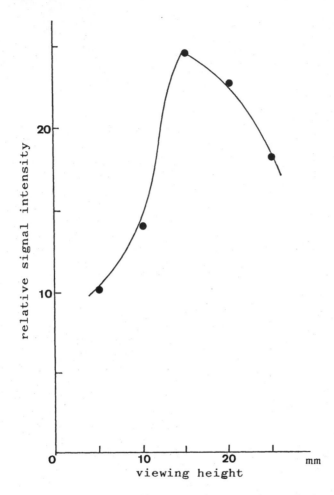

Fig. 19 Effect of viewing height on Cu emission signal intensity at 324.75 nm. Mobile phase, toluene; flow rate, 16 μl/min; sample, Cu-DDTC (150 ng of Cu).

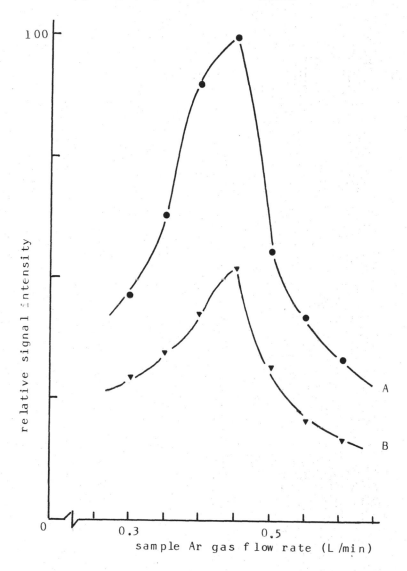

Fig. 20 Effect of mobile phase flow rate on Cu emission signal intensity at 324.75 nm. Mobile phase, toluene; flow rate (A) 16 μl/min, (B) 4 μl/min; sample, Cu-DDTC.

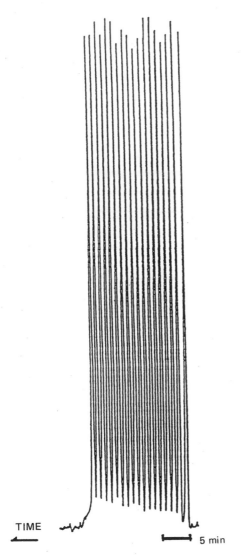

TIME

5 min

Fig. 21 Reproducibility of Cu emission signal intensity with toluene as the mobile phase. Mobile phase flow rate, 8 µl/min; sample, Cu-DDTC; detection, Cu 324.75 nm. (Copyright Dr. Alfred Huethig Verlag, 1985.)

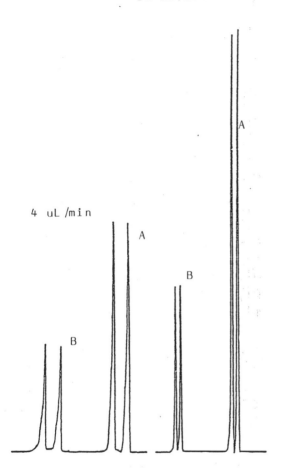

16 uL /min

4 uL /min

Fig. 22 Effect of mobile phase flow rate on Cu emission signal intensity at 324.75 nm. Sample Ar gas flow rate (A) 0.45 liter/min, (B) 0.60 liter/min; sample, Cu-DDTC. (Copyright American Chemical Society, 1985.)

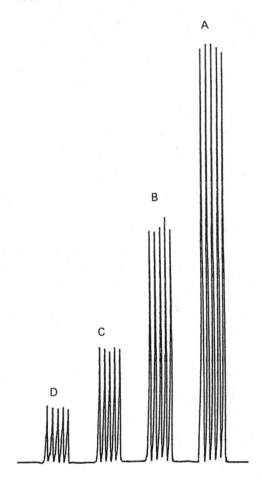

Fig. 23   Relationship between emission signal intensity and amount of copper.   Sample, Cu-DDTC; (A) 500 ng of Cu; (B) 250 ng of Cu; (C) 125 ng of Cu; (D) 63 ng of Cu.

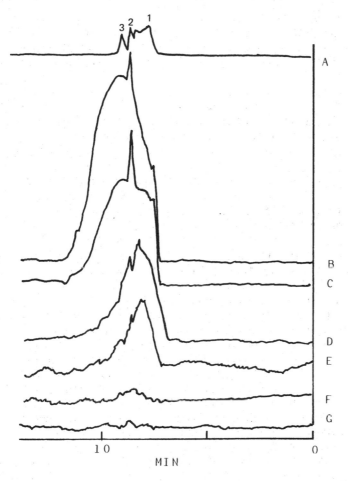

Fig. 24 SEC chromatograms of NBS SRM-1580 shale oil. Column, 0.53 mm i.d. × 600 mm long packed with TSK Gel 1000-H; mobile phase, chloroform; flow rate, 8 μl/min. (A) UV chromatogram of polystyrene standards monitored at 254 nm; molecular weights: (1) 900, (2) 370, (3) 266; (B) UV chromatogram of shale oil monitored at 254 nm; (C) UV chromatogram of shale oil monitored at 300 nm; (D) ICP chromatogram of shale oil monitored at 279.55 nm magnesium emission line; (E) ICP chromatogram of shale oil monitored at 238.20 nm iron emission line; (F) ICP chromatogram of shale oil monitored at 309.31 nm vanadium emission line; (G) ICP chromatogram of shale oil monitored at 396.15 nm aluminum emission line. (Copyright American Chemical Society, 1985.)

fused silica capillary packed with TSK Gel 1000-H; Tosoh, Tokyo) that excludes compounds of molecular weight larger than 1000 was used, because it has been reported that most interesting organometallic compounds in petroleum samples have molecular weights less than 900 [39].

Figure 24 shows the SEC UV chromatograms of polystyrene standards and of NBS SRM-1580 shale oil monitored by the UV and ICP detectors. It indicates that the shale oil contains mainly compounds with molecular weights of a few hundred. A sharp peak in the UV chromatograms is observed around molecular weight 400 calibrated with polystyrene. The ICP element-specific detection indicated to us that iron- and magnesium-containing compounds distribute widely in a range of molecular weights larger than 500, while vanadium-containing compounds distribute in a range of molecular weights smaller than those of other metal compounds. The elution time of the sharp peak appearing in the UV chromatograms is almost consistent with that of the vanadium peak. This result can be explained by the fact that vanadyl porphrin compounds have molecular weights less than 600, although complexation in the asphaltene fraction of the oil could drastically increase their apparent molecular weight [38]. Even though the chromatogram monitored at the aluminum emission line is not clear because of its low sensitivity, aluminum-containing compounds also seem to be present at molecular weights around 300. This example strongly implies that the microcapillary LC-ICP system can provide a molecular weight categorization of organometallic compounds present in shale oil. With this technique, identification of organometallic compounds in petroleum samples can be effectively attempted, since the method provides information on molecular weight and spectroscopic characterization along with elemental composition. Thus, it is expected to widen application, especially in the fields of oil chemistry and biochemistry, as many important substances have metal atoms in their functional sites.

CONCLUSION

In the coupling of microcolumn LC to ICP, the distance between the microcolumn and the plasma flame must be shortened to prevent peak broadening and sensitivity loss due to the loss of nebulized sample aerosols in the ICP sample introduction part. With the no-spray chamber, the ICP system has been demonstrated to be a reliable detector for microcolumn LC without changing the spectrometric characteristics of the ICP system.

The results of the instrumental investigations described in this chapter indicate that the technique is a viable approach for selective detection of trace metals in complex mixtures such as biological and petroleum samples. Further improvements and optimization of the microcolumn LC and ICP systems may offer better detection limits than those available with conventional ICP systems.

REFERENCES

1. D. R. Jones and S. E. Manahan, *Anal. Chem. 48*, 502 (1976).
2. J. F. Dankow and G. E. Janauer, *Anal. Chim. Acta 69*, 47 (1974).
3. J. C. Van Loon, B. Radziuk, N. Kahn, F. J. Fernandez, and J. D. Kerber, Pittsburgh Conference on Analytical Chemistry and Applied Spectroscopy, February 28—March 4 (1977).
4. J. C. Van Loon, B. Radziuk, N. Kahn, J. Lichwa, F. J. Fernandez, and J. D. Kerber, *At. Absorpt. Newsl. 16*, 79 (1977).
5. K. I. Irgolic, E. A. Woolson, R. A. Stockton, R. D. Neuman, N. R. Bottina, R. A. Zingaro, P. C. Kearney, R. A. Pyles, S. Maeda, W. J. McShane, and E. R. Cox, *Environ. Health Perspect. 19*, 61 (1977).
6. V. A. Fassel and R. N. Knisely, *Anal. Chem. 46*, 1110A (1974).
7. V. A. Fassel and R. N. Knisely, *Anal. Chem. 46*, 1155A (1974).
8. D. M. Farely, D. A. Yates, and S. E. Manahan, *Anal. Chem. 51*, 2225 (1979).
9. M. Morita, T. Uehiro, and K. Fuwa, *Anal. Chem. 51*, 349 (1980).
10. C. H. Gast, J. C. Kraak, H. Poppe, and F. J. M. J. Maessen, *J. Chromatogr. 185*, 549 (1979).
11. R. S. Whaley, K. R. Snable, and R. F. Browner, *Anal. Chem. 54*, 162 (1982).
12. W. S. Gardner, P. F. Laundrum, and D. A. Yates, *Anal. Chem. 54*, 1196 (1982).
13. D. W. Hausler and L. T. Taylor, *Anal. Chem. 53*, 1223 (1981).
14. D. R. Heine, M. B. Deuton, and T. D. Schlabach, *Anal. Chem. 54*, 81 (1982).
15. M. Morita, T. Uehiro, and K. Fuwa, *Anal. Chem. 53*, 1806 (1981).
16. M. Morita and T. Uehira, *Anal. Chem. 53*, 1997 (1981).
17. K. Yoshida, T. Hasegawa, and H. Haraguchi, *Anal. Chem. 55*, 2106 (1983).
18. K. Yoshida and H. Haraguchi, *Anal. Chem. 56*, 2580 (1984).
19. K. E. Lawrence, G. W. Rice, and V. A. Fassel, *Anal. Chem. 56*, 292 (1984).

20. M. W. Dong and J. L. DiCesare, *J. Chromatogr. Sci.* *20*, 49
     (1982).

21. J. L. DiCesare, M. W. Dong, and F. L. Vandemach, *Am. Lab.*
     (Fairfield, Conn.) *13*, 52 (1981).

22. D. Ishii, K. Asai, K. Hibi, T. Jonokuchi, and M. Nagaya,
     J. Chromatogr. *144*, 157 (1977).

23. R. P. W. Scott and P. Kucera, *J. Chromatogr.* *125*, 251 (1976).

24. F. L. Yang, *J. Chromatogr. Sci.* *20*, 241 (1982).

25. Y. Hirata and K. Jinno, *J. High Resolut. Chromatogr.*
     Chromatogr. Commun. *6*, 196 (1983).

26. K. Jinno, Y. Hiyoshi, and Y. Hirata, *J. High Resolut.*
     Chromatogr. Chromatogr. Commun. *5*, 102 (1982).

27. K. Jinno and H. Tsuchida, *Anal. Lett.* *15*, 427 (1982).

28. K. Jinno, H. Tsuchida, S. Nakanishi, Y. Hirata, and
     C. Fujimoto, *Appl. Spectrosc.* *37*, 258 (1983).

29. K. Jinno and S. Nakanishi, *J. High Resolut. Chromatogr.*
     Chromatogr. Commun. *6*, 210 (1983).

30. K. Jinno, S. Nakanishi, and T. Nagoshi, *Anal. Chem.* *56*,
     1977 (1984).

31. K. Jinno, S. Nakanishi, and T. Nagoshi, *Chromatographia 18*,
     437 (1984).

32. K. Jinno, S. Nakanishi, and C. Fujimoto, Proceedings of the
     6th International Symposium on Capillary Chromatography,
     Riva del Garda, Italy (May 1985).

33. K. Jinno, S. Nakanishi, and C. Fujimoto, *Anal. Chem.* *57*,
     2229 (1985).

34. P. Barrett and E. Pruszkowska, *Anal. Chem.* *56*, 1927 (1984).

35. R. F. Browner and D. D. Smith, *Anal. Chem.* *55*, 374 (1983).

36. V. L. McGuffin and M. Novotny, *Anal. Chem.* *55*, 2296 (1983).

37. R. H. Fish and J. J. Komlenic, *Anal. Chem.* *56*, 510 (1984).

38. R. H. Fish and J. J. Komlenic, *Anal. Chem.* *56*, 2452 (1984).

39. J. M. Sugiyama, J. F. Branthaver, G. Y. Wu, and
     C. Weatherbee, *Prepr. Am. Chem. Soc. Div. Pet. Chem.* *15*,
     C5 (1970).

7

Practice and Applications of On-Line Multidimensional Chromatography Using Micro–HPLC and Capillary GC

HERNAN J. CORTES / Dow Chemical Company, Midland, Michigan

INTRODUCTION

In recent years the efficiencies of chromatographic separations have increased steadily, with extensive use of capillary columns for gas chromatography (GC) [1], trends toward smaller particle packings [2], and more efficient columns in liquid chromatography (LC) [3–5]. Davis and Giddings [6] indicated through statistical theory that the possibilities of peak overlap in complex samples are much greater than generally recognized, pointing out that a chromatogram for a mixture "must be 95% vacant to insure a 90% probability that a given peak will not be overlapped by another component." Although the assumption of random distribution of peaks is generally not valid, the theory presented indicates the need for advances yet to be made in order to obtain true "high-resolution chromatography."

One approach commonly used to obtain greater resolution of complex mixtures is the technique of multidimensional chromatography. In this method, selected components separated on one particular system are transferred in some manner to a second separation system, where further resolution can be obtained. Multidimensional systems are useful when complex samples are analyzed and the resolution obtained on one system is not sufficient.

Multidimensional separations can be conducted in two ways: off-line and on-line. In off-line applications, fractions eluting from one separating system are collected, either manually or by a fraction collector, and can be concentrated, derivatized, or treated in some manner prior to introduction into a second separating system. Off-line techniques are widely used because of the relative simplicity of collection of effluents from a chromatographic system, especially in the liquid phase, and they also allow the removal of incompatible solvents, if necessary, prior to introduction into the second chromatographic system. Off-line techniques, however, are generally time-consuming, operator-intensive, and difficult to automate and reproduce. More important, the possibility of inadvertently losing or changing the sample in a detrimental manner is present.

In on-line applications, two or more chromatographic systems are connected in series and selected fractions eluting from the first system are introduced into the second by means of appropriate valves or pressure control. On-line techniques offer the advantages of automation simplicity and in general are less time-consuming and more reproducible. However, the initial setup of a multidimensional system requires more expense, and loss in sensitivity can occur by dilution effects if on-column concentration does not take place.

Numerous examples of multidimensional chromatographic separations exist. One of the earliest ones is still used in thin-layer chromatography [7], where a single plate is developed conventionally, then subsequently developed in the second dimension by rotating the plate 90° and using a developing solvent with different characteristics. Gas chromatographic multidimensional separations have been accomplished using both packed and capillary columns [8–10], and various instruments are commercially available.

Liquid chromatographic multidimensional separations have also been employed, utilizing size exclusion chromatography [11], ion chromatography [12], and normal-phase [13,14] and reversed-phase [15] system combinations.

LC offers some advantages over GC; some separations are more readily attainable because both the mobile and stationary phases can influence the separation strongly, compared to only the stationary phase in conventional GC. A variety of phases are available and separations are usually conducted at low temperatures, which lead to better resolution of thermally labile samples. In addition, non-volatile, ionizable, and very polar compounds can be routinely chromatographed. On the other hand, GC offers a wide range of more sensitive and selective detectors, greater resolving power, and can benefit from small sample requirements.

The coupling of liquid chromatography with gas chromatography offers a different perspective on multidimensional separations.

Because of the basic differences in the techniques, selectivities that are difficult to obtain with multidimensional systems using either liquid or gas chromatography alone can in principle be obtained using the wide range of mobile phases, stationary phases, temperature profile combinations, and detector systems that are possible by combining these two dissimilar techniques. The aim of this chapter is to review the on-line coupling of LC and GC with emphasis on the more recent use of micro-LC columns in multidimensional chromatography utilizing capillary GC columns as the secondary separation mode.

INTERFACES

Autosamplers

The first reported on-line coupling of liquid chromatography and gas chromatography was by Majors et al. [16,17], who connected a conventional LC to an autosampler. In this application, the effluent from a liquid chromatographic detector is directed to a flow-through syringe and goes to a waste container, the syringe being continuously flushed with liquid chromatographic solvent. A signal from the electronics module initiates the injection cycle; the syringe is lifted from the waste container and lowered into the GC injector port, where the syringe plunger is depressed to make the injection. The connection between the LC detector and the side-arm syringe is made via a 0.009-in.-i.d. Teflon tube (Fig. 1). The system was used successfully for the determination of atrazine in sorghum fodder, as illustrated in Fig. 2.

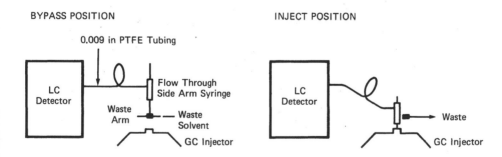

BYPASS POSITION INJECT POSITION

0.009 in PTFE Tubing

LC Detector — Flow Through Side Arm Syringe — Waste Arm — Waste Solvent — GC Injector

LC Detector — Waste — GC Injector

Fig. 1 Schematic diagram of the interface for an automated on-line LC-GC multidimensional system. (Reproduced with permission from Ref. 17. Copyright Preston Publications, Inc.)

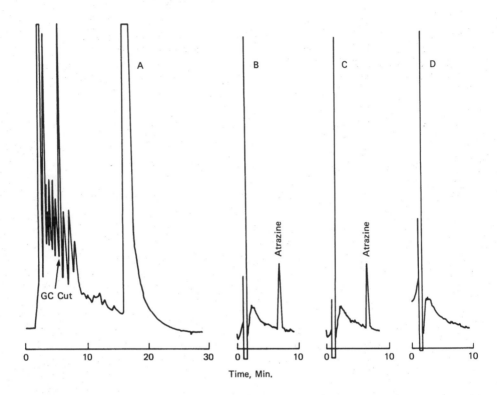

Fig. 2 Chromatograms of atrazine in sorghum: (A) HPLC chromatogram of sorghum extract containing atrazine (15 cm × 4 mm column containing cyano bonded phase, 2 cm^3/min flow rate, 2% isopropanol in hexane mobile phase; column was flushed for 1 min with
100% isopropanol at 18 min mark to remove strongly retained components). (B) GC chromatogram of HPLC/GC cut from chromatogram
A (25 m OV-101 glass capillary column, 8 μl splitless injection, 200°C
isothermal column temperature, nitrogen selective thermionic detector). (C) GC chromatogram of atrazine standard carried through
entire HPLC/GC procedure (0.2 ppm atrazine, conditions same as
for chromatogram B). (D) GC chromatogram of sorghum control
carried through entire HPLC/GC procedure (conditions same as for
chromatogram B). (Reproduced with permission from Ref. 17.
Copyright Preston Publications, Inc.)

More recently, Apffel and McNair [18] utilized a similar approach for the fractionation of gasoline and diesel fuel, where saturates, olefins, and aromatics were separated from each other by LC and the peak maximum of each family of compounds was introduced into a capillary GC (Fig. 3).

Although the utilization of an autosampler as described for the introduction of effluent from a liquid chromatograph to a gas chromatograph can be applied in some cases, a major limitation of this approach is the small amount of liquid volume that can be injected into a capillary GC column, an obvious incompatibility when using conventional LC columns that are operated at flow rates of 0.4 to 3.0 ml/min. Because of this limitation, only a small fraction of a selected peak can be sampled for introduction into the GC column and the injections made represent only a portion of the LC peak; thus, "it would be impossible to take one sample for GC analysis that is representative of the LC fraction" [18]. In addition, the reproducibility of the analyses can suffer if sampling does not occur at exactly the same place in the peak profile on subsequent injections, and quantitative results can be difficult.

Retention Gap

The introduction of relatively large volumes of sample into a capillary GC column has been studied for some time. For this technique to be successful, various criteria must be met: No overloading (solvent or concentration) should take place, and the sample components of interest should not be distorted or broadened.

Grob and Grob [19] described focusing in the stationary phase, a process in which large amounts of sample can be introduced into a capillary GC column without overloading by maintaining the column temperature at "100°C or more below the boiling point of the most volatile solute," the stationary phase acting to focus the components of interest at the head of the analytical column. This process has been described as the solvent effect [20—22], where a dilute vapor or liquid sample forms a film at the inlet of the system which subsequently evaporates, transferring the focused solutes to the column. This approach is applicable, however, to relatively small injection volumes when utilizing on-column injections, as illustrated in Fig. 4.

When the solute is not effectively focused at the head of the capillary GC column, a split peak pattern is observed. This is created by the large amount of condensed solvent spreading out the solute at the column inlet. This effect has been described as "band broadening in space" by Grob, and the concept of the "retention gap" was introduced to overcome this problem [23,24].

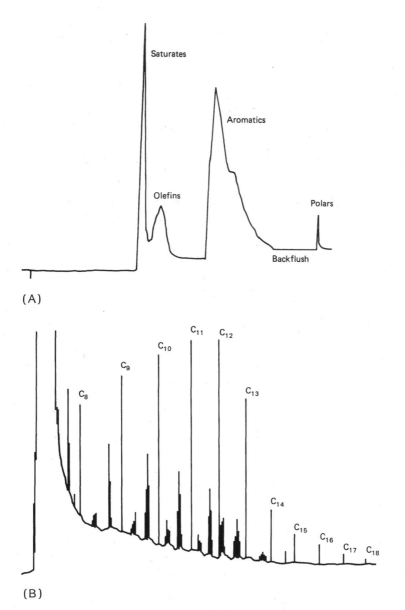

Fig. 3 (A) HPLC fractionation of Texaco unleaded gasoline. Conditions: column, 32 × 4.4 mm i.d., 5-μm silica; mobile phase, hexane, 0.5 ml/min; refractive index detection at ×32 attenuation; sample, 10 μl. (B) LC-GC analysis of saturates in Texaco unleaded

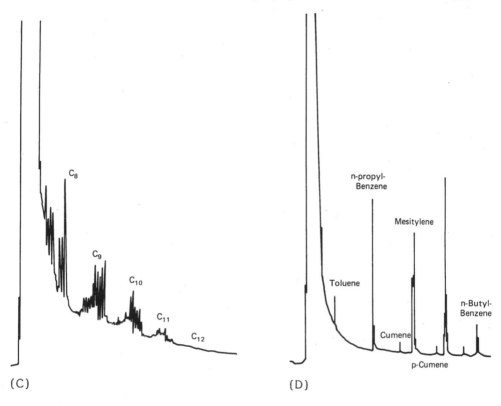

(C)

(D)

Fig. 3 (Continued)

gasoline. LC conditions as in (A). GC conditions: column, 33 m × 0.25 mm i.d. WCOT SP-2100; carrier gas, helium at 2 ml/min; temperature program, 40°C for 15 min, 2°C/min to 220°C; flame ionization detection, 8×10^{-11} AUFS; 2-µl splitless injection; sample diluted 1:10 in hexane; LC sampled at 6.81 min. (C) LC-GC analysis of olefins in Texaco unleaded gasoline. LC sampled at 9.05 min. (D) LC-GC analysis of aromatics in Texaco unleaded gasoline. LC sampled at 12.95 min. (E) LC-GC analysis of aromatics in Texaco unleaded gasoline. LC sampled at 14.95 min. (Reproduced with permission from Ref. 18. Copyright Elsevier Scientific Publishing Company.)

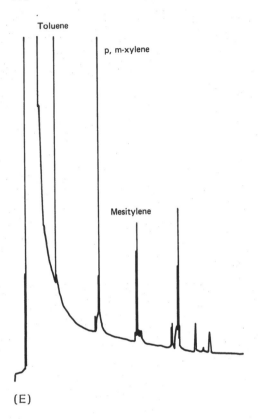

(E)

Fig. 3   (Continued)

In the retention gap approach, an inlet section of the column containing no stationary phase, and therefore having minimal reten- tion of the solutes of interest when compared to the separating sec- tion, is used, allowing reconcentration of the solutes of interest at the front of the film of the stationary phase. The mechanism is depicted in Fig. 5.

The use of a retention gap to introduce LC effluent into a capil- lary GC was described [25]. In this application, a conventional LC column (100 × 3 mm i.d.) was packed with Spherisorb S-5W and op- erated with a nonpolar eluent (cyclohexane) at a flow rate of 0.4 ml/ min. The LC effluent was introduced into a capillary GC to qual- itatively identify the components in a methanol extract of tooth- paste. The components of interest were anethole and guajazulene. No information was supplied on the concentrations of the components present in the toothpaste. The chromatograms obtained are depicted

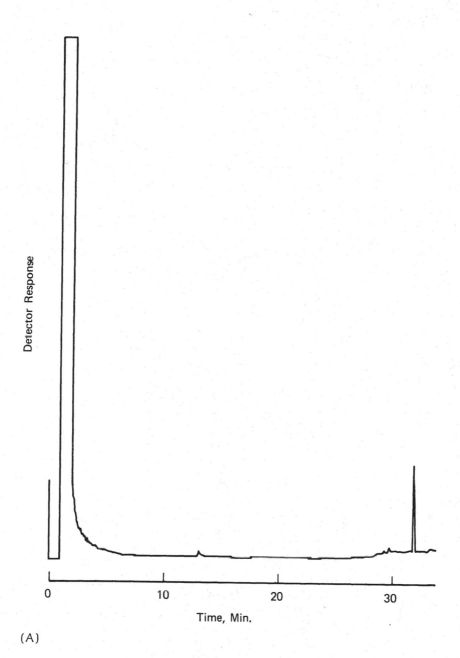

(A)

Fig. 4   Effect of injection volume on peak shape for on-column injection.   (A) 1.0 μl of 1.0 μg/ml 3,5-dichlorobiphenyl. Conditions: column, 30 m × 0.25 mm i.d. Supelcowax 10, 0.25-μm film; program, 115°C for 7 min, 5°C/min to 240°C; carrier, helium at 68 cm/s; detector, flame ionization; attenuation, ×4. (B) 10.0 μl injection; attenuation, ×16; other conditions as in (A).

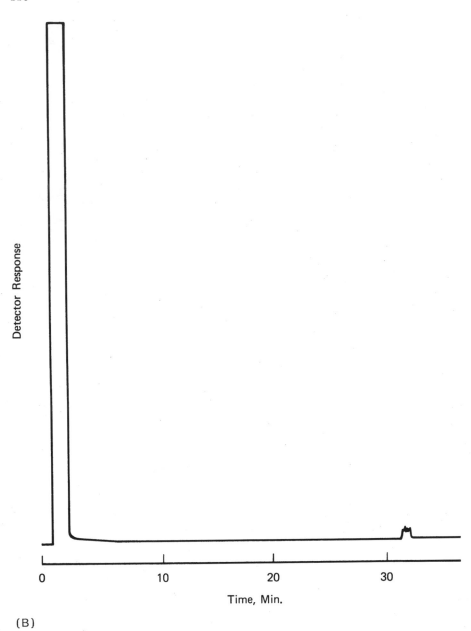

(B)

Fig. 4 (Continued)

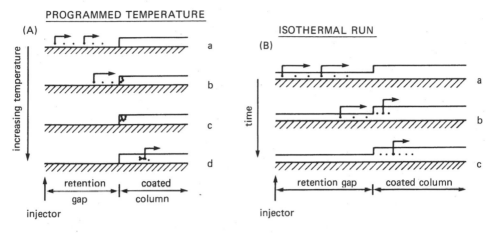

Fig. 5 Reconcentration mechanism of the retention gap: summary of the two explanations given previously, both of which are simplified and account for one aspect of the phenomenon each. (A) Reconcentration in a temperature-programmed run. (a) The sample is spread out in the retention gap section. (b) At a certain column temperature a sample component is enabled to migrate in the section of low retention. However, the material is unable to migrate in the regularly coated column. (c) The component is blocked at the beginning of the regular coating. (d) At an increased temperature, the reconcentrated band of the compound begins chromatography. (B) A hypothetical isothermal run; reconcentration by an insufficient retention gap with still half of the film thickness of the stationary phase (or half the retention) compared to the regularly coated part of the column. Accordingly, a component migrates through the retention gap zone at double the speed in the normally coated column. Assume that the first molecules of the band of, e.g., 50 cm width have reached the beginning of the regular coating. The last molecules are 50 cm back. Where are the first molecules at the moment when the last molecules have reached the beginning of the regular coating situation (c)? As the forward molecules migrated at only half the speed of the rear molecules, they advanced only 25 cm. Hence the bandwidth is reduced to half. An effective retention gap, however, should reduce the bandwidth by a factor of 10 to 100 (depending on the conditions). Thus the film thickness of the stationary phase needs to be reduced accordingly. (Reproduced with permission from Ref. 24. Copyright Elsevier Scientific Publishing Company.)

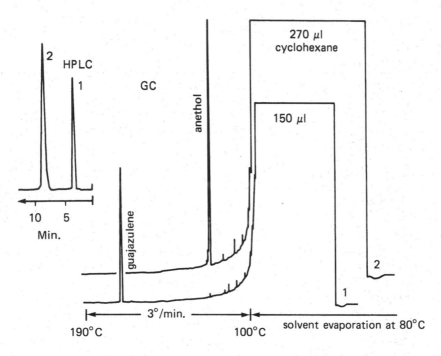

Fig. 6 LC-GC of two solutes visible on the HPLC trace (left), run for the analysis of azulene dyestuffs in a toothpaste. Peak 1 was transferred with 150 μl of eluent and coeluted in the GC system with guajazulene. Peak 2 was identified as anethole by LC-GC-MS. (Reproduced with permission from Ref. 25. Copyright Elsevier Scientific Publishing Company.)

in Fig. 6. The LC chromatogram in this application is fairly clean, and the components are well resolved, making the identification by GC retention times simple. It would seem ideal to have introduced both components into the capillary GC simultaneously, in order to obtain the required information in one chromatographic run. However, a 50-m retention gap was used, implying that the effluent volume may have been too large to allow the introduction of both components in one run.

MICROCOLUMN LC-GC COUPLING*

The reduction of the internal diameter of the LC column utilized for multidimensional chromatographic applications to microcolumn

---

*The method and apparatus described are the subjects of pending patent applications.

dimensions (0.1—2.0 mm) introduces various advantages to the technique: The volume of eluent used is significantly reduced, which means the solutes of interest are diluted in much less eluent. Much larger sections of the LC chromatogram can be introduced into the capillary GC, allowing quantitative transfer of the components of interest, resulting in greater reproducibility and better opportunity for quantitative analyses. Because microcolumns can be effectively prepared at lengths greater than the conventional 25 cm, greater total column efficiencies can be obtained.

The first application of multidimensional LC-GC utilizing microcolumns and reversed-phase eluents was described in 1985 [26]. A block diagram of the system used is shown in Fig. 7. A complex hydrocarbon sample (coal tar extract) was analyzed for the presence of two polychlorinated biphenyls. The LC system consisted of a fused silica capillary (95 cm × 250 μm i.d.) packed with Zorbax ODS of 7-μm particle diameter. Acetonitrile was used as the eluent. The length of the retention gap was 4 m and the volume introduced into the capillary GC was 40 μl. The chromatograms obtained are presented in Fig. 8. The polychlorinated biphenyls of interest are resolved from all other components in the mixture, allowing a separation that was not possible with HPLC or GC alone.

In another application [27], the technique was used to analyze a sample of fuel oil for the presence of various chlorinated benzenes. In this case, a normal-phase eluent was used in a capillary fused silica column packed with Zorbax silica of 7-μm particle diameter. The liquid chromatograph effectively removed all interfering components from the sample, allowing quantitation of the components of interest. The detection limits obtained in this application ranged from 8 to 17 μg/g with a flame ionization detector (FID). The chromatograms obtained are shown in Fig. 9.

The determination of 2-chloro-N-isopropylacetanilide (Propachlor* herbicide) in soil was accomplished using a reverse-phase system with an eluent of methanol/water (90:10). A 50-g sample of soil was extracted with methylene chloride, filtered, and subsequently evaporated. The residue was dissolved in eluent, filtered, and injected into the LC system using a fused silica column of 102 cm × 250-μm i.d. packed with Spherisorb ODS of 5-μm particle diameter. Detection limits obtained were estimated as 3.6 μg/ml using an FID. The separations obtained are presented in the chromatograms in Fig. 10.

Chlorpyrifos [O,O-diethyl O-(3,5,6-trichloro-2-pyridyl) phosphothioate[in rodent feed used for toxicological studies was determined using a normal-phase system with an eluent consisting of heptane/methyl-t-butyl ether (95:5). Fifteen grams of rodent feed was extracted with 100 ml of benzene, the solvent was allowed to evaporate under a nitrogen stream, and the residue was dissolved

---

*Trademark of Dow Chemical Company.

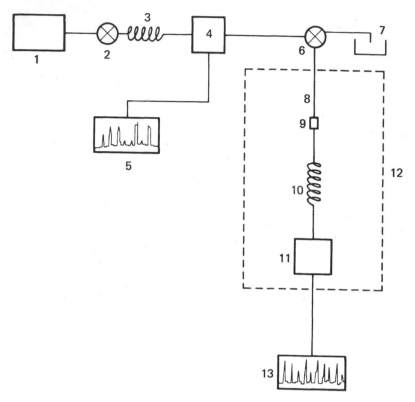

Fig. 7   Block diagram of on-line packed capillary LC-capillary GC
system.   (1) Pump; (2) injection valve; (3) packed capillary LC
column; (4) detector; (5) recorder; (6) switching valve; (7) effluent waste; (8) retention gap; (9) butt connector; (10) capillary GC
column; (11) detector; (12) gas chromatograph oven; (13) recorder.
(Reproduced with permission from Ref. 26.   Copyright Dr. Alfred
Huethig Publishers.)

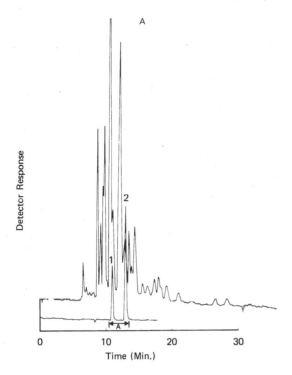

Fig. 8 (A) Chromatogram of coal tar on packed capillary LC system. Column, 96 cm × 250-μm i.d. packed with Zorbax ODS, dp = 7 μm; eluent, acetonitrile; flow, 8.0 μl/min; detector, Jasco Uvidec II at 220 nm, 0.01 AUFS; injection size, 60 nl. (1) 2,4'-Dichlorobiphenyl, (2) 2,3,4-trichlorobiphenyl. A = Section injected into capillary GC column. (B) Chromatogram of coal tar section from on-line packed capillary LC system. Column, 30 m × 0.25 mm i.d. Supelcowax 10, 0.25 μm film; oven, 115°C for 7 min, 5°C/min to 240°C; retention gap, 4 m × 0.25 mm i.d. fused silica; carrier, helium at 68 cm/s; injection size, 40 μl. (C) Chromatogram of coal tar section containing PCBs from packed capillary LC system. Conditions as in (B). (1) 2,4'-Dichlorbiphenyl, (2) 2,3,4-trichlorobiphenyl. (Reproduced with permission from Ref. 26. Copyright Dr. Alfred Huethig Publishers.)

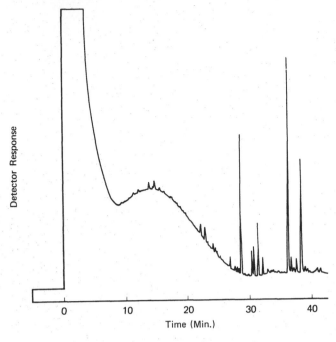

(B)

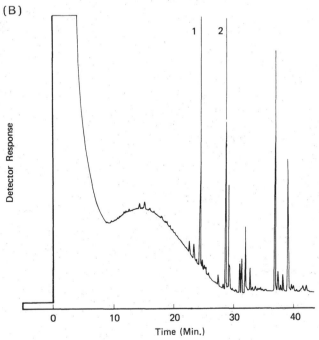

(C)

Fig. 8   (Continued)

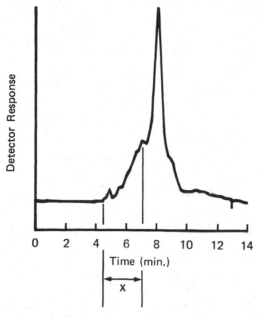

(A)

Fig. 9 (A) Micro-LC chromatogram of fuel oil sample. Column, 105 cm × 250 µm i.d. packed with Zorbax silica, d_p = 7 µm; mobile phase, heptane; flow, 10.6 µl/min; detector, Jasco Uvidex II at 214 nm; pressure, 3800 psig. Sample prepared to give concentration of 1.00 g/10 ml in heptane. X = section introduced into the gas chromatograph. (B) Capillary gas chromatogram of fuel oil from LC. Column, 30 m × 0.25 mm i.d. Supelcowax 10 (d_f = 0.25 µm); retention gap, 15 m × 0.25 mm fused silica; oven temperature, 105°C for 9 min, program to 245°C at 5°C/min; carrier, helium at 70 cm/s; makeup gas, nitrogen at 30 ml/min; detector, FID at 275°C; injected volume, 21 µl. Retention times of chlorobenzenes of interest are indicated. (1) Chlorobenzene, (2) 1,2-dichloro-benzene, (3) 1,2,4,5-tetrachlorobenzene, (4) 1,2,3,4-tetrachloro-benzene, (5) pentachlorobenzene, (6) hexachlorobenzene. (Reproduced with permission from Ref. 27. Copyright Elsevier Scientific Publishing Company.)

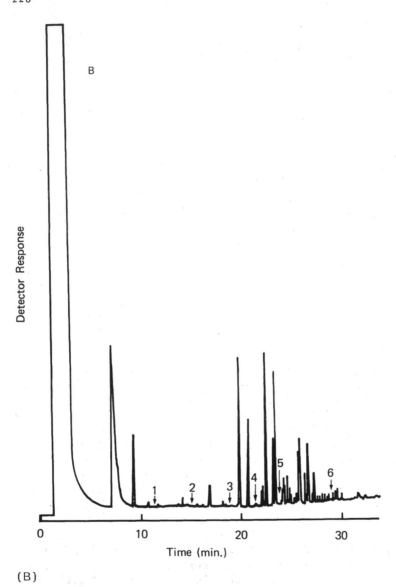

(B)

Fig. 9   (Continued)

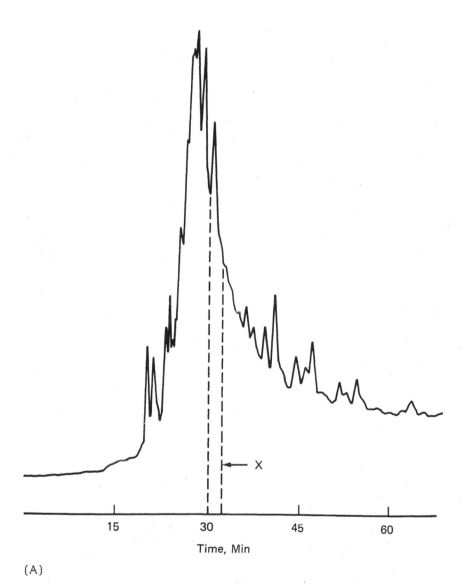

(A)

Fig. 10 Chromatograms of soil extract. (A) Micro-LC. Column, 105 cm × 250 μm i.d. fused silica packed with Spherisorb ODS (d_p = 5 μm); mobile phase, methanol/water (90:10); flow rate, 3.0 μl/min; detector, Jasco Uvidex II at 214 nm; injection size, 200 nl. X = Section introduced into the gas chromatograph. (B) Capillary GC. Column, 30 m × 0.25 mm J&W Carbowax (d_f = 0.5 μm); retention gap, 10 m × 0.25 mm i.d. fused silica; oven temperature, 100°C for 10 min, program to 230°C at 5°C/min; carrier, helium at 38 cm/s; makeup gas, nitrogen at 30 ml/min; detector, FID at 270°C. (1) 2-Chloro-*N*-isopropylacetanilide (14 μg/g).

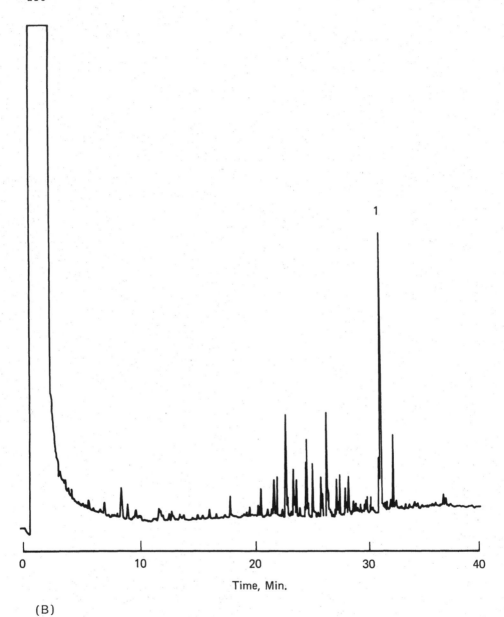

Time, Min.

(B)

Fig. 10 (Continued)

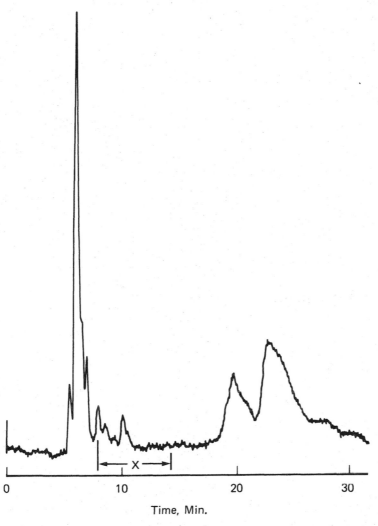

(A)

Fig. 11   Chromatogram of rodent feed.   (A) Micro-LC.   Column,
100 cm × 250 μm i.d. fused silica packed with Zorbax ODS (d_p =
7 μm); mobile phase, heptane/methyl-t-butyl ether (95:5); flow
rate, 12 μl/min; detector, Jasco Uvidex II at 214 nm; injection
size, 500 nl.   X = Section introduced into the capillary GC.   (B)
Capillary GC.   Column, 50 m × 0.20 mm i.d. 5% phenyl/methyl
silicone (d_f = 0.25 μm); retention gap, 10 m × 0.25 mm i.d. fused
silica; oven temperature, 130°C for 5 min, program to 250°C at 5°C/
min; carrier, helium at 35 cm/s; makeup, nitrogen at 30 ml/min;
detector, electron capture at 270°C.   (1) Chlorpyrifos (1.1 μg/g).

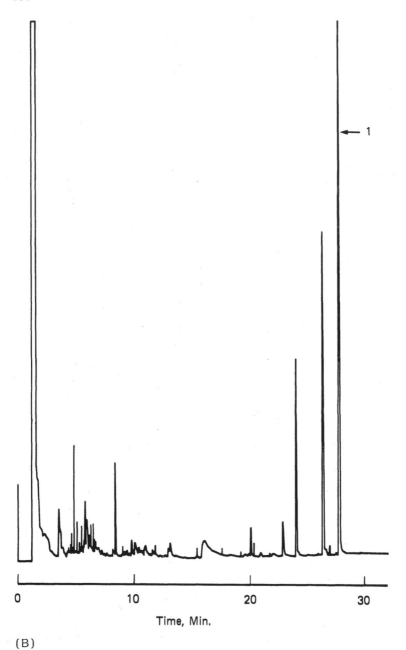

(B)

Fig. 11   (Continued)

in 5.0 ml of eluent. The resulting solution was filtered and injected
into the micro-LC system using a 110 cm × 250-μm i.d. fused silica
column packed with Zorbax silica of 7-μm particle diameter. The
section of the LC chromatogram known to contain the component of
interest was introduced into the capillary GC as previously described.
Detection limits obtained in this application were estimated as 20
parts per billion with an electron capture detector. The chromatograms obtained are presented in Fig. 11.

ADVANTAGES AND CONSTRAINTS

The multidimensional technique described offers various advantages
over previously reported multidimensional techniques. These include speed of analysis, minimal sample preparation, automation,
sensitivity, quantitation, and reproducibility. Some of the constraints of the technique are component volatility, interface dura-
bility, the type of GC columns that can be used, and eluent restrictions. These points are discussed in more detail in the following
sections.

Speed of Analysis

The LC system used is a highly efficient means of effectively removing undesirable interferences in the sample matrix in a relatively
short time. Manual extractions, preseparations using LC columns,
and tedious collection of cuts can be eliminated, especially when
analyzing samples that are liquids, in which a dilution may be all
the preparation necessary.

Sample Manipulation

Since pretreatment of the sample is often not needed, the integrity
of the sample is maintained, minimizing the possibilities of sample
loss through handling or changing the components of interest in a
detrimental manner.

Automation

A multidimensional system is easily automated once the appropriate
chromatographic conditions are established. An autoinjector is
needed to introduce the sample into the LC system. Utilizing time
control, the eluting sections of interest can be introduced repro-
dubibly into the GC. This can easily be accomplished with any
integration device equipped with external relay controls. The GC

can be modified to initiate the temperature program at a signal from the integrator.

Sensitivity

The components of interest are retained at the head of the GC column by utilization of a retention gap and judicious initial temperature selection. It is possible to make multiple injections into the GC system to increase the amount of material to be detected, as long as the initial temperature is maintained below the boiling points of the solutes of interest. By using inherently sensitive detectors, such as an electron capture, sensitivities in the low parts-per-billion range are attainable.

Quantitation

The use of microcolumns for LC allows the quantitative transfer of the total component(s) of interest into the GC, in contrast to only sections of a peak. There is therefore less dependence on absolute reproducibility of retention times in the LC. This advantage allows better quantitation of the material since all the component eluted from the LC column is transferred to the GC.

Reproducibility

In a well-designed micro-LC system, the chromatographic reproducibility is comparable to that obtained in a conventional LC system. Retention time variability of 1% is attainable and adequate. Validation data obtained in our laboratory yielded relative precisions at the 95% confidence level between 9.2 and 17.2% at concentrations of 0.9 parts per million. Recoveries of known quantities of the components of interest added to the matrix were found to range between 90 and 110%. These data are comparable to those obtained on conventional chromatographic systems.

Some consideration must be given to the areas in which the technique described may not be applicable and also to the experimental parameters that should be kept in mind when utilizing this technology. These constraints are discussed in the following sections.

Volatility

The components of interest must be retained at the head of the GC column at the initial temperature selected. This means that, in many cases, only compounds with boiling points above that of the solvent can be effectively trapped and quantitated. In addition, only compounds that can be chromatographed in the gas phase are amenable to the technique.

Interface Durability

For effective chromatography to take place, the retention gap used must have minimal retention for the components being determined. The introduction of components that are nonvolatile may not be avoided in some cases, and these components can collect in the retention gap to a point where they can act as a stationary phase, broadening the components of interest. By monitoring changes in peak shape and retention times, the time when the retention gap must be cleaned or replaced can be determined.

Capillary GC Columns

Because relatively large volumes of solvent pass through the GC column, the stationary phase of the column must be stable so as not to be stripped off by the eluent. With the increasing commercial availability of "bonded" or cross-linked phases that routinely withstand severe conditions such as those encountered when using supercritical fluids, this limitation becomes less significant but should still be considered.

Eluent Restrictions

The eluents used in the LC section of the technique have included acetonitrile, methanol, heptane, water, and others. It is expected that most other common mobile phases will behave in a satisfactory manner. However, nonvolatile buffer salts, which may be required in some applications, can create problems by deposting in the retention gap, and in severe cases plugging can occur.

CONCLUSIONS

Multidimensional chromatography, especially utilizing the combination of micro-LC and capillary GC, is a technique likely to become important in the analysis of complex matrices when a high degree of resolution is required. The use of a microcolumn for LC offers a desirable approach, helping to overcome some of the problems associated with the introduction of large volumes into a capillary GC.

Recent advances in the preparation and use of microcolumns have made the multidimensional approach feasible, as demonstrated by various examples in this chapter. Although limitations still exist, especially in the types of components that can be chromatographed in the gas phase, many unique analytical opportunities can easily be envisioned. Routine analyses where tedious manual cleanup is required can be automated and the time savings can be significant.

Continued research in microcolumn technology and interface performance and design may be expected to produce further improvements in this interesting area of chemical analysis.

REFERENCES

1. M. L. Lee, F. J. Yang, and K. D. Bartle, *Open Tubular Column Gas Chromatography, Theory and Practice*, Wiley, New York (1984).
2. J. Dicesare, M. Dong, and L. Ettre, *Chromatographia 14*, 257 (1981).
3. F. J. Yang, *J. Chromatogr. 236*, 265 (1982).
4. M. Novotny, V. McGuffin, A. Hirose, and J. Gluckman, *Chromatographia 17*, 303 (1983).
5. D. Ishii and T. Takeuchi, *J. Chromatogr. 285*, 97 (1984).
6. J. M. Davis and J. C. Giddings, *Anal. Chem. 55*, 418 (1983).
7. E. Stahl, ed., *Thin Layer Chromatography*, Springer-Verlag, New York (1969).
8. D. R. Deans, *Chromatographia 1*, 18 (1968).
9. H. Schomburg, K. Kotter, and F. Hack, *Anal. Chem. 45*, 1236 (1973).
10. W. Bersch, W. Anderson, and A. Holzerg, *Chromatographia 10*, 449 (1977).
11. J. N. LePage, W. Lindner, G. Davis, D. E. DeSeitz, and B. L. Karger, *Anal. Chem. 51*, 433 (1970).
12. T. Z. Bishay, *Anal. Chem. 44*, 1087 (1972).
13. R. G. Dolphin, R. W. Willmott, A. D. Mill, and L. J. Hoogeven, *J. Chromatogr. 122*, 259 (1976).
14. J. F. K. Huber, I. Fogy, and C. Fioresi, *Chromatographia 13*, 408 (1980).
15. G. C. Davis and P. T. Kissinger, *Anal. Chem. 51*, 1960 (1979).
16. R. E. Majors, E. L. Johnson, S. P. Cram, and A. E. Brown, III, Pittsburgh Conference on Analytical Chemistry and Spectroscopy, Cleveland, Ohio, Paper 115 (1979).
17. R. E. Majors, *J. Chromatogr. Sci. 18*, 571 (1980).
18. J. P. Apffel and H. McNair, *J. Chromatogr. 279*, 139 (1983).
19. K. Grob and G. Grob, *J. Chromatogr. Sci. 7*, 584 (1969).
20. F. J. Yang, A. C. Brown, and S. P. Cram, *J. Chromatogr. 158*, 91 (1978).
21. K. Grob and K. Grob, Jr., *J. High Resolut. Chromatogr. Chromatogr. Commun. 1*, 57 (1978).

22. W. Bersch, C. S. G. Phillips, and V. Pretorius, *J. High Resolut. Chromatogr. Chromatogr. Commun.* 6, 234 (1982).
23. K. Grob, Jr., *J. Chromatogr.* 237, 15 (1982).
24. K. Grob, Jr. and R. Muller, *J. Chromatogr.* 244, 185 (1982).
25. K. Grob, Jr., D. Frolich, B. Schilling, P. Neukom, and P. Nageli, *J. Chromatogr.* 295, 55 (1984).
26. H. J. Cortes, C. D. Pfeiffer, and B. E. Richter, *J. High Resolut. Chromatogr. Chromatogr. Commun.* 8, 469 (1985).
27. H. J. Cortes, C. D. Pfeiffer, B. E. Richter, and D. E. Jensen, *J. Chromatogr.* 349, 55 (1985).

8

Capillary Supercritical Fluid Chromatography: Practical Aspects

KARIN E. MARKIDES and MILTON L. LEE / Brigham Young University, Provo, Utah

DOUGLAS W. LATER / Lee Scientific, Inc., Salt Lake City, Utah

INTRODUCTION

Mobile phases in chromatography can be either gases, liquids, or supercritical fluids. For a given chromatographic system (same column and mobile phase), the optimum plate heights are identical, but the optimum mobile phase flow rates are highest for gases, lowest for liquids, and intermediate for supercritical fluids [1]. Therefore, analysis speeds (or efficiencies per unit time) are expected to increase in the sequence of liquid chromatography (LC), supercritical fluid chromatography (SFC), and gas chromatography (GC). This is a result of the decreasing densities and, hence, increasing solute diffusion coefficients for these mobile phases, respectively.

Mobile phase density also dictates the column type, packed or capillary, that is most practical to use. While capillary columns are preferred in GC because they offer excellent efficiency, packed columns are necessary in LC in order to facilitate mass transfer in the mobile phase because diffusion is too slow. This severely limits the total chromatographic efficiency that can be obtained in LC. In addition, there is an inherently greater pressure drop across a packed column than across a capillary column, and only short

packed columns can give acceptable flow rates at tolerable pressures. In comparison, capillary columns can be used in SFC, and
much greater total chromatographic efficiencies are obtainable. It
should be noted, however, that if high efficiency is not needed,
greater analysis speeds (plates per unit time) are achievable with
packed column SFC as compared to both LC and capillary SFC [2].

The superiority of capillary column GC for separations, in general, stems from (a) its unsurpassed total chromatographic efficiency
and (b) its compatibility with a wide variety of selective and nonselective detectors. Its limitation is that it is applicable only to
compounds that are thermally stable and that have sufficient volatility that they can be analyzed in the gas phase. Unfortunately,
the vast majority of organic compounds do not meet these criteria
and must be analyzed by other methods. While liquid chromatography eliminates the need for thermal stability and volatility, total
chromatographic efficiency and detector compatibility are compromised. In comparison, capillary SFC possesses many of the advantages of both GC and LC. Nonvolatile and thermally labile compounds can be chromatographed using a wide range of detectors
with efficiencies that approach those obtainable by capillary GC.

Capillary SFC was first demonstrated in 1981 using polycyclic
aromatic hydrocarbons as test solutes [3]. Since that time, the
technique has found application in such separations as fossil fuel
heavy-ends, drugs, pesticides, polymers, surfactants, dyes, fats,
carbohydrates, and antibotics. As instrumentation and technical
know-how mature, the influence of this new technique in separation
science will continue to increase. In this chapter, the practical
aspects of capillary SFC are outlined with emphasis on mobile phases,
column technology, instrumentation, and operating conditions.

MOBILE PHASES

Criteria for Selection

In gas chromatography, the mobile phase serves only as a carrier
and contributes little, if at all, to the solvation of the analyte or
selectivity of the separation. Variations in selectivity are generally
limited to modifications in the stationary phase. In both LC and
SFC, the mobile phase contributes to the selectivity because of its
solvating power, which is a result of its higher density as compared to gases. It seems reasonable to expect that as the density
increases, the solvating power and, hence, selectivity of the mobile
phase should also increase. By controlling the density of a supercritical fluid mobile phase via pressure control, the solvating ability

of the mobile phase can be adjusted. Pressure or density programming has become a widely applied control technique in SFC and is analogous to gradient programming in LC [4]. Density programming in SFC is duscussed in more detail later in this chapter.

The properties that should be considered in the selection of supercritical mobile phases for chromatography are critical pressure, critical temperature, density range, viscosity, selectivity, detector compatibility, stability, reactivity, and toxicity. Table 1 lists a number of supercritical fluids along with their critical parameters and corresponding liquid densities. Fluids with low critical temperatures have been the most preferred mobile phases because of their usefulness in the analysis of thermally labile compounds. Many large nonvolatile compounds that can be analyzed by SFC are thermally unstable at elevated temperatures, and it is therefore preferable to perform the separations at the lowest possible temperature. For this reason, in addition to its other favorable properties, carbon dioxide has been the fluid of choice in most studies to date.

Fluids with low critical pressures are also preferred because of the pressure limitations of available pumping systems. Currently, the syringe pumps typically used for capillary SFC have design limitations that restrict their use to pressures below approximately 400 atm. Therefore, fluids with critical pressures somewhat lower are required in order to provide a suitable pressure range for density programming. The wider the operating pressure range between the critical pressure and maximum attainable pump pressure, the greater is the density range and, hence, the greater the utility and selectivity achievable by density programming of the mobile phase. For example, supercritical carbon dioxide (at 40°C) has a usable density range of 0.19 g/ml (72 atm) to 0.96 g/ml (400 atm), while n-pentane (at 210°C) has a more limited density range of 0.09 g/ml (30 atm) to approximately 0.5 g/ml (400 atm).

The viscosities of supercritical fluids near their critical points are approximately 10^{-4} to 10^{-3} g/cm·s. This is only slightly higher than gas viscosities and is a factor of 10 to 100 times lower than liquid viscosities. Because of this, pressure drops across the column length are much smaller for supercritical fluids than those observed with liquid mobile phases. However, the effects of a pressure drop in supercritical fluid systems are more adverse than in liquids because of the density gradient caused by the pressure drop across the column. Such a density gradient would have obvious deleterious effects on chromatographic performance. Column pressure drops have been shown to decrease the selectivity and, hence, resolution in chromatographic systems using supercritical fluid mobile phases [2]. Open-tubular (capillary) columns and low-viscosity mobile phases can help eliminate this problem.

TABLE 1   Physical Parameters of Selected Supercritical Fluids

Fluid	T_c (°C)	P_c (atm)	ρ_c (g/ml)[a]	$\rho_{400\ atm}$ (g/ml)[b]	ρ_L [a,c] (g/ml)	
CO_2	31.3	72.9	0.47	0.96	0.71	(63.4 atm, 25°C)
N_2O	36.5	72.5	0.45	0.94	0.91	(sat., 0°C)
					0.64	(59 atm, 25°C)
NH_3	132.5	112.5	0.24	0.40	0.68	(sat., −33.7°C)
					0.60	(10.5 atm, 25°C)
$n\text{-}C_5$	196.6	33.3	0.23	0.51	0.75	(1 atm, 25°C)
$n\text{-}C_4$	152.0	37.5	0.23	0.50	0.58	(sat., 20°C)
					0.57	(2.6 atm, 25°C)
SF_6	45.5	37.1	0.74	1.61	1.91	(sat., −50°C)
Xe	16.6	58.4	1.10	2.30	3.08	(sat., 111.75°C)
CCl_2F_2	111.8	40.7	0.56	1.12	1.53	(sat., −45.6°C)
					1.30	(6.7 atm, 25°C)
CHF_3	25.9	46.9	0.52	—	1.51	(sat., −100°C)

[a]Data taken from Matheson Gas Data Book (1980) and from CRC Handbook of Chemistry and Physics, CRC Press (1984).

[b]$T_r = 1.03$.

[c]Calculated from compressibility data from G. N. Lewis and M. Randall, Thermodynamics (revised by K. S. Pitzer and L. Brewer), McGraw-Hill, New York (1961), pp. 605−629, according to J. C. Fjeldsted, Ph.D. dissertation, Brigham Young University (1984).

The effect of viscosity on solute diffusion is also important. In both gases and liquids, viscosity has been shown to have significant effects on diffusion; diffusion coefficients decrease with increasing viscosity [5,6]. Chromatographic efficiency is, therefore, improved in SFC by using low-viscosity mobile phases.

The choice of mobile phase can significantly affect the selectivity offered by SFC, similar to what is observed in LC. For example, supercritical carbon dioxide was shown to be a better solvent for purines than supercritical ammonia, whereas the latter was a better solvent for carbohydrates and nucleosides [7]. Selectivity is dependent on solute-mobile phase interactions, for which few data related to SFC are presently available. However, it is to be expected that the selectivity contribution of the mobile phase density of modifiers to supercritical mobile phases can change the selectivity of the separation. For example, addition of methanol or 2-propanol to n-pentane improved the selectivity in the separation of nonpolar polymers in packed column SFC [8]; however, data are just now becoming available on the use of mixed phases in capillary column SFC.

One of the advantages of capillary SFC is its compatibility with a wide variety of detectors. This subject is treated extensively in a subsequent chapter. However, it should be mentioned here that not all supercritical fluids are compatible with all detectors, and this aspect should be part of the mobile phase selection criteria. For instance, while carbon dioxide and nitrous oxide give little or no response in the flame ionization detector, n-pentane or carbon dioxide doped with methanol cannot be used with this universal detector. It has recently been found that supercritical ammonia gives a low but sufficient response in the flame ionization detector that it cannot be used during density programming [9]. Carbon dioxide gives some emission at the 365 nm emission wavelength for sulfur in the flame photometric detector, which decreases the sensitivity for sulfur compounds by an order of magnitude [10]. Most of the desirable supercritical mobile phases, however, can be used with UV absorbance and fluorescence detectors.

The stability and reactivity of a supercritical fluid are also important in deciding whether or not it would be appropriate as a mobile phase. One prerequisite is that it must be chemically stable under the pressure and temperature conditions to which it will be subjected. A number of possible mobile phases have been evaluated in this regard by Asche [11]. Halogenated hydrocarbons usually give acidic decomposition products, and acetonitrile can react with the support surface to give off ammonia. Supercritical ammonia

reacts with silica, and care must be taken in selecting column materials when supercritical ammonia is used as a mobile phase. Even
supercritical carbon dioxide will react with primary amines, and
another mobile phase must be selected when attempting to chromatograph such compounds. When using mixed mobile phases, both
must be miscible and chemically compatible. Carbon dioxide and
ammonia cannot be mixed because of the formation of urea. It would
seem unwise to mix ammonia and nitrous oxide in the pump; ammonia
is flammable, nitrous oxide is an oxidant, and a mixture of the two
is potentially explosive.

In a number of cases, toxicity precludes the use of a fluid in
SFC. For example, sulfuryl fluoride has some interesting properties
which might prove useful in SFC, but its extreme toxicity (5 ppm
threshold limit value) limits even experimentation with the fluid.

In conclusion, a number of supercritical fluids have been investigated for use as mobile phases in chromatography. Most of these
experiments have been performed with packed columns, and many
fluids were eliminated for various reasons as discussed above. In
comparison, few fluids have been evaluated for use in capillary SFC.
Since void volumes and flow rates are so much smaller in capillary
column systems than in packed column systems, and capillary column materials are much less active (and, therefore, more chemically
inert) than packed column materials, the restrictions for mobile
phase selection are not as great. At present, carbon dioxide and
n-pentane have been used most extensively in capillary SFC. Nitrous
oxide is only slightly more polar than carbon dioxide and offers
little advantage over carbon dioxide except for the analysis of compounds such as the primary amines, for which carbon dioxide is
reactive. Ammonia has great potential as a polar mobile phase,
but work still must be done to develop instrumentation and column
materials that are resistant to this fluid. From the present activities
of several research groups, it is expected that a number of new
supercritical fluids, including mixed fluids, will soon be added to
the list of useful mobile phases for chromatography.

COLUMN TECHNOLOGY

Open-tubular columns are, both theoretically and practically, the
best choice for high-performance supercritical fluid chromatography.
Efficient capillary columns give the advantages of high resolution
and high sensitivity. Knowledge about column materials, dimensions,
stationary phases, and methods for column preparation is important
for using and treating such columns properly and for optimizing
the performance of supercritical fluid chromatographic systems.

Materials

Synthetic fused silica is preferred as capillary tubing material for the same reasons that it has become popular in GC. Fused silica columns of 100 μm i.d. or less can withstand pressures of over 400 atm without breaking. The low metal content in the fused silica, less than 1 ppm, prevents adsorption and degradation of polar and chemically unstable sample solutes. In addition, hydrogen-bonding silanol functionalities on the fused silica surface can be deactivated by using controlled reactions to form a stable surface that gives no contribution to the retention of sample solutes.

Column Dimensions

Column internal diameters of 50 to 100 μm are primarily used in capillary SFC. From a practical standpoint, the 100-μm-i.d. columns are much easier to prepare than 50-μm-i.d. columns, but the latter are superior for high-resolution separations. The effects of different column diameters and lengths on resolution have been treated theoretically [2]. Reasonable analysis times for high-resolution separations were found to be possible only when the internal diameters of the columns were 50 μm or less. Over 10^5 effective theoretical plates could then be achieved in less than 2 h.

The outer diameter of the capillary tubing is not critical for analytical performance, but it determines extracolumn effects at connecting fittings and butt connections, and it influences the strength of the material as well as its heat capacity. Thin-wall columns are more sensitive toward temperature differences in the chromatographic oven.

The required column length is determined by the theoretical plates needed for the specific separation. The maximum length is limited by the pressure drop across the column. Acceptable maximum lengths for 50- and 75-μm-i.d. columns were calculated from theory to be 40 and 100 m, respectively [12]. However, complex mixtures often can be sufficiently resolved on 10 to 15-m columns at higher speeds and milder conditions.

Pretreatments

Silanol groups on the fused silica surface become active sites in the capillary column if they are not deactivated. Sample solutes that are sensitive to hydrogen bonding interact with these sites, resulting in reversible and/or irreversible adsorption and poor chromatographic peak shape. In addition, polar modifiers in the mobile phase adsorb at these sites and lead to problems including inhomogenity of the stationary phase. Several methods are known for

deactivation of active sites on the inner capillary surface. A deactivation method based on a dehydrocondensation reaction has
proved its usefulness in deactivation of 50-μm-i.d. capillaries [13].
Silicon hydride groups in an organohydrosiloxane polymer react
with surface silanols, yielding hydrogen gas as the only by-product.
The reaction can be carried out at about 100°C lower than other
known deactivation methods, leaving the protective polyimide coating
chemically intact and still transparent.

A stationary phase with a given surface tension will wet any
surface that has a higher critical surface tension. Untreated fused
silica is wettable by any stationary phase because it has a very
high critical surface tension. However, chemical reactions of the
fused silica capillary wall with deactivation reagents can drastically
change its critical surface tension and thereby its wettability. Different organic substituents can be incorporated into the deactivation
reagents to ensure proper critical surface tensions of the chemically
modified surfaces. Combined deactivations and surface modifications
for 50% phenyl- and 50% cyanopropyl-substituted stationary phases
have been reported [14,15].

In addition to deactivation of the capillary surface, the pretreatment can also improve the hydrolytic stability of the stationary
phase. Hydrolytic instability in the bonds between an organic
polymer and a mineral surface is well known [16]. A highly cross-
linked organofunctional silane resin of several monolayers thickness
is recommended as surface modifier for hydrolytic stability. This
has been achieved by dehydrocondensation of silicon hydride reagents with the fused silica surface. Figure 1 shows chromatograms
of a polar test mixture on a nonpolar and a polar deactivated column, resepectively, demonstrating the high quality of the dehydrocondensation deactivation.

Stationary Phases

To date, mainly nonpolar and polarizable stationary phases have
been used in capillary SFC. Contradictory explanations have been
presented concerning the actual role of stationary phase selectivity
in retention and resolution in capillary SFC. Some consider that
variation of mobile phase composition provides the best approach to
maximizing sample resolution, while others believe that selectivity
can best be achieved with different bonded stationary phases. In
actuality, neither should be looked on as more or less important;
a consideration of solute solubility in both phases seems to be
necessary.

Requirements for stationary phases in open-tubular SFC are
somewhat different from what normally is true for phases in capillary
gas chromatography. A high degree of thermostability is usually

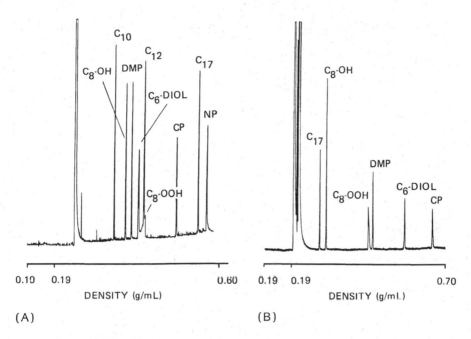

Fig. 1   Capillary supercritical fluid chromatograms of a polarity
test mixture using carbon dioxide mobile phase at 50°C.   Conditions:
(A) 15 m × 50 μm i.d. column deactivated with polymethylhydrosil-
oxane and coated with SE-54 stationary phase (d_f = 0.25 μm).   (B)
12 m × 50 μm i.d. column deactivated with cyanopropyl methyl-
hydrosiloxane and coated with 50% cyanopropyl methylpolysiloxane
stationary phase (d_f = 0.25 μm).   Peak identifications: C_8-OH,
1-octanol; C_{10}, n-decane; DMP, 2,6-dimethylphenol; C_6-DIOL,
1,6-hexanediol; C_{12}, n-dodecane; C_8-OOH, 1-octanoic acid; CP,
2,4,5-trichlorophenol; C_{17}, n-heptadecane; and NP, 4-nitrophenol.

not essential in SFC, whereas resistance to dissolution is critical.
The necessary combination of low solubility in the mobile phase and
high solute diffusion in the stationary phase can be achieved only
by using properly designed synthesis procedures.   It is known that
a polysiloxane chain forms a stable backbone for the stationary
phase polymer, and it provides favorable solute diffusion properties.
The polymer should be relatively free from branching and should
consist of relatively high-molecular-weight material with a narrow
distribution range.   These long straight polymer chains should be
substituted with at least two types of functional groups: the major
portion with substituents that determine the selectivity of the phase,

and the minor portion with groups that are susceptible to cross-linking for immobilization of the phase after it is coated on the capillary wall.

Experience, to date, with stability and selectivity of different stationary phases for SFC is somewhat modest. Gum methyl polysiloxane stationary phases have led to the highest efficiencies (e.g., >4000 plates/m). Another nonpolar stationary phase, a 50% n-octyl polysiloxane, was shown to have excellent properties for SFC [17]. An efficient, chemically stable, and insoluble film can easily be achieved with this phase, and the n-octyl functional groups provide better partitioning than methyl groups. While it is easy to coat stable and efficient films of nonpolar stationary phases, polarizable phases, such as the 50% phenyl polysiloxane, are difficult to coat as uniform films in 50-μm-i.d. capillary columns. Lower efficiencies are obtained for polarizable phases as compared to methyl phases. However, the advantage of the 50% phenyl and 25% biphenyl phases is that they demonstrate better selectivity than the methyl phases for isomeric compound classes such as the polycyclic aromatic hydrocarbons. The biphenyl phase is preferred over the phenyl phase in SFC because it is easier to cross-link (contains more methyl groups) and the larger, more polarizable biphenyl group provides greater interaction with solutes. Polar stationary phases in SFC, such as the 50% cyanopropyl polysiloxane stationary phase, have the advantage that polar molecules have increased solubilities in these phases, resulting in high sample capacities for these solutes. This is not the case for polar compounds in nonpolar phases.

The previously discussed stationary phases can be coated with film thicknesses ranging from approximately 0.05 to 1.5 μm in 50-μm-i.d. capillary columns. The thinner films favor analysis speed and column efficiency, while the thicker films favor sample capacity and column neutrality. Fields et al. [18] showed that film thicknesses up to 1 μm could be used in capillary SFC with less than a 10% loss in column efficiency. Several considerations are important when selecting the film thickness. Because it is more difficult to elute solutes from a thick stationary phase film, and efficiencies are lower, a thinner film is generally more advantageous. However, the sample capacity required should also be considered. Finally, it has been observed that selected stationary phases can swell in certain supercritical fluids. For example, methyl polysiloxane phases swell to ∿3 times their normal thickness in supercritical hydrocarbon fluids, while little swelling is observed for the same phases in supercritical CO_2 [19].

Preparation of Capillary Columns

Although it is expected that few SFC users will actually prepare their own capillary columns, it is important for them to know how the columns are prepared. This knowledge can help in proper column handling and treatment and in understanding the chromatographic behavior of certain samples during analysis.

Several basic precautions are essential to prevent plugging when working with 50-μm-i.d. capillaries. It is very important to filter all column preparation solutions before they are introduced into the column. This is true during column preparation as well as for injection of sample solutes. The ends of the fused silica column should always be cut off with a diamond-tipped pen or silicon wafer after the column is inserted through a ferrule or rubber septum. The basic steps used in our laboratories for column preparation are described in the following paragraph.

Hydrolysis of silicon tetrachloride during manufacturing of synthetic fused silica capillary tubing leaves trace amounts of volatile acidic residues and surface silanols in the final capillary material. An inert gas purge at elevated temperature easily removes most of the impurities. The deactivation reagent for capping the silanol groups (e.g., an organohydropolysiloxane, MW = 1000–5000, containing 25–50% hydride functionality) is dissolved in an inert solvent and dynamically coated on the pretreated capillary surface. The capillaries are sealed after solvent evaporation and heated to a reaction temperature of 250 to 350°C for 2 to 15 h, depending on the organofunctionality in the reagent. Hydrogen gas, being the only by-product formed in the dehydrocondensation reaction, can be detected as a slight overpressure in the deactivated column. Excess reagent is thoroughly rinsed out of the column, leaving a thin resin layer on the capillary wall. The stationary phase is coated using a static coating procedure at a slightly elevated temperature [20]. After coating, the stationary phase film is immobilized by a free-radical mechanism. Free radicals are initiated using azo-*t*-butane [21], and covalent carbon–carbon bonds are formed between the stationary phase polymer chains.

Proper Use and Treatment of Columns

Capillary columns should be stored with ends sealed when they are not being used. After installation of a column in the chromatograph, and prior to routine chromatographic analyses, the column should be conditioned with repeated pressure programs at the proper operating temperature until column bleed has been stabilized. All solvents and sample solutions should be filtered before being introduced into the capillary column in order to prevent plugging. Injection of samples that are not soluble in the mobile phase and injection of large volumes of water, methanol, and buffered solutions are not recommended because they will shorten the column lifetime. Impure modifiers in the mobile phase, such as chloroform, or reactive mobile phases, such as ammonia, can also destroy the stationary phase. Finally, when pressure or density ramps are used in the analysis method, moderate depressurization rates from the upper pressure/density limit to the initial setpoints are advised. Too rapid decompression (greater than 50 atm/min) from high pressures can have a destructive effect on the chemically bonded stationary phase, resulting in reduced efficiency and increased column bleed.

INSTRUMENTATION

A supercritical fluid chromatograph has five major components: (a) a high-pressure pumping system for delivering a fluid mobile phase under supercritical pressures, (b) an injector, (c) a precision oven equipped with a capillary column, (d) a detection system, and (e) a microprocessor-based control system. The fifth component can be either an on-board microprocessor or a microcomputer, but it should have the capability of controlling pump and oven functions, including pressure and/or density ramping and temperature regulation of the capillary column oven. A schematic diagram of a capillary SFC instrument is shown in Fig. 2.

Until recently, most of the instrumentation used for SFC research and analysis has been home-built. Systems have been constructed by modifying and combining gas and liquid chromatographic pumps and ovens with microprocessors and personal computers for electronic control. Recently, several complete SFC systems have been introduced by instrument companies and are now commercially available. In this section, the current state of development of SFC instrumentation is discussed.

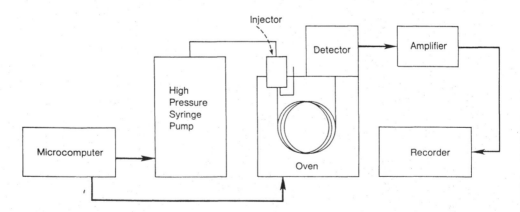

Fig. 2   Schematic diagram of capillary SFC instrumentation.

Pumping Systems

Because of the low flow rates (several microliters per minute) required for capillary SFC and the necessity of having precision control of a nonfluctuating fluid supply, syringe pumps have been the predominantly used type of pump in SFC. Syringe pumps have the capability of delivering a pulseless fluid flow to the capillary column under either pressure or flow control. Supplying an even, hysteresis-free fluid flow is critically important to the optimized chromatographic performance of the SFC system.

Currently, pressure is the most common control parameter in capillary SFC. Using a high-precision pressure transducer as the control component, a pumping system can be pressure-ramped during an analysis to increase the density and, hence, the mobile phase solvent strength. Because solvent strength is more accurately correlated to fluid density, particularly at temperatures near the critical point, computerized conversion of pressure to density has been used to operate SFC pumps in a density program mode. To date, flow-controlled pumping has not been widely applied in capillary SFC.

A wide range of ramping rates, in either the pressure or density mode, is advantageous. Typically, SFC analyses are performed at relatively low ramp rates (e.g., pressure, 0.1–10.0 atm/min; density, 0.001–0.050 g/ml·min), particularly if a complex mixture is being analyzed. However, higher ramp rates (e.g., pressure, 10–150 atm/min; density, 0.050–0.50 g/ml·min) can be used to significantly shorten analysis times if simple mixtures are analyzed or a selective detector is employed. Figure 3 demonstrates the level of

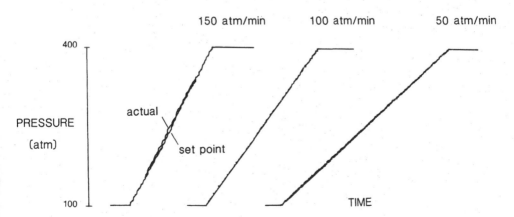

Fig. 3 Control accuracy of an SFC syringe pump during density programming operation; CO_2 mobile phase at 40°C.

control that can be achieved in the fast pressure ramp mode using a computer-controlled syringe pump. No deviation is noticeable between the actual and setpoint pressures at 50 and 100 atm/min, and very little deviation is observed at 150 atm/min.

The operating pressure range of the pump is also an important design criterion that affects the density range for the mobile phase fluids being used in the system. As discussed earlier in this chapter, the wider the pressure range, the greater the utility and selectivity obtainable in density programming with a variety of fluids. Currently, SFC pumps are available with pressure ranges of approximately 1 to 400 atm (15 to 6000 psi). In general, these are most commonly syringe pumps, which are powered by electrical stepping motors and gear trains with gearing ratios of 1/20 to 1/35. Sufficient torque and speed are achievable with this type of pump to perform analyses within the pressure limits and ramp rates cited. Pneumatic amplifier syringe pumps have also been used in SFC. Advantages of this type of pump include rapid pressure control, quick filling times, and, in general, higher pressure ranges (up to approximately 700 atm or 10,000 psi).

Another important feature of the syringe pumps currently used for SFC instrumentation is their volume. The volume of a syringe pump determines the period in which the SFC system can perform uninterrupted analyses between fill cycles. Generally, total volumes greater than 50 ml are preferred. Most commercially available SFC pumps range from 100 to 250 ml in total cylinder volume. If the system is leak-free and the split vent is closed between injections, this volume range is normally sufficient for several days of operation between fills. Such systems take from 5 to 10 min to fill. With system computer control, fill cycles can be automatically performed between chromatographic runs. One manufacturer offers a dual-piston syringe pump, each piston having a volume capacity of 25 ml; synchronized operation of this system allows continuous operation (one piston supplies fluid while the other cylinder fills). Both pistons can also be pumped simultaneously, extending the capabilities of the pumping system to gradient SFC operation as performed in LC. The pneumatically driven syringe pumps are generally of lower total volume (10 ml or less) but can be filled in a few seconds.

Finally, the material from which the syringe pump cylinder, piston components, and seals are fabricated must be chemically inert and resistant to corrosion and wear with all fluids being used in the system. Series 300 stainless steels are most frequently used in SFC pumps. Fluids ranging in chemical properties from carbon dioxide (noncorrosive) to ammonia (extremely corrosive in the supercritical state) have been used successfully with 316 stainless steel. Seals must also have a high level of chemical inertness while

maintaining good resistance to wear and long lifetime expectancies.
In general, Teflon, graphite, and carbon fiber-filled Teflons have
been predominantly used for the manufacture of the high-pressure
seals used in SFC pumping systems.

Chromatographic Oven

In general, current GC oven technology has been adequately developed for capillary SFC. Ovens from most GC vendors are suitable
for use with SFC instruments. As with any oven used for capillary
chromatography, the oven should be designed so that thermal
gradients between any two points in the column zone are less than
±0.1°C. If larger thermal gradients exist, peak splitting is likely
to occur. This is crucial in SFC for two reasons. First, capillary
columns in SFC are typically 50 μm i.d. and are more subject to
adverse thermal effects than their larger-bore counterparts used
for GC. Second, slight changes in temperature can have a pro-
nounced effect on the fluid density of the mobile phase, an effect
which is not particularly pertinent in either GC or LC.

Injection Systems

Various injector configurations for SFC instrumentation have been
investigated. By far the most widely used injector has been the
high-pressure, internal sample loop, rotary microvalve [3,22].
Such a valve can be actuated manually or automatically. Furthermore, the inlet to the capillary column can be equipped with an
adapter for split injections, or the rotary valve can be actuated
rapidly under electronic control for operation in the nonsplit mode.
Such injection systems are quite adequate for introducing a small
band of sample onto the head of the chromatographic column.

   Figure 4 is a schematic diagram of an automatic rotary valve
and demonstrates one system that can be used in both the split and
nonsplit modes. In the split mode, only a fraction of the injected
sample is allowed to enter the head of the column. Typical split
ratios range from 1:10 to 1:50. As with split injections in capillary
GC, the major disadvantages of split injections in SFC are sample
component discrimination and poor quantitative reproducibility. If
the split tee and restrictor are removed from the injection system,
operation in the nonsplit mode is possible. Electronic control of
the valve's rotation has enabled fast pulse rotations, on the order
of 10–50 ms. At these speeds, nanoliter (nl) injection volumes
have been achieved with quantitative reproducibility (5% RSD at
1 nl for hydrocarbon solutes).

   An injection system component not depicted in Fig. 4 is the
cooling jacket. Most SFC analyses reported to date have used

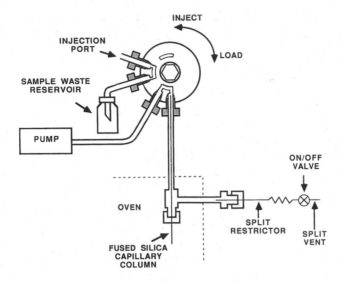

INJECT

INJECTION
PORT

LOAD

SAMPLE WASTE
RESERVOIR

PUMP

ON/OFF
VALVE

OVEN

SPLIT
RESTRICTOR

SPLIT
VENT

FUSED SILICA
CAPILLARY
COLUMN

Fig. 4   Schematic diagram of the split/splitless valve injector.

injection systems equipped with subambient cooling devices.   The
purpose of the cooling is to maintain the injector, which is in close
proximity to the heated oven compartment, at a temperature below
the boiling point of the sample carrier solvent.   One of the limitations of cooled injection systems is that the components that are
introduced onto the column must be soluble in the liquid solvent
carrier.   Of course, one of the major advantages of using supercritical fluids for mobile phases is their enhanced solvating properties.   Hence, the ideal injection system would involve injection of
the sample using the supercritical fluid being used as the mobile
phase.   Preliminary work with supercritical fluid extraction/injection
techniques has already been reported [23,24] and will most likely
dominate future approaches to improving injection systems for
capillary SFC.

Detectors

One of the major advantages of capillary SFC is that both LC and
GC detectors are readily adapted for use as SFC detection systems.
Virtually all of the GC detectors, including the selective detectors,
and the spectrophotometric LC detectors have been successfully
interfaced and used as SFC detectors [25].   In addition, several
groups have reported hyphenated detection systems including SFC-
MS and SFC-FTIR.   There have been a few reports of using less

familiar detection systems such as the ion mobility detector, supersonic jet spectroscopy, and the inductively coupled plasma. Even
more exotic detectors based on thermal lens absorption, light-
scattering measurements, and electrochemical reactions appear to
hold promise for future SFC detector innovations. The subject of
detectors for SFC is discussed in detail in a following chapter and,
therefore, is not treated further here.

Computer Control

The fifth major component of a capillary SFC instrument is the
microprocessor-based control device used for operation of the pump
and oven. Functions that are convenient to have under either microprocessor or microcomputer control include pump and oven operating
functions and status displays, injection system operation, thermal
regulation of all heated and cooled zones, pressure and density
ramping of the pumping system, and, finally, acquisition, storage,
and manipulation of the detector signal output data.

In general, the microcomputer system has provided the most
flexibility, particularly for data processing and pressure/density
ramping of the SFC system with fluids that have markedly different
critical properties. For example, using equations of state and compressibility data, the microcomputerized system can be programmed
to generate and operate in the various density ramping modes for
virtually any fluid by simply entering the fluid variables, namely
critical pressure, critical temperature, molecular weight, and the
accentric factor. In addition, the operating software for microcomputer systems can be designed to be adequately user-friendly,
thus rendering the operation of the capillary SFC system as routine
as either GC or LC instrumentation. For these reasons, the microcomputer appears to be emerging as the optimum system control
device. However, the microprocessor approach is currently used
either alone or in conjunction with microcomputer control systems
and should experience further development, particularly for on-line
and dedicated SFC analyses.

PRACTICE OF SFC

The supercritical fluid chromatograph is a sophisticated combination
of electronic, mechanical, and chemical systems that must function
together to yield the anticipated separation. Thorough knowledge
and care of the supercritical fluid chromatograph will be rewarded
by high performance and long lifetime. Capillary SFC from a
practical standpoint is the final topic covered in this chapter. Ad-
vice for setting up the system will be followed by guidelines on how
to select conditions for different types of analytical separations.

Setting Up the System

When setting up a supercritical fluid chromatograph, there are several things that, if not considered, can cause unnecessary problems. Mobile phase purity is of considerable importance to successful operation. The mobile phase should generally be free from particles, traces of organic compounds, and water. The cost of the mobile phase is generally not a limiting factor, even for the highest purity, because extremely low consumption rates are experienced in capillary SFC. The high-purity mobile phase should be transported from the high-pressure cylinder or glass bottle to the head of the chromatographic column without introducing any contamination in the process. Gaseous (at room temperature) mobile phases, with and without modifiers, are commercially available. Aluminum cylinders are recommended over steel cylinders for SFC mobile phases because aluminum cylinders have lower particulate levels. A dip tube in the high-pressure cylinder facilitates transfer of fluids such as liquid CO_2 to a cooled syringe pump and results in a more complete fill of the system. A helium headspace over the CO_2 in the cylinder provides a higher pressure in the cylinder. All surfaces in the system with which the mobile phase comes in contact (e.g., seals, valves, pressure relief devices, split devices, and pump cylinder) must be free from lubricants and extensively cleaned before the system is filled with mobile phase. All seals, connections, and valves in the system should be made of inert materials, such as stainless steel, Teflon, or gaphitized Teflon. Tubing and connections from the mobile phase cylinder or solvent reservoir to the pump and from the pump to the injector should be made of stainless steel material. Tubing ends should be deburred before the ferrules are mounted and the connections made. A series of solvents, when used in the order given, will remove all foreign material from stainless steel parts: methyl ethyl ketone or acetone, isopropanol, distilled water, nitric acid (50%) for 5 min, and distilled water (followed by a final methanol rinse and low-temperature, 100—120°C, baking). After the pumping system has been cleaned and dried with oil-free gas, it can be filled with the mobile phase. Several purge cycles in which the fluid is purged through all the connecting lines and fittings are recommended to remove any residual cleaning solvents or fluid-soluble contaminants. This procedure is also necessary when changing from one fluid to another. Most supercritical fluid mobile phases have special requirements, some of which are listed below.

High-Pressure Fluids

The pump cylinder should be cooled to around 0°C to ensure efficient transfer of the fluid from the high-pressure cylinder to the pump as a liquid. Precautions must be taken to avoid a possible

failure in the closed, pressurized system in which the pressure
could increase drastically, leading to a system rupture. For example, if a line which has been used to transfer fluid between the
pressurized cylinder and pump is isolated immediately after use, extremely high pressures can be generated as the line reequilibrates
to room temperature. Use of a pressure relief valve in the transfer
line would obviate this potential hazard. In addition, a check valve
in the transfer line would prevent inadvertent contamination of the
pressure tank fluid from a pump that either is not completely clean
or has been used with another fluid.

Noncorrosive High-Pressure Fluids

These fluids should be filtered through a 2-μm-pore metal filter
between the high-pressure cylinder and the pump. Carbon dioxide
is available in high purity, neat or with a wide variety of added
modifiers. Although SF_6 is an inert, noncorrosive fluid, SO_2 and
HF are formed when it is combusted in flame detectors. Gold plating
of the collector is therefore recommended.

Corrosive High-Pressure Fluids

These fluids should be filtered through a 0.2-μm replaceable, in-line
Teflon filter to protect the syringe pump from scoring and the restrictors from plugging by particles. Below 140°C, NH_3 is compatible with Teflon, stainless steel, and fused silica materials.
Above this temperature, gold or platinum is recommended.

Room Temperature Liquids

These fluids should be filtered through a 0.2-μm Teflon filter and
degassed before filling in the pump reservoir or pump cylinder.
Gas leak detectors should be placed in the chromatographic oven
when combustible mobile phases such as hydrocarbons (e.g.,
n-butane and n-pentane) are being used. The pump should never
be pressurized when it is partly filled with air.
    It is very important to avoid poor connections that will lead to
extracolumn effects between the injector and detector when the
capillary column is being installed. Whether or not a splitter is
used, the column or retention gap must be connected directly to
the sample loop outlet of the injector valve.
    Many detectors, such as flame detectors, operate at atmospheric
pressure, and the mobile phase must be depressurized through a
restrictor just before it enters the zone of detection. The connection between column and restrictor must be made, in this case,
using a butt connection with very low dead volume. Often, the
column and restrictor do not have matching outer diameters and/or
the butt connector has a larger i.d. than the o.d. of the capillaries

to be connected. A sleeve of fused silica can be used to line up
the ends with each other. The ends must be cut at straight angles
for this connection. A dead volume at this point could produce
poor chromatographic peak shapes during the chromatographic run.

The supercritical fluid system should be pressurized and all
connections leak-tested before the chromatographic oven is closed
and brought up to temperature. After installation, the column
should be conditioned until the baseline and the noise level stabilize.
A program rate of 0.007 g/ml·min from 0.19 to 0.65 g/ml, with a
depressurizing period of 20 min, can be used as typical for such a
procedure. The mobile phase linear velocity is usually regulated by
the length, diameter, and thermostatting of the restrictor at the
end of the column. A compromise between theoretical optimum
velocity ($\sim$0.2 cm/s) and analysis time has resulted in a recommended
linear velocity of $\sim$2 cm/s.

Selecting Conditions

Capillary SFC offers the user a wide variety of chromatographic
conditions that can be modified and controlled to optimize the analytical separation. It is important to understand how the choice of
operating conditions will affect the chromatographic performance.
Both chemical and instrumental parameters must be properly selected
for a specific separation. The chemical parameters include mobile
and stationary phase composition, while the instrumental parameters
include column dimensions, temperature, and mobile phase density
control during the analysis. The selection of these SFC conditions
is based on knowledge of the sample molecular weight, complexity,
chemical composition, reactivity, and thermal stability.

Mobile Phase Composition

The basic requirements for the mobile phase were described in previous sections. Most applications in capillary SFC have been performed with unmodified mobile phases. However, it is also possible
to use a modified mobile phase by either adding a constant percentage of modifier to the fluid or using a gradient pumping system
where the solvent strength is continuously changed by adding increasing amounts of modifier during the analysis.

Stationary Phases

Stationary phases with selectivities based on nonpolar, polarizable,
and polar solute-stationary phase interactions are available for use
in capillary SFC. As experienced in both gas and liquid chromatography, nonpolar stationary phases give the highest efficiencies,
and they are recommended for most SFC separations. A polarizable

phase gives improved separation of isomeric polar compounds, while a polar stationary phase gives better solvation of the same compounds. For example, free carboxylic acids are best analyzed on a polar stationary phase, for which the higher column capacity results in a more efficient separation than is possible on a nonpolar stationary phase. For the analysis of trace and/or polar compounds, such as pesticides and pharmaceuticals, well-deactivated columns are required with which there is no chance for adsorption, reaction, rearrangement, or degradation of the sample solutes during the analysis. In Fig. 5, a chromatogram of a polar penicillin standard compound on a deactivated column coated with a polarizable biphenyl polysiloxane stationary phase is shown.

Temperature

It is generally recommended that the highest temperatures the solutes can withstand be used for the SFC analysis [26]. By so doing, the volatilities of the solutes will contribute as much as possible to their chromatographic migration and they will elute at lower mobile phase densities, resulting in higher chromatographic efficiencies. However, at elevated elution temperatures, the mobile phase linear velocity increases much more rapidly with density than at lower temperatures [27], and lower maximum densities are attained at the maximum pump operating pressure (Table 2). Figure 6 compares chromatograms of a polyol oligomer mixture obtained at two different temperatures. Analysis at the higher temperature made it possible to achieve higher resolution.

Low temperature separations, on the other hand, are recommended for thermally labile solutes. Figure 7 shows an example of the analysis of thermally labile nitrated polycyclic aromatic compounds in a diesel particulate extract on a well-deactivated column coated with a 5% phenyl methylpolysiloxane stationary phase [28].

Density

The importance of controlling mobile phase density during SFC analysis is based on the fact that the solvating properties of supercritical fluids are a direct function of their densities at constant temperature. Mobile phase density programming changes the solvating strength of the mobile phase during the analysis. In most cases, nonlinear density programming provides the best resolution of mixture components when density programming is employed. For example, the retention times of members of a homologous series are a logarithmic function of density. Therefore, asymptotic density programming can be used to obtain even spacing of compounds in such samples.

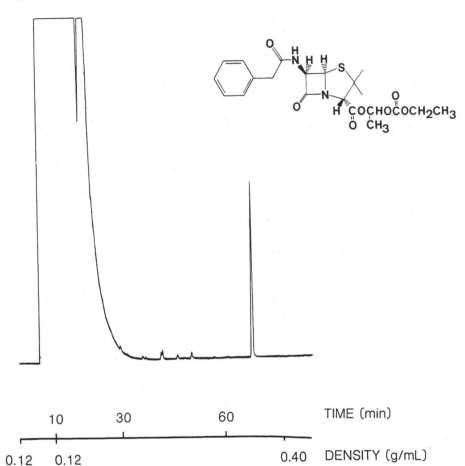

Fig. 5 Capillary supercritical fluid chromatogram of benzylpeni-cillin 1-ethoxycarbonyloxyethyl ester. Conditions: 12 m × 50 μm i.d. deactivated column coated with 50% cyanopropyl methylpoly-siloxane (d_f = 0.25 μm).

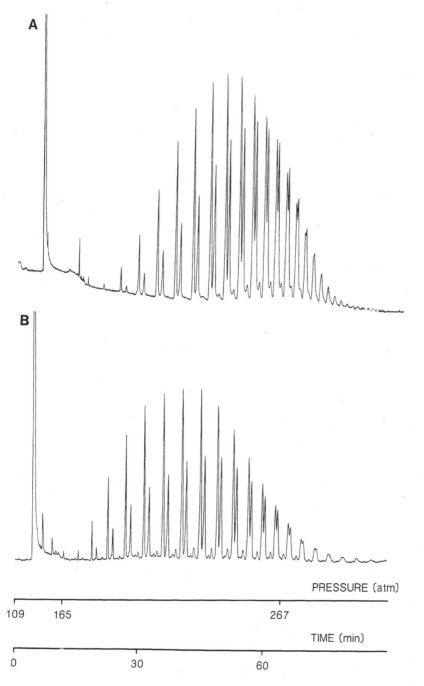

Fig. 6   Capillary supercritical fluid chromatograms of nonvolatile polyglycol oligomers with carbon dioxide mobile phase at (A) 120°C and (B) 160°C.   Conditions: 20 m × 50 μm i.d. column coated with SE-33 stationary phase (d_f = 0.50 μm).

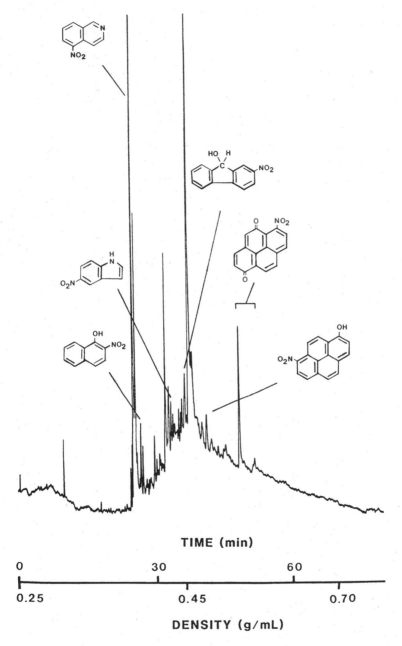

TIME (min)

0	30	60
0.25	0.45	0.70

DENSITY (g/mL)

Fig. 7   Capillary column supercritical fluid chromatogram of a
diesel particulate extract.   Conditions: 15 m × 50 μm i.d. fused
silica capillary coated with SE-54 (d$_f$ = 0.25 μm).   Carbon dioxide
mobile phase at 101°C, density programmed from 0.25 to 0.70 g/ml
at 0.0075 g/ml·min after an initial 10-min isoconfertic period.
TID-1-N$_2$ detector (300°C).   (From Ref. 28.)

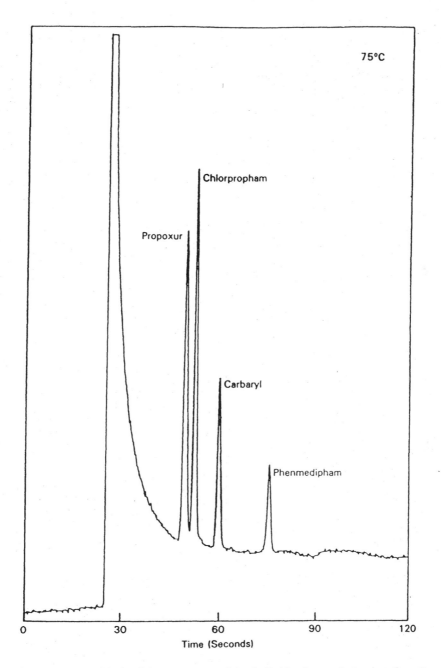

Fig. 8 Rapid capillary supercritical fluid chromatogram of carbamate pesticides. Conditions: 0.9 m × 0.25 μm i.d. fused silica capillary coated with SE-54 stationary phase (d_f = 0.20 μm). Carbon dioxide mobile phase at 75°C, programmed from 75 atm at 100 atm/min immediately after injection. (From Ref. 29.)

TABLE 2   Dependence of Density
on Temperature and Pressure for
Supercritical CO_2

T (°C)	P (atm)	ρ (g/ml)[a]
40	72	0.22
	400	0.96
60	72	0.17
	400	0.90
80	72	0.14
	400	0.82
100	72	0.13
	400	0.76
120	72	0.12
	400	0.70
140	72	0.11
	400	0.64

[a]Calculated from compressibility
data from G. N. Lewis and
M. Randall, *Thermodynamics* (revised by K. S. Pitzer and
L. Brewer), McGraw-Hill, New
York (1961), pp. 605–629, according to J. C. Fjeldsted, Ph.D.
dissertation, Brigham Young
University (1984).

For applications in which speed of analysis is more important
than high efficiency, high density programming rates can be utilized.
Figure 8 demonstrates a fast separation (<80 s) of a carbamate
pesticide test mixture.

REFERENCES

1. L. G. Randall, *Ultrahigh Resolution Chromatography*, *ACS Symp.
    Ser. 250* (S. Ahuja, ed.), p. 135, American Chemical Society,
    Washington, D.C. (1984).
2. P. A. Peaden and M. L. Lee, *J. Chromatogr.* *259*, 1 (1983).

3. M. Novotny, S. R. Springston, P. A. Peaden, J. C. Fjeldsted, and M. L. Lee, *Anal. Chem. 53*, 407A (1981).
4. J. C. Fjeldsted, W. P. Jackson, P. A. Peaden, and M. L. Lee, *J. Chromatogr. Sci. 21*, 222 (1983).
5. J. C. Giddings, *Dynamics of Chromatography*, p. 233, Marcel Dekker, New York (1965).
6. P. A. Peaden and M. L. Lee, *J. Liquid Chromatogr. 5*(Suppl. 2), 1979 (1982).
7. J. C. Giddings, N. M. Myers, L. McLaren, and R. A. Keller, *Science 162*, 67 (1968).
8. J. E. Conaway, J. A. Graham, and L. B. Rogers, *J. Chromatogr. 16*, 102 (1978).
9. J. C. Kuei, K. E. Markides, and M. L. Lee, unpublished results.
10. K. E. Markides, E. D. Lee, R. Bolick, and M. L. Lee, *Anal. Chem. 58*, 740 (1986).
11. W. Asche, *Chromatographia 11*, 411 (1978).
12. S. M. Fields, R. C. Kong, J. C. Fjeldsted, and M. L. Lee, *J. High Resolut. Chromatogr. Chromatogr. Commun. 7*, 312 (1984).
13. C. L. Woolley, K. D. Bartle, K. E. Markides, and M. L. Lee, *J. High Resolut. Chromatogr. Chromatogr. Commun. 9*, 506 (1987).
14. K. E. Markides, B. J. Tarbet, C. M. Schregenberger, J. S. Bradshaw, and M. L. Lee, *J. High Resolut. Chromatogr. Chromatogr Commun. 8*, 741 (1985).
15. K. E. Markides, B. J. Tarbet, C. L. Woolley, C. M. Schregenberger, J. S. Bradshaw, and M. L. Lee, *J. High Resolut. Chromatogr. Chromatogr. Commun. 8*, 378 (1985).
16. E. P. Plueddeman, *Silane Coupling Agents*, Plenum, New York (1982).
17. J. C. Kuei, B. J. Tarbet, W. P. Jackson, J. S. Bradshaw, K. E. Markides, and M. L. Lee, *Chromatographia 20*, 25 (1985).
18. S. M. Fields, R. C. Kong, M. L. Lee, and P. A. Peaden, *J. High Resolut. Chromatogr. Chromatogr. Commun. 7*, 423 (1984).
19. S. R. Springston, P. David, J. Steger, and M. Novotny, *Anal. Chem.*, submitted for publication.
20. R. C. Kong, S. M. Fields, W. P. Jackson, and M. L. Lee, *J. Chromatogr. 289*, 105 (1984).
21. B. E. Richter, J. C. Kuei, N. J. Park, S. J. Crowley, J. S. Bradshaw, and M. L. Lee, *J. High Resolut. Chromatogr. Chromatogr. Commun. 6*, 371 (1983).
22. J. C. Fjeldsted and M. L. Lee, *Anal. Chem. 56*, 619A (1984).
23. W. P. Jackson, K. E. Markides, and M. L. Lee, *J. High Resolut. Chromatogr. Chromatogr. Commun., 9*, 213 (1986).

24. K. Sugiyama, M. Saito, T. Hondo, and M. Sonda, *J. Chromatogr.* *332*, 107 (1985).
25. M. Novotny, *J. High Resolut. Chromatogr. Chromatogr. Commun.* *9*, 561 (1985).
27. S. M. Fields and M. L. Lee, *J. High Resolut. Chromatogr. Chromatogr. Commun. 349*, 305 (1985).
28. W. R. West and M. L. Lee, *J. High Resolut. Chromatogr. Chromatogr. Commun. 3*, 161 (1986).
29. B. W. Wright and R. D. Smith, *J. High Resolut. Chromatogr. Chromatogr. Commun. 1*, 8 (1985).

9

Detection Systems for Capillary Supercritical Fluid Chromatography

HERBERT H. HILL, JR. and CHRISTOPHER B. SHUMATE /
Washington State University, Pullman, Washington

INTRODUCTION

In the early days of SFC, when chromatography was conducted on
packed columns, component detection was one of the most significant
and challenging instrumental problems facing supercritical fluid separations [1]. High mobile phase flow rates, large internal detector
volumes, and low exit pressures contributed to the early technical
challenges of interfacing SFC to sensitive chromatographic detection
systems. The advent of capillary SFC in 1981 [2] (coupled with
the experience that had been gained over the years from the development of high-resolution capillary gas chromatography and low-
volume, low-flow systems for liquid chromatography) facilitated the
interfacing of a variety of sensitive and selective detection methods.
Because SFC is a separation technique with properties between those
of gas chromatography and liquid chromatography, it is compatible
with many detection methods that are traditionally employed with
either gas or liquid chromatography.

Three general approaches to detection after SFC exist. In the
first approach the mobile phase is maintained at the same temperature
and pressure as the column so that it remains a supercritical fluid
while passing through the detector. In the second method, the

mobile phase is cooled to form a liquid prior to entering the detector cell. And in the third approach, the mobile phase is decompressed and heated so that detection occurs in the gas phase. Whether supercritical fluid, liquid, or gas, the transfer characteristics of the mobile phase from the column to the detector are critical for optimum component response. Detection of components under supercritical fluid conditions has the advantage that no change of state occurs between separation and detection and the disadvantage that the detector cell must be compatible with high pressure. Moreover, under supercritical fluid conditions the response of many detectors is sensitive to fluctuations in temperature and pressure. Stringent temperature and pressure requirements on the detector can be reduced if the column effluent is converted to either a liquid or gas prior to entering the detection cell. The major difficulty associated with liquid detection after capillary SFC is the small dead volume that is required to maintain the integrity of the chromatographic separation. Decompression of the fluid coupled with heating produces a gas or aerosol that can be monitored by standard high-resolution gas chromatographic detection methods, but the process often is plagued by deposition of nonvolatile components at the column exit.

Whether liquid, supercritical fluid, or gas phase, the selection of detection methods that can be used with SFC continues to grow as successful adaptations are made from many common liquid and gas chromatographic detectors. From construction and operation perspectives, it is convenient to discuss SFC detectors as a function of the pressure under which they are operated. For LC-type detectors, SFC column effluents are detected as either liquids or supercritical fluids at pressures that range from ambient to high, whereas with GC-type detectors the fluid must be converted to a gas at pressures in the range from ambient to low. The following sections describe the similarities and differences of high-pressure, ambient-pressure, and low-pressure detection systems after supercritical fluid chromatography.

HIGH-PRESSURE DETECTION

High-pressure detectors for capillary SFC are normally modified liquid chromatographic detectors. Depending on whether the detector is heated or not, the mobile phase may exist as either a liquid or a supercritical fluid during the detection process. For capillary SFC, high-pressure detection has been reported with ultraviolet detectors (UVD), fluorescence detectors (FD), and Fourier transform infrared (FTIR) detectors.

Ultraviolet Detection (UVD)

For packed column SFC and, to a certain extent, capillary column SFC, UV absorption is the most common method of detection. Ultraviolet detectors used in liquid chromatography can be directly interfaced to a packed column supercritical fluid chromatograph, but for capillary SFC the detector's design must be modified to reduce the cell volume. Detectors used in packed column liquid chromatography have cell volumes that are too large, and much of the resolution gained in the capillary column would be lost in the detector.

One approach for reducing the detector cell volume was to construct a UVD cell from a zero dead volume union as shown in Fig. 1 [3]. This particular cell had a 3-mm light path length and an internal diameter of 0.8 mm, giving a cell volume of 1.5 μl. Ultraviolet light was transmitted to and from the heated detection cell via fiber optics and the optical configuration of a Tracor model 960 ultraviolet detector was modified to fit the new detector. In order to transfer the effluent from the column to the detector, supercritical conditions were maintained by heating the transfer line and the detector. The major difficulty with this approach is that resolution can be lost during the transport process from the column to the

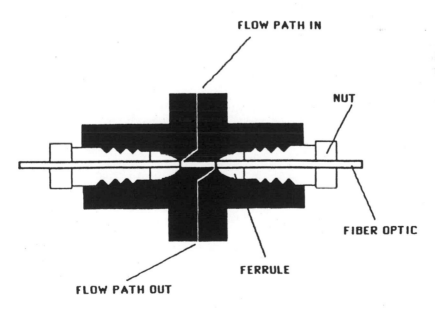

Fig. 1 Fiber-optic UVD. Constructed from a Valco zero dead volume union; cell volume, 1.5 μl. (From Ref. 3.)

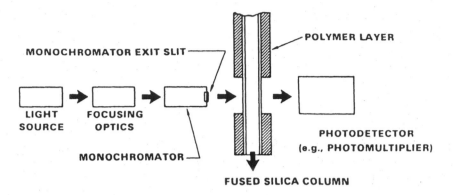

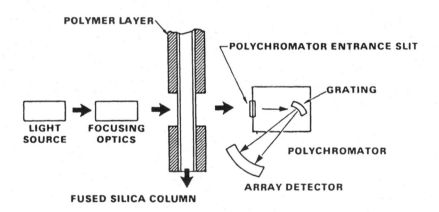

Fig. 2   On-column UVD.   Schematic diagram for on-column detection in conjunction with fused silica open-tubular column.   (From
Ref. 4.)

detector.   Also, conditions required to maintain sufficient solvating
power of the supercritical fluid during the transfer process while
maintaining optimal flow characteristics for the transfer of the solute to off-column high-pressure detectors are not yet well understood.
        The most common method used to avoid the necessity of effluent
transfer between column and detector is an on-column detection approach in which the capillary near the end of the column becomes
the flow-through cell.   This design is patterned after an on-column
detector for use with microbore LC columns [4].   About 1 cm of
the polyimide external coating is stripped from the column and the
bare fused silica is used as the sample cell window by placing it
between the slits of a standard UVD.   Figure 2 depicts a schematic
of an on-column design which has been constructed from a fused

silica open-tubular column. Focusing optics may be used to enhance the sensitivity of the detector, as the path length in these detectors is limited by the internal diameter of the capillary column.

Because it is inconvenient to heat the sample cell, these on-column UV detectors are ideal for use as the first detector of a series, especially if decompression is to occur with the second detector.

Fluorescence Detection (FD)

The lack of high sensitivity of the UV detectors after capillary SFC points to the need for other methods of detection. As in liquid chromatography, certain compounds of interest have a sufficient quantum yield to be monitored by fluorescence. For example, many polycyclic aromatic hydrocarbons exhibit strong fluorescence and are particularly suited for SFC separation. Again, as with the UV detector, special design considerations must be taken into account to ensure minimum resolution loss in the detector cell.

An on-column flow cell design can be constructed which is similar to that of the UV detector [5]. Because the photomultiplier in this design must be positioned at a right angle to the excitation beam, light scattering from the capillary wall creates a larger background than is normally observed with fluorescence techniques. Although this background is acceptable for single-wavelength monitoring, a different detector design is necessary to enable fluorescence scanning [6].

To reduce light scattering from capillary walls, the excitation light beam is introduced into the detector in a manner similar to that of the UV detector, but the emission from the compounds of interest is captured within the flow cell and conducted to the photomultiplier through a 145-μm optical fiber, which is positioned end to end with the exit of the analytical column. In the design shown in Fig. 3 a 0.3-mm-i.d. capillary served as a sleeve around both the optical fiber and the analytical column to maintain the cell volume as small as possible, typically 180 nl.

With this design the minimum detectable amount for a rapidly eluting (about 3.5 min) peak of pyrene was found to be on the order of 10 pg, using a constant excitation wavelength of 335 nm with a 10-nm bandwidth and an emission wavelength of 385 nm with a 20-nm bandwidth. By choosing the proper excitation or emission wavelength to monitor, selective detection of mixtures can be accomplished.

The real advantage of the internal optical fiber flow cell design is the ability to obtain qualitative information by scanning the peak on-the-fly. Figure 4 shows single 1-s scans of the excitation spectra from 300 to 420 nm of benzo[a]pyrene taken on both the leading and trailing edges of the chromatographic peak. Because of the dynamic nature of chromatography, it is important to take

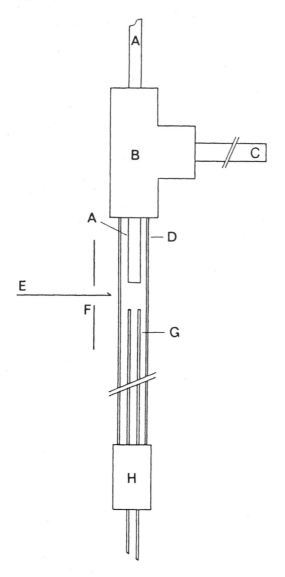

Fig. 3   Flow cell of scanning fluorescence detector.   (A) 145-μm-o.d. optical fiber; (B) 1/16-in. stainless steel tee connector; (C) 10-μm-i.d. fused silica capillary restrictor; (D) 300-μm-i.d. fused silica capillary; (E) excitation light; (F) slit for reduction of stray light; (G) analytical column; (H) stainless steel connecting union. (From Ref. 6.)

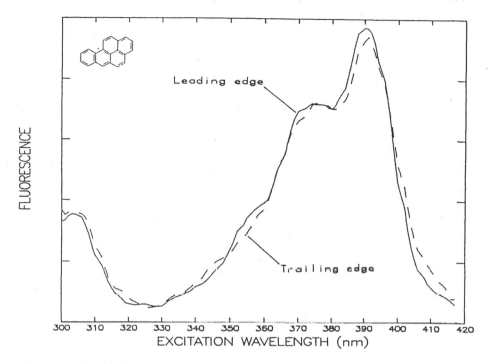

Fig. 4   Excitation spectra of benzo[a]pyrene from the leading and trailing edges of the chromatographic peak. Scanned from 300–400 nm (1 s) with 10-nm slit width. (From Ref. 6.)

scans as rapidly as possible to ensure that skewing of the spectra does not occur due to concentration changes during the scan. Since SFC peaks with reasonable capacity factors typically have widths that are minutes wide, 1-s scans are sufficient to obtain reproducible results. Well-defined excitation spectra have been obtained for polycyclic aromatic hydrocarbons at amounts as low as 1 ng/peak.

On-Line Fourier Transform Infrared Spectrometry (FTIR)

Although scanning fluorescence detection provides moderate qualitative information for a few fluorescing compounds, the need for a more universal identification method following separation by SFC is clearly evident. Valuable structural information on components in complex mixtures can be obtained from Fourier transform infrared

spectrometry after either liquid or gas chromatographic separation.
In LC, however, many common mobile phases absorb infrared radiation in regions that interfere with spectra, increasing the detection
limit and making very short path lengths necessary in order to observe IR spectra. In gas chromatography, mobile phases can be
selected which do not interfere with component spectra, and FTIR
spectra obtained after GC are generally more sensitive than those
obtained after LC.

On-line supercritical fluid chromatography/Fourier transform
infrared spectrometry can produce easily identifiable spectra for a
variety of compounds [7]. The use of FTIR after SFC is particularly appealing because in supercritical fluids the gaslike characteristics of the mobile phase increase sensitivity, whereas the liquid-
like characteristics facilitate the introduction of larger compounds.

Because it is advantageous to retain the mobile phase in the
supercritical condition for FTIR detection, high-pressure, heated
flow-through detection cells of the type shown in Fig. 5 are

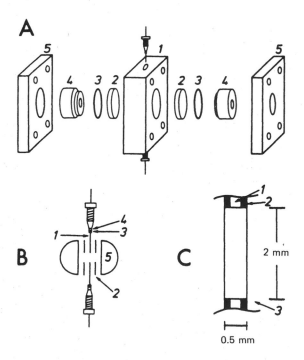

Fig. 5   High-pressure flow cell for on-line FTIR. (A) 1, Cell
body; 2, window; 3, O-ring; 4, support ring; 5, faceplates.
(B) 1, Capillary column; 2, indium gasketing; 3, indium washers;
4, Teflon tubing; 5, lead spacers. (C) 1, Capillary column;
2, indium washers; 3, lead spacer. (From Ref. 8.)

necessary. This 1-µl flow-through cell consists of a stainless steel body with lead spacers and indium gasketing to maintain pressures up to 3000 psi [8]. The IR-transparent windows are constructed of a hot-pressed polycrystallized zinc sulfide called IRTRAN 2. IRTRAN 2 is insoluble in water, has a high modulus of rupture of 14,100 lb/in.^{2} at 250°C, and is transparent to IR in the range 5000 to 690 cm^{-1}. It is the relatively high modulus of rupture of the IRTRAN 2 that allows the windows on this high-pressure cell to be only 3 mm thick.

The chromatographic effluent is introduced directly into the flow cell from the fused silica column and exits the column through a fused silica transfer line, which also serves as a flow restrictor. The entire flow cell is placed inside an FTIR (IBM IR/85) spectrometer with a narrow-bandwidth mercury cadmium telluride (MCT) photodetector. A special beam condenser is used to reduce the beam to approximately 2 mm^2.

Figure 6 is a nonselective tracing of a test chromatogram containing a mixture of alkylphenols, aromatic ketones, and aromatic aldehydes, which was obtained by comparing real-time interferograms to background interferograms through the use of Gram-Schmidt algorithms [9]. Basically, these algorithms determine when the interferograms vary from interferograms of the background. The primary objective of the FTIR technique is to obtain on-the-fly spectra of the separated mixture. Figure 7 illustrates the transmittance spectrum for 2 µg of benzaldehyde obtained from the average of several interferograms taken during the elution of the middle portion of the chromatographic peak. Although this spectrum is clear, it is obvious from the chromatogram in Fig. 6 that background correction is required because the baseline changes as a function of the pressure program.

Change in background as a function of pressure is one of the major problems in SFC/FTIR and is caused by interferences from the mobile phase. Carbon dioxide is one of the most common mobile phases employed in SFC. As a liquid, supercritical fluid, and gas it has large transparent windows in the mid-IR spectra, but it does exhibit a variation in intensity of two absorbance bands, 1282 and 1385 cm^{-1}, as the pressure is programmed from ambient to supercritical. In the liquid phase this variation is not as severe as when CO_2 is a supercritical fluid, and although CO_2 exhibits more absorption as a liquid than as a supercritical fluid, it is recommended that, when pressure programming, FT/IR should be carried out with CO_2 as a liquid. Propane, dichlorodifluoromethane, and xenon [10] have also been investigated as mobile phases for SFC/FTIR. Both propane and dichlorodifluoromethane become almost completely opaque at pressures commonly used in SFC. Xenon, however, has some very promising characteristics. Not only are the chromatographic properties of supercritical xenon adequate, but it is more transparent than CO_2.

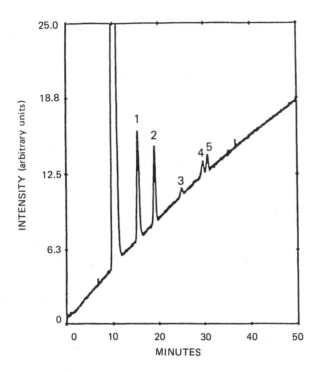

Fig. 6 Nonselective chromatogram obtained by FTIR. Gram-Schmidt real-time chromatogram of test mixture: (1) benzaldehyde; (2) o-chlorobenzaldehyde; (3) 2,6-di-t-butylphenol; (4) 2-naphthol; (5) benzophenone. Chromatographic conditions: oven temperature, 40°C; initial linear flow rate, 3.0 cm/s; pressure programmed from 78.2 to 105.5 atm at 0.54 atm/min; 2 μg of each compound injected; CO_2 used as a mobile phase. (From Ref. 8.)

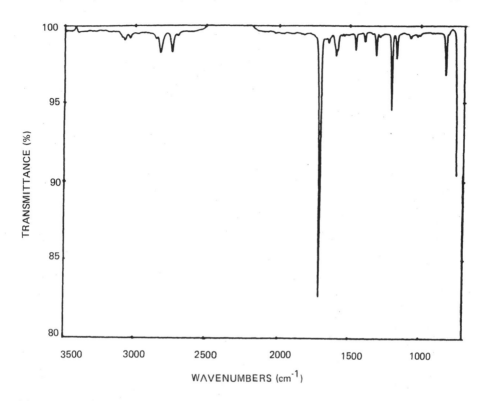

Fig. 7   Infrared spectrum in transmittance of 2 μg of benzaldehyde. Taken on-the-fly during chromatographic run where CO_2 was used as the mobile phase. (From Ref. 8.)

AMBIENT-PRESSURE DETECTION

Off-Line Fourier Transform Infrared Spectrometry

Interference from the mobile phase has so far appeared to be the primary problem with obtaining data by FTIR after SFC. Absorption by the mobile phase limits not only the reliability and ultimate sensitivity of the spectra but also the number of solvents that can be used for the separation. To date, only CO_2 and xenon have been used successfully with on-line FTIR detection. With a mobile phase that is gaseous at room temperature, nonvolatile compounds can be separated from the mobile phase by deposition on a moving track after decompression to atmospheric pressure. This moving track, with the compound of interest adsorbed to the surface, subsequently passes through the sample chamber of an FTIR spectrometer [3,11,12]. Continuous deposition of the chromatographically

separated compounds on a moving strip not only provides a means
of separating the mobile phase from the solute but also provides for
more sensitive identification. For highest sensitivity the sample strip
can be slowed or stopped to average a number of scans.

   With off-line SFC/FTIR both diffuse reflectance and transmittance FTIR are possible. With diffuse reflectance the moving track
is typically a finely ground and packed KCl bed that is 1 mm wide,
5 mm deep, and 300 mm long. The moving track for transmittance
FTIR is a ZnSe window 19 mm wide, 2 mm thick, and 39 mm long.
Using a microscope adapter, the 39-mm length of the ZnSe window
strip is sufficient for a 2-h SFC run. With both diffuse reflectance
and transmittance, identifiable spectra for about 50 ng of test compounds have been obtained. Although both methods still limit the
choice of mobile phases to those that are gaseous at standard temperature and pressure, they do permit the use of modifiers and of
pressure programming without background interference. The disadvantage, of course, is increased complexity of the apparatus and
the need to prepare or clean new windows for each chromatographic
run.

Flame Ionization Detection (FID)

Since its development in 1957, the FID has served as the workhorse
of gas chromatography detectors. The sensitivity and ubiquity of
its response have been the envy of liquid chromatographers for
years. Although schemes such as moving belts, bands, and ribbons
have been used to interface the FID with LC, the general complica-
tion of these approaches has limited successful operation. When CO_2
and other nonionizable substances are used as supercritical fluid
mobile phases, the FID can be readily interfaced to SFC to provide
a detection method with universal response characteristics. Because
the FID is covered in detail elsewhere, we will not repeat the information here. The separation and detection of small quantities of
nonconjugated high molecular weight compounds has been one of the
most significant contributions of SFC/FID.

Hydrogen Atmosphere Flame Ionization
Detection (HAFID)

The ability to interface SFC to ambient-pressure detectors that have
traditionally been used for GC detection provides a tremendous
arsenal of possible detection methods for SFC. Simple modifications
of a standard FID, for example, can lead to some unique and selective detection methods.

   One of the first mixtures of compounds to be separated by SFC
was a mixture of metal porphyrins [13]. Selective detection of

metal-containing compounds can be sensitively accomplished by a modification of a flame ionization detector to produce a flame that burns in a hydrogen atmosphere [14]. Although this flame has a number of requirements for proper operation, multiparameter optimization of the detector can produce detection limits in the low picogram and high femtogram range for many organometallic compounds. Initial investigations of this detector with respect to SFC have indicated that organometallics can be selectively detected after separation by SFC [15]. Figure 8 is a chromatogram of a mixture of three ferrocenes used to demonstrate the metal-sensitive nature of the hydrogen atmosphere flame ionization detector. For this work the chromatographic system consisted of a CO_2 mobile phase, which was delivered by an Altex 100A dual-piston pump. The piston heads were cooled with dry ice to prevent vapor lock, and a 15-ft coil of 1/8-in.-i.d. stainless steel tubing was placed between the pump and the injector to damp pressure pulses from the piston. Separation was accomplished on a 10 m × 100 μm i.d. fused silica capillary column with a chemically bonded and cross-linked SPB-1 stationary phase with a film thickness of 0.2 μm. The selectivity of the HAFID for metal-containing species to hydrocarbons ranged from 10,000 to 40,000. Selectivity of the detector can be adjusted significantly by varying the collector electrode height. Modified from a Hewlett-Packard FID (model 5830A), the detector used for SFC is shown schematically in Fig. 9. A stainless steel chimney containing an ignitor and electrode was fitted onto the base of the commercial FID. The collector electrode, which consisted of a co-axial bulkhead fitting, was mounted in a hollow stainless steel tube so that the electrode height could be adjusted relative to the flame. The original FID jet tip was bored out to an internal diameter of 1.5 mm and the SFC restrictor was terminated flush with the end of the jet tip. Detector gases used in the HAFID were reversed from those normally used in an FID: O_2 and N_2 were used as make-up gases and introduced through the jet tip, while the hydrogen, which was doped with small quantities of silane, was introduced directly into the detector housing. Thus the detector flame is an air diffusion flame burning in a hydrogen atmosphere. The potential of a sensitive and selective metal detector for SFC is of interest because of the large number of metal-containing compounds that are either not volatile or thermally labile. More detailed investigations of the HAFID as a detector for SFC are needed before the practicality of its application to SFC can be fully assessed.

Thermionic Detection (TID)

Another gas chromatographic detector that has been successfully adapted to SFC is the thermionic detector [16—18]. First discovered

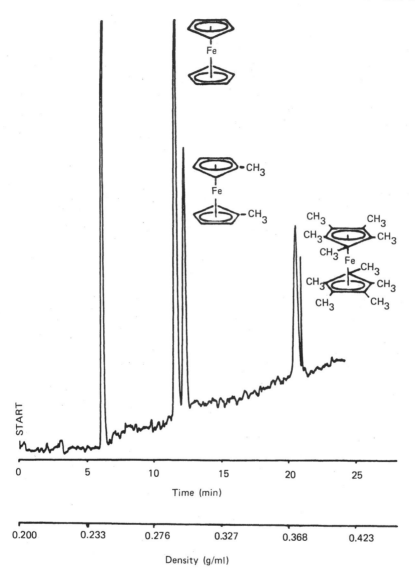

Fig. 8   Selective detection of organometallics by hydrogen atmosphere flame ionization.   (From Ref. 15.)

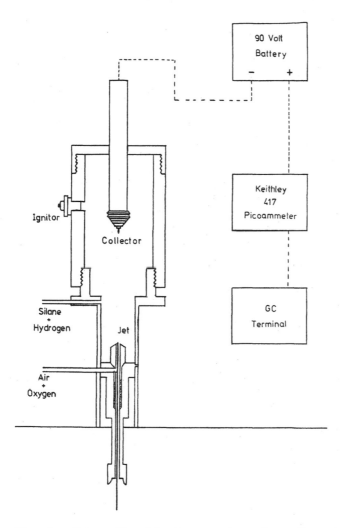

Fig. 9   Schematic of hydrogen atmosphere flame ionization detector used with SFC.   (From Ref. 15.)

by observing enhanced ionization in a common flame ionization detector when the flame was contaminated with alkali salt, it rapidly became an important detector for organophosphate pesticides and residues in the environment. The selectivity of its response permitted the detection of trace quantities of nitrogen- and phosphorus-containing compounds in the presence of a wide variety of contaminating hydrocarbon species. Since that time, the thermionic detector has undergone several radical design changes as its ionization mechanism has become better understood.

Under certain conditions this unique ionization phenomenon can be optimized for nitrogen- and phosphorus-containing compounds. The determination of biologically important organonitrogen compounds from complex mixtures has continually demanded more and more attention. One class of compounds of specific interest for SFC methods is nitrated polycyclic aromatics. Several compounds in this class decompose or are irreversibly adsorbed onto the walls of the injector, column, or transfer line during GC separation. Figure 10 shows an example of SFC separation and selective detection of a number of nitrated polycyclic aromatic compounds (PAC) [16]. The selective detection advantage of producing a less complicated tracing for data processing should be obvious. Also, relative response ratios between the TID and FID can provide qualitative information about sample components.

Although most thermiomic detectors require combustion, Fig. 11 shows a general schematic of a TID design that has a nonflame operation mode. When the detector is operated in this mode, a source with a high surface layer concentration of cesium is warmed to a relatively low temperature of 400—600°C in a nitrogen atmosphere. Under these conditions electronegative compounds such as nitrated polycyclic aromatic hydrocarbons capture an electron from the surface to form a negative ion, which is collected at a positive-biased electrode, producing the detector's response. For the nitro-PAC compounds this mode of operation is the most sensitive. Figure 12 shows a capillary column supercritical fluid chromatogram of an NBS 1650 diesel particulate extract with nitro-selective detection. The numbered peaks in the chromatogram were tentatively identified as nitro-containing compounds by retention time comparison with standards.

Flame Photometric Detection (FPD)

As demonstrated in the previous discussions, gas chromatographic flame detectors are readily adapted to supercritical fluid chromatography. A particularly important one has been the flame photometric detector, in which sulfur- and/or phosphorus-containing compounds are detected selectively and with high sensitivities. Modification of

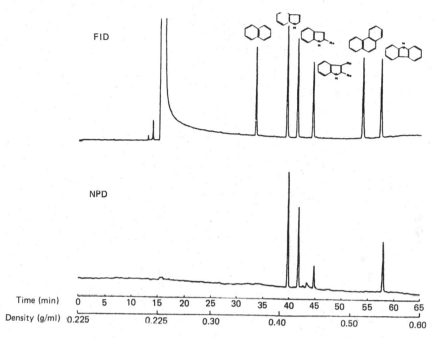

Fig. 10 Thermionic detection of aromatic nitrogen compounds. An FID and an NDP tracing of a supercritical fluid chromatographic separation of a mixture of aromatic hydrocarbons showing the selective detection of the nitrogen-containing compounds. Chromatographic conditions: 34 m × 50 μm i.d. fused silica column; SE-54 stationary phase with 0.25 μm film thickness and cross-linked with azo-*t*-butane; mobile phase CO_2 at 40°C; density programmed from 0.225 to 0.60 g/μl at 0.0075 g/μl after an initial isoconfertic period of 15 min. (From Ref. 16.)

commercially available designs of this detector provides an efficient detector for sulfur-containing compounds after SFC separation.

The basic mode of operation of the flame photometric detector has been known for years. When compounds containing P or S atoms burn in a hydrogen atmosphere flame, the reducing atmosphere of the flame leads to the formation of small amounts of S_2 or HPO. A small fraction of these combustion products is formed in the excited state and emits light at characteristic wavelengths to produce the measured response of the detector. A single hydrogen atmosphere flame is sufficient to produce selective responses for S and P, but response factors may differ between compounds. A double flame is often preferred to a single flame because the primary flame produces chemically equivalent precursors and destroys many

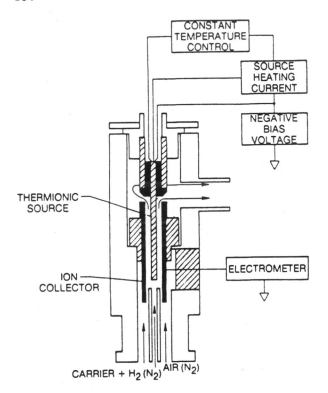

Fig. 11 Schematic of a nonflame thermionic detector. When operated under an inert atmosphere of nitrogen, this detector is highly sensitive to electronegative compounds such as nitro-containing compounds. The mode of operation is through electron transfer from a surface with a low work function to a sample molecule, forming a gas-phase negative ion. (From Ref. 18.)

of the sample interferences; the secondary flame is then the analytical flame. Thus, the dual-flame photometric detector reduces the difference in response factors from compound to compound (making the flame more quantifiable) and experiences less interference from non-sulfur- or non-phosphorus-containing compounds (making the detector more selective).

Figure 13 shows a schematic of a typical GC dual-flame photometric detector that has been modified for operation with SFC [19]. Only minor modification was required; the detector base was shortened so that the restrictor could be completely inserted into the detector and positioned just below the primary flame. A makeup gas line was added above the primary air inlet (number 7 in Fig. 13) so

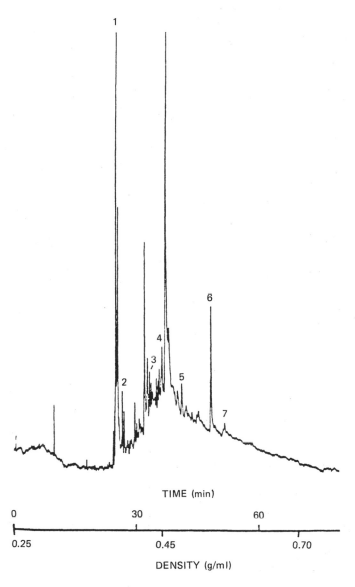

TIME (min)

0   30   60

0.25   0.45   0.70

DENSITY (g/ml)

Fig. 12   Thermionic selective detection of nitro-containing compounds in NBS 1650 diesel particulate extract. (1) 5-Nitroisoquinoline; (2) 1-hydroxy-2-nitronaphthalene; (3) 5-nitroindole; (4) 9-hydroxy-2-nitrofluorene; (5) 1-hydroxy-8-nitropyrene; (6) 1-hydroxy-6-nitropyrene; (7) 1-hydroxy-3-nitropyrene.   SFC conditions: column, 15 m × 50 μm i.d. fused silica capillary coated with SE-54 (d_f = 0.25 μm); mobile phase, CO_2 at 101°C; density programmed from 0.25 to 0.70 g/μl at 0.0075 g/μl after an initial 10-min isoconferic period; initial linear flow rate, 2.8 cm/s.   (From Ref. 17.)

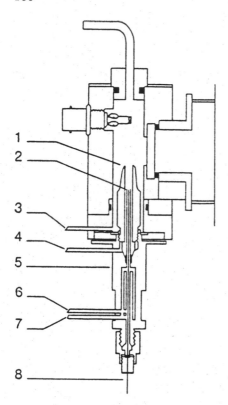

Fig. 13   Dual-flame photometric detector. Schematic diagram of
the detector after modification for supercritical fluid chromatography.
(1) Secondary flame jet; (2) primary flame jet; (3) secondary air
inlet; (4) H_2 gas inlet; (5) detector base; (6) makeup gas inlet;
(7) primary air inlet; (8) column (restrictor). (From Ref. 19.)

that 30 to 40 cm^3/min of helium could be added to help sweep the effluents into the primary flame.

Although initial results with this FPD indicated that the detector can be used with SFC, detection limits were found to be considerably higher than those commonly found after gas chromatography. The detection limit for a sulfur-containing compound, benzo[b]thiophene, was reported to be 25 ng. These detection limits were determined when using carbon dioxide as the mobile phase. It is not clear whether the low sensitivity in the sulfur mode was a function of CO_2 emission, CO_2 quenching, or a combination of the two effects. It was noted, however, that because of the emission of CO_2 at 365 nm, baseline rise was experienced as a function of density programming.

In the phosphorus mode light emission is monitored at 530 nm, so the emission from CO_2 at 394 nm can be discriminated against; thus, no interferences are observed from the CO_2 mobile phase and density programming can be used for the separation and detection of phosphorus-containing compounds by SFC/FPD. For the phosphorus-containing compound methomyl the detection limit was 0.5 ng. This value is still somewhat higher than that expected in gas chromatography, but it is considerably better than that observed in the sulfur mode of operation.

Figure 14 provides an example of a supercritical fluid chromatogram of a standard mixture of sulfur- and phosphorus-containing pesticides showing selectivity as a function of sulfur (365 nm) or phosphorus (530 nm) operating mode. Note that these test compounds, parathion, chlorpyrifos, and Larvin, contained carbon, hydrogen, oxygen, and nitrogen. In addition to these common elements, parathion contains phosphorus and sulfur atoms; chlorpyrifos contains three chlorines, a phosphorus, and a sulfur atom; and Larvin contains only sulfur. In the sulfur mode all three pesticides are detected, while the solvent, which contains neither sulfur nor phosphorus, produces only a small detector response. In the phosphorus mode, Larvin, which does not contain phosphorus, is completely discriminated against. It is important to note that the conditions used to obtain these chromatograms were considerably milder (lower temperature) than those required in gas chromatography to produce a similar separation.

Ion Mobility Detection (IMD)

Next to the flame ionization detector, the ion mobility detector may prove to be one of the most useful detection methods for SFC. Although this detector is not yet commercially available, it has the potential of providing a variety of responses that otherwise would require a number of dedicated chromatographic detectors [20]. Chromatographic traces similar to those obtained with a flame

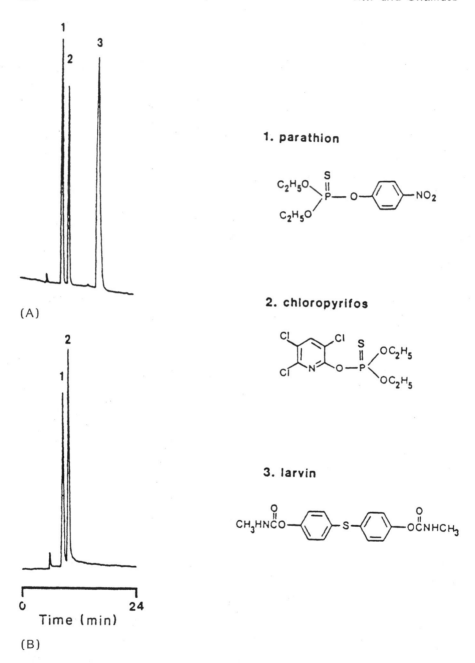

Fig. 14   Flame photometric detection of pesticides.   Supercritical fluid chromatogram of a standard mixture of sulfur- and phosphorus-containing pesticides.   Flame photometric detection in (A) sulfur

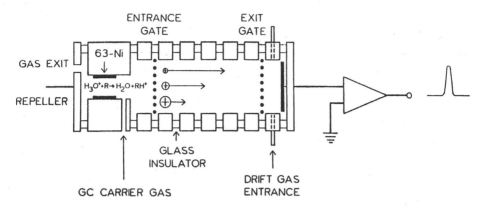

ENTRANCE GATE   EXIT GATE

GAS EXIT   63-Ni

$H_3O^+ \cdot R \rightarrow H_2O \cdot RH^+$

REPELLER

GLASS INSULATOR

GC CARRIER GAS

DRIFT GAS ENTRANCE

Fig. 15   Ion mobility spectrometer.   (From Ref. 20.)

ionization detector, an electron capture detector, a photoionization detector, and a variety of selective GC detection methods can be obtained with an ion mobility detector by simply changing operating conditions.

Figure 15 shows a schematic diagram of an ion mobility detector in which ambient-pressure ions are generated in an ion chamber. To date, two types of ionization sources have been employed for this detector: a [63]Ni beta-emitting foil and a photoionization lamp. With the [63]Ni source, a standing background of ions called reactant ions is produced which, in a chemical ionization manner, transfers charge to neutral compounds as they enter the detector. These new ions are called product ions. The entire chamber in which the ions are produced is under an electric field that causes migration of the ions in a given direction. By introducing ion pulses into the drift region of the detector, reactant and product ions can be separated according to their average velocities under a given electric field at ambient pressure. Ion mobility spectrometry is not a high-resolution ion separation technique, but sufficient separation occurs for many product ions to be separated from other product ions and from the reactant ions. Thus two types of information can be obtained with this detection method. The first mode of operation

mode using 365 nm and (B) phosphorus mode using 530 nm. Conditions: 15 m × 75 μm i.d. fused silica column; *n*-octyl polysiloxane stationary phase (1.0 μm film thickness) cross-linked with azo-*t*-butane; CO_2 mobile phase at 120°C and isoconfertic at 0.5 g/ml. (From Ref. 19.)

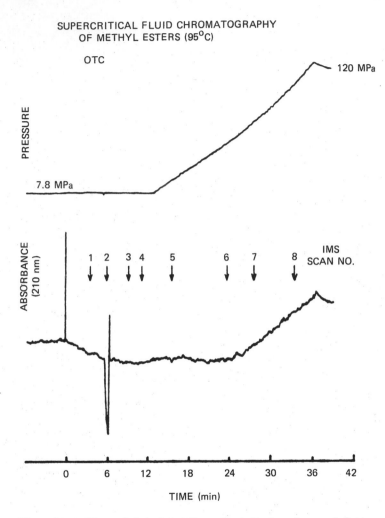

Fig. 16   Ultraviolet detection of methyl esters of fatty acids after SFC separation.   (From Ref. 21.)

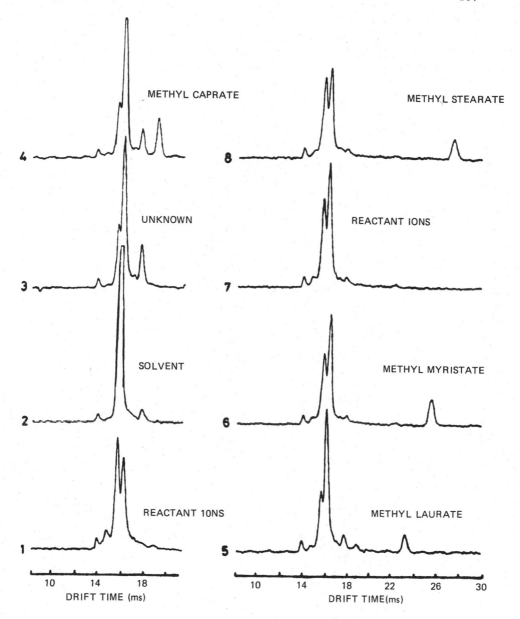

Fig. 17   Ion mobility spectra of fatty acid methyl esters after SFC. (From Ref. 21.)

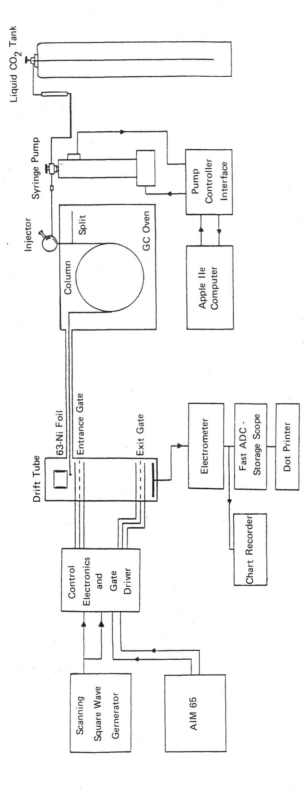

Fig. 18   Schematic diagram of SFC FT-IMD system.   (From Ref. 22.)

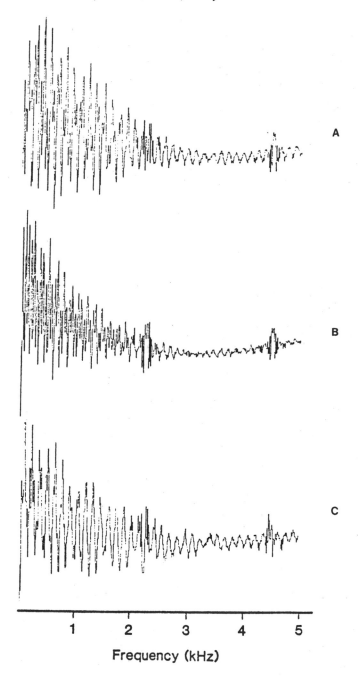

Fig. 19 Mobility interferograms of Triton X-100. (A) Tenth component, (B) fifth component, and (C) third component. (From Ref. 22.)

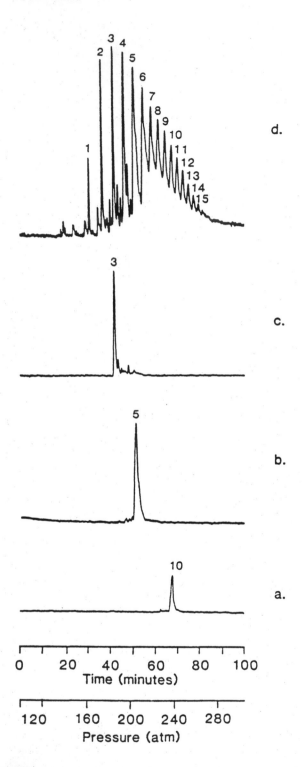

provides qualitative information by scanning the arrival times of all the ions to produce an ion mobility spectrum. The second mode of operation continuously monitors only one ion transit time. In this mode, the detector can be used as a quantitative selective detection method for compounds that produce ions which migrate with the transit time being monitored. Figure 16 shows a supercritical fluid chromatographic separation of a mixture of methyl esters at concentrations too low to detect by on-column UV-visible absorbance detection. The ion mobility scans taken throughout this chromatographic run are shown in Fig. 17. Distinct ion mobility spectra can be seen for the reactant ions, the solvent, an unknown, methyl caprate, methyl laurate, methyl myristate, and methyl stearate [21].

Signal averaging was used to obtain the ion mobility spectra shown in Fig. 17. In this method the entire spectrum is taken as a waveform a number of times and then averaged to increase the signal-to-noise ratio. An alternative method of taking ion mobility spectra is with the entrance and exit gates opened and closed simultaneously while the frequency at which they are opened and closed is increased over time [22]. The instrumentation required for this mode of operation, known as Fourier transform ion mobility spectrometry (FTIMS), is shown in Fig. 18. Figure 19 illustrates the signal obtained from FTIMS with frequency scans called mobility interferograms. These interferograms were obtained for the third, fifth, and tenth components in a mixture of Triton X-100 which was separated by SFC. The Fourier transformation of these interferograms produces normal ion mobility spectra with peaks for the reactant ions and signal product ions for these individual components. Continuous monitoring of the selected product ions provided the selective chromatograms shown in Fig. 20.

LOW-PRESSURE DETECTION

Mass Spectrometry

Low-pressure detectors primarily involve the variety of mass spectrometers and mass spectrometric methods that can be used with the separation and introduction method of supercritical fluid chromatography. With gas chromatography, mass spectrometry has become a completely integrated combination for detection and identification after high-resolution separation. For the highly polar, nonvolatile, or thermally labile compounds typically separated by liquid

---

Fig. 20 Selective detection of individual oligomers of Triton X-100 after SFC separation. (a) Selective result detecting tenth component, (b) selective result detecting fifth component, (c) selective result detecting third component, and (d) nonselective result.

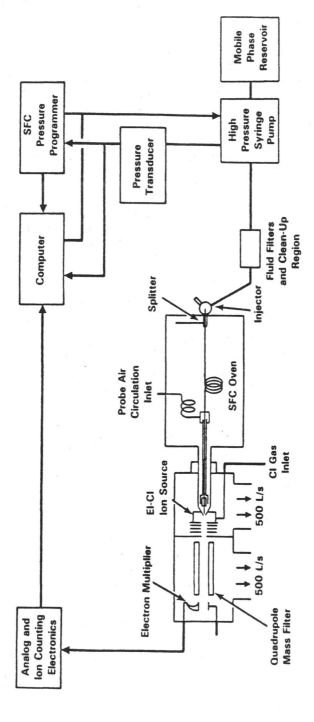

Fig. 21   Schematic diagram of capillary SFC/MS instrumentation.   (From Ref. 27.)

chromatography rather than gas chromatography, the success of mass spectral application has been less spectacular, although significant gains in LC/MS have been made in recent years. However, SFC can be more readily interfaced to mass spectrometers than HPLC and promises to offer high-resolution separation of compounds which are not suited for gas chromatography. A detailed discussion of the principles and applications of SFC with mass spectrometric detection is given in Chapter 10.

A schematic diagram of a typical two-stage SFC/MS is shown in Fig. 21. In this design the capillary SFC column is inserted through a probe which is located in the oven so that its temperature remains the same as that of the oven. From the probe the sample passes through a steep pressure gradient in which it is decompressed from the supercritical pressure of several hundred atmospheres to vacuum conditions of around 10^{-5} torr in the ionization chamber of the mass spectrometer. Electron impact and chemical ionization are the two most common ionization mechanisms used with SFC introduction. The advantages and disadvantages of each of these ionization methods are discussed later in this chapter. The ions are accelerated from the ion source region into an independently pumped quadrupole mass analyzer, where they are separated and recorded with all of the options that have become available for analytical mass spectrometers. Both the supercritical fluid chromatograph and the mass spectrometer are unmodified except for the interface region. As with all of the detection methods that have been discussed in this chapter, the most important single component of the SFC/MS combination is the interface.

The Interface

The traditional approach to creating a dynamic high- to low-pressure interface has been to permit expansion of the mobile phase through a nozzle, followed by production of a collimated beam using molecular beam skimmers. Applications of this molecular beam approach have been demonstrated [23,24], but the complexity of this multiparameter operation has restricted its use in SFC/MS.

A primary objective of the SFC/MS interface is to reduce solvent-solute clustering and prevent solute precipitation during the decompression process. As the supercritical fluid expands, the solvating capacity of the mobile phase diminishes and solvent–solvent clusters can be formed because of the cooling effect of the adiabatic expansion process. The purpose of the nozzle-skimmer combination was to permit decompression to occur in stages, reducing the effects of clustering and establishing a molecular beam of the solute which was free of solvent–solvent and solute–solute interactions. The nozzle-

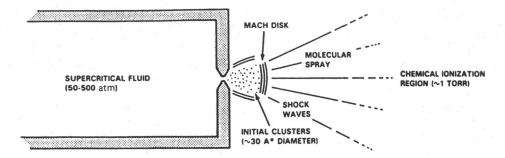

Fig. 22 Schematic illustration of the direct fluid injection process. The Mach disk is the shock wave perpendicular to the expansion axis resulting from expansion into the lower-pressure region. (From Ref. 28.)

skimmer approach demonstrated that low-volatility solutes could be injected successfully into the mass spectrometer without substantial clustering or clogging from precipitation, but it did not provide a practical method for interfacing supercritical fluid chromatographs to mass spectrometers and it exhibited relatively poor sensitivities with detection limits in the microgram range.

With capillary SFC, the mobile phase flow rates are so low that cooling and clustering can be avoided by external heating and solvent molecules can easily be pumped away. Thus, the approach of direct fluid injection (DFI), which was first described for SFC/MS in 1982 by Smith et al. [25] simply requires that a flow restrictor be placed between the SFC column and the ion source of the mass spectrometer. The ideal restrictor for DFI provides three functions: (a) it maintains the proper mobile phase flow rate through the capillary column, (b) it maintains the pressure in the column (and transfer line) as close to the exit as possible, and (c) it produces a jet of vapors into the ionization source in which the compounds of interest are entrained.

Figure 22 is a schematic illustration of the ideal direct fluid injection process. In this illustration the interface between the supercritical fluid conditions of the chromatographic column and the vacuum conditions of the mass spectrometer is a vanishingly small orifice in an infinitely thin barrier. Under these conditions, the mobile phase maintains its integrity as a supercritical fluid until it passes through the pinhole leak into the vacuum regions.

Immediately on entering the low-pressure region, adiabatic expansion begins and solvent—solvent and solvent—solute clustering is observed. When expansion occurs into a vacuum with a finite background pressure (as is the case with chemical ionization mass spectrometry), the expanding gas interacts with the background gases to produce a standing shock wave system as illustrated in Fig. 22. The shock wave, which is formed perpendicular to the jet axis, is called the Mach disk and serves the·purpose of heating and breaking up the clusters formed in the expansion process.

In practice, an ideal restrictor interface cannot be constructed, but several restrictor designs have proved adequate for SFC/MS operation. Pinhole designs have been attempted with laser-drilled orifices in stainless steel diaphragms. For capillary columns of 50 μm i.d., an orifice of 0.25 μm is needed; for columns of 100 μm, 0.50 μm is used. One of the major difficulties of the pinhole orifice approach is the construction of the restrictor. Another major problem with using such a small orifice is plugging. These orifices can and do readily become plugged from unfiltered particles in the line. Also, the thin stainless steel diaphragm can deform from the pressure, changing the size of the orifice.

An alternative to the pinhole leak is the capillary restrictor approach. A short length of narrow-diameter fused silica tubing is connected with a zero dead volume butt connector to the exit of the separation column. For columns of 50 to 100 μm, a 5-μm-i.d. length of tubing serves as a convenient restrictor. The length of the capillary restrictor is varied to adjust the flow of the mobile phase in the column. The major disadvantage of the capillary restrictor is that decompression occurs over the entire length of the tube, decreasing the solvent strength of the mobile phase and forcing the solute out of solution before it reaches the restrictor exit. For solutes with some volatility, heating the restrictor aids in transferring the compounds into the mass spectrometer.

A practical compromise between the ideal pinhole restrictor and the capillary restrictor is the tapered capillary restrictor. Similar to those first developed for interfacing SFC to the FID, these tapered restrictors offer improved performance. The pressure gradient in tapered restrictors is much sharper than that in capillary restrictors and they are easier to construct than laser-drilled pinhole restrictors.

Ion Sources

The two most common analytical ion sources for mass spectrometry, electron impact (EI) and chemical ionization (CI), have also been investigated for use with capillary SFC. Direct introduction of the column effluent into an electron impact ionization source might seem to be the most desirable method of operation, since the EI source is the most widely used mass spectrometry source and fragmentation patterns of a variety of compounds have been investigated. Many of the compounds which require the use of SFC for separation and MS introduction do not have reliable EI spectra established in the

literature.   Standards of extremely labile compounds often exhibit
some thermal degradation in the direct probe approach, and high-
molecular-weight standards are difficult to purify without chromatographic separation.

· Unfortunately, the use of EI has been limited because with SFC
the sensitivity of the method is decreased and, as a result of solvent—solvent and solvent—solute interaction, the EI spectra are
often contaminated with chemical ionization spectra.   The primary
reason for the loss in sensitivity of the source is apparently the
increase in pressure that is associated with the expansion of the
SFC mobile phase into the ionization region.   Even with an open EI
source the effective pressure of the region increases, reducing the
range and efficiency of the ionizing electrons by increasing their
frequency of collision with background gases.

The problems associated with the EI source can be partially
alleviated with an "extremely" open source.   Smith et al. [26] have
been able to maintain the ion source chamber pressure in the range
of $(1-5) \times 10^{-5}$ torr.   Under these conditions minimum detectable
amounts as low as 5 pg and mass spectra of reasonable quality with
few CI contributions were obtained.

Because it normally operates at an elevated pressure relative to
the EI source, the chemical ionization source is ideal for coupling
with SFC.   It has the flexibility of using different reagent gases
and exhibits excellent sensitivity.   The CI reagent gases which are
commonly used with SFC are methane, isobutane, and ammonia.   Detection limits, of course, depend on the compounds being ionized
and the reagent gas used, but in general CI can be expected to be
capable of obtaining spectra in the subnanogram range and of detecting chromatographic peaks by single-ion monitoring in the low
picogram range.

Examples

Figure 23 is the total-ion tracing obtained from the capillary supercritical fluid separation of the polycyclic aromatic hydrocarbon fraction of a marine diesel fuel (MDF) sample.   The chromatography was
accomplished on a 15 m × 50 μm fused silica column coated with a
highly cross-lined 50% phenyl polymethylphenylsiloxane stationary
phase.   A capillary restrictor approximately 20 mm × 5 μm i.d. was
used to form the interface between the column and the CI source.
Carbon dioxide, being the most convenient fluid for capillary SFC,
was used as the mobile phase, and the operating temperature was
selected to be 60°C, 1.1 times the critical temperature of carbon
dioxide.   The isobutane chemical ionization reagent used with this
analysis formed primarily $(M + 1)^+$ ions with the polycyclic aromatics
to produce normal CI spectra for this type of compound.   Figure 24

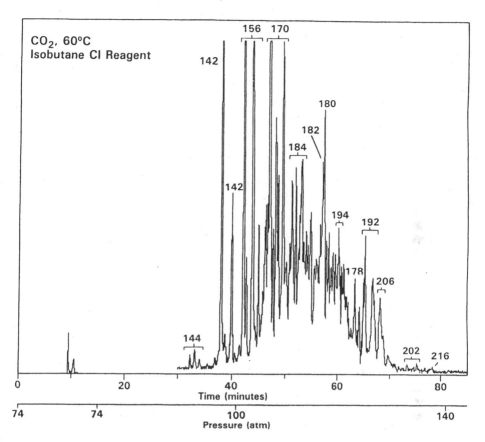

Fig. 23 Total-ion chromatogram from capillary SFC/MS analysis of the polycyclic aromatic hydrocarbon fraction of MDF 81-6. Peak labels refer to the actual molecular weights of the compounds. (From Ref. 27.)

shows the typical isobutane CI mass spectra obtained for selected peaks from this SFC/MS analysis.

To illustrate the separation and MS identification of higher-molecular-weight compounds, Fig. 25 shows the supercritical separation of several triacylglycerols using CO_2 at 100°C. Figure 26 is the methane CI spectrum of tristearoylglycerol having a molecular weight of 890. The $(M + 1)^+$ ion was clearly present, as was the $(MH—RCOOH)^+$ ion. Currently, a molecular weight of about 1000 seems to be the size limit for detection by SFC/MS. As the challenge of yet higher-molecular-weight compounds continues to beckon

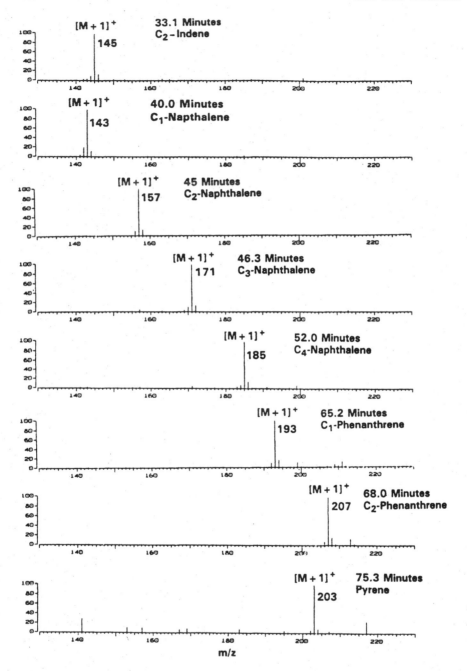

Fig. 24   Typical isobutane CI mass spectra for selected peaks from the SFC/MS analysis in Fig. 23. (From Ref. 27.)

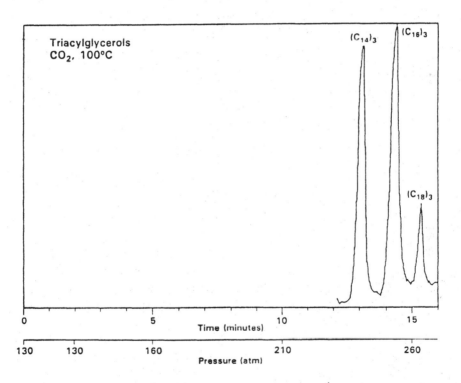

Fig. 25 Capillary SFC/MS separation of a mixture of three tri-acylglycerols. (From Ref. 29.)

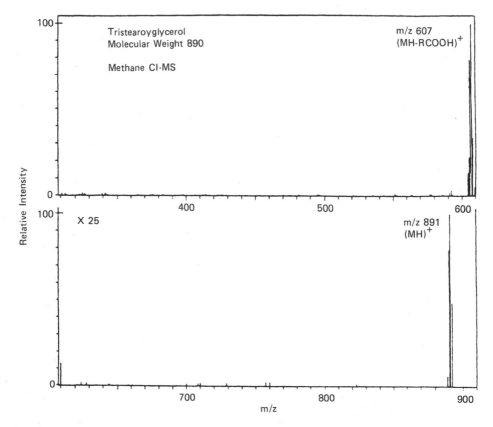

Fig. 26   Methane chemical ionization mass spectrum of tristearoyl-glycerol from SFC/MS analysis in Fig. 25.   (From Ref. 29.)

the supercritical fluid chromatographer, investigations will inevitably be conducted with magnetic sector-type instruments, increasing the potential molecular weight range of this promising analytical combination.

SUMMARY

Novel designs for on-column UV absorbance detection and for fluorescence detection in which the column exit is monitored have extended the use of traditional LC detectors to capillary SFC. More important, however, is the ease with which SFC can be interfaced to gas chromatographic detection systems. The high sensitivity and selectivities of detectors such as the ion mobility detector, the nitrogen-phosphorus thermionic detector, and the flame photometric detector offer a variety of unique methods for the determination of compounds that are normally separated by liquid chromatography. Moreover, the flame ionization detector offers a universal sensitivity which is currently impossible to achieve with LC detectors. With SFC, high-molecular-weight compounds without chromophors can be separated and detected using the FID. Also, through a direct fluid interface (DFI), SFC-compatible compounds can be efficiently introduced into mass spectrometers for either detection or identification.

As demonstrated above, SFC has shown exceptional promise as an efficient method for introducing nonvolatile or thermally labile compounds into mass spectrometers. Moreover, the compatibility of capillary SFC with the wide variety of chromatographic detection methods described earlier in this chapter provides separation/detection combinations that can complement those currently available with capillary gas chromatography and high-performance liquid chromatography. Although unique interfacing challenges still exist for each detection system employed, the multiplicity of detection methods that are now available (or have been demonstrated to be feasible) attests to the ultimate potential of capillary SFC as a versatile chromatographic method.

REFERENCES

1. J. C. Giddings, *Science 162*, 67 (1968).
2. M. Novotny, S. R. Springston, P. A. Peaden, J. C. Fjeldsted, and M. L. Lee, *Anal. Chem. 53*, 407A (1981).
3. K. H. Shafer, S. L. Pentoney, Jr., and P. R. Griffiths, *Anal. Chem. 58*, 58 (1986).
4. F. J. Yang, *J. High Resolut. Chromatogr. Chromatogr. Commun. 4*, 83 (1981).
5. P. A. Peaden, J. C. Fjeldsted, M. L. Lee, S. R. Springston, and M. Novotny, *Anal. Chem. 54*, 1090 (1982).
6. J. C. Fjeldsted, B. E. Richter, W. P. Jackson, and M. L. Lee, *J. Chromatogr. 279*, 423 (1983).
7. K. H. Shafer and P. R. Griffiths, *Anal. Chem. 55*, 1939 (1983).
8. S. V. Olesik, S. B. French, and M. Novotny, *Chromatographia 18*, 489 (1984).
9. J. A. de Haseth and T. L. Isenhour, *Anal. Chem. 49*, 1977 (1977).
10. S. B. French and M. Novotny, *Anal. Chem. 58*, 164 (1986).
11. K. H. Shafer, S. L. Pentoney, Jr., and P. R. Griffiths, *J. High Resolut. Chromatogr. Chromatogr. Commun. 7*, 707 (1984).

12. S. L. Pentoney, Jr., K. H. Shafer, and P. R. Griffiths, *J. High Resolut. Chromatogr. Chromatogr. Commun.* 9, 168 (1986).

13. E. Klesper, A. H. Corwin, and D. A. Turner, *J. Org. Chem.* 27, 700 (1962).

14. D. G. McMinn, R. L. Eatherton, and H. H. Hill, Jr., *Anal. Chem.* 56, 1293 (1984).

15. M. Morrissey and J. H. Hill, Jr., unpublished work.

16. J. C. Fjeldsted, R. C. Kong, and M. L. Lee, *J. Chromatogr.* 279, 449 (1983).

17. W. R. West and M. L. Lee, *J. High Resolut. Chromatogr. Chromatogr. Commun.* 9, 161 (1986).

18. P. L. Patterson, R. A. Gatten, and C. Ontiveros, *J. Chromatogr. Sci.* 20, 97 (1982).

19. K. E. Markides, E. D. Lee, R. Bolick, and M. L. Lee, *Anal. Chem.* 58, 740 (1986).

20. H. H. Hill, Jr. and M. A. Baim, *Trends Anal. Chem.* 1, 232 (1982).

21. S. Rokushika, H. Hatano, and H. H. Hill, Jr., *Anal. Chem.* 59, 8 (1987).

22. R. L. Eatherton, M. Morrissey, W. F. Siems, and H. H. Hill, Jr., *J. High Resolut. Chromatogr. Chromatogr. Commun.* 9, 154 (1986).

23. J. C. Giddings, M. N. Myers, and A. L. Wahrhaftig, *Int. J. Mass Spectrom. Ion Phys.* 4, 9 (1970).

24. L. G. Randall and A. L. Wahrhaftig, *Anal. Chem.* 50, 1703 (1978).

25. R. D. Smith, J. C. Fjeldsted, and M. L. Lee, *J. Chromatogr.* 247, 231 (1982).

26. R. D. Smith, H. R. Udseth, and H. T. Kalinoski, *Anal. Chem.* 56, 2971 (1984).

27. B. W. Wright, H. R. Udseth, R. D. Smith, and R. N. Hazlett, *J. Chromatogr.* 314, 253 (1984).

28. R. D. Smith and H. R. Udseth, *Anal. Chem.* 55, 2266 (1985).

29. B. W. Wright, H. T. Kalinoski, H. R. Udseth, and R. D. Smith, *J. High Resolut. Chromatogr. Chromatogr. Commun.* 9, 145 (1986).

10

Principles and Applications of Supercritical Fluid Chromatography with Mass Spectromic Detection

RICHARD D. SMITH and BOB W. WRIGHT / Batelle Pacific Northwest Laboratory, Richland, Washington

INTRODUCTION

The combination of chromatography with mass spectrometric detection provides the most widely applicable analytical approach for characterization of organic mixtures. In the past decade gas chromatography—mass spectrometry (GC—MS) has been established as the method of choice for the characterization of amenable compounds. The rapid rise to dominance of GC—MS has fostered continuing research aimed at providing a similar analytical technology for the large majority of compounds not sufficiently volatile or stable for GC separation. The extensive advances in liquid chromatography during the past two decades and the variety of commercially available instrumentation have naturally focused attention on development of LC—MS.

The three commercially available LC—MS interfaces are based on the moving belt [1], direct liquid introduction [2], and, most recently, thermospray [3]. Interfaces based on particle beams generated from nebulized liquids are also available, and are best suited for compounds that may be effectively volatilized at ion source temperatures. Thermospray is clearly the superior approach in other instances, since highly labile and nonvolatile compounds, including many not suitable for the other methods, can often be analyzed. Despite these successes, limitations still exist with the thermospray interface;

optimization is often required for each compound, the range of mobile
phases is limited, ionization and detection efficiency can be poor and
vary widely between compounds, and application is currently impractical without the use of "make-up" streams below flow rates of ~ 0.2 ml/
min.

The current interest in supercritical fluid chromatography (SFC)
and SFC—MS arises from two major considerations. First, the more
favorable diffusion rates and viscosities of supercritical fluids allow
faster and more efficient separations with the potential of addressing
many problems not amenable to gas chromatography. Second, the
properties of supercritical fluids allow the use of conventional liquid-
phase detectors and, after decompression by expansion through a
restrictor, many gas-phase detectors. The operation of many GC-
type detectors (flame ionization, electron capture, photoionization)
is possible with SFC when compatible mobile phases are utilized
(e.g., CO_2), but only the mass spectrometric detector promises
truly universal on-line applicability [4—10].

In this chapter the basic principles and current practice of SFC—
MS are described. The fundamental processes allowing the instrumental hybrid are considered, with emphasis on expansion of the
supercritical fluid through a restrictor. Implications of the expansion phenomena are discussed with respect to SFC—MS interface development. While this chapter emphasizes capillary column SFC, the
relative merits of capillary and packed column (1—4 mm diameter) SFC
and the practical considerations of operation for both approaches are
reviewed. Since both SFC and SFC—MS are relatively new and still
developing techniques, the current interface methods are described in
detail. Examples are presented illustrating both the wide range of applicability of SFC—MS and its practical application to problems which
would be difficult or impossible by GC—MS or LC—MS. Finally, the
current limitations of SFC and SFC—MS (since they are largely the
same) and the promise for much wider applicability are discussed.

SUPERCRITICAL FLUIDS AS MOBILE PHASES

Properties of Supercritical Fluids

The ability to combine SFC and mass spectrometry by the "direct
fluid injection" interfacing method [4] rests on the properties of
supercritical fluids [11]. Above the critical pressure and temperature of a substance, the single fluid phase has properties that are
intermediate between those of the gas and liquid phases and are
dependent on the fluid composition, temperature, and pressure [12—
14]. The density of a supercritical fluid is typically 10^2 to 10^3
times greater than that of the gas, at somewhat lower pressures,

and approaches (and sometimes exceeds) that of the liquid. Molecular interactions are similar to those in liquids because of short intermolecular distances. The "liquidlike" behavior of a supercritical fluid results in greatly enhanced solubilizing capabilities compared to the "subcritical" gas, but with higher diffusion coefficients and lower viscosities compared to corresponding liquids. The variable solvent properties for a number of fluids have been demonstrated using the solvatochromic method [15]. Polymers with molecular weights in excess of several thousand are known to be soluble in supercritical carbon dioxide, which has a critical temperature of only 31°C, and with molecular weights of over several hundred thousand (for polystyrene) in supercritical pentane [16].

The critical properties of the most commonly used supercritical fluid solvents for SFC are given in Table 1. In addition to pure solvents, fluid mixtures can be advantageously employed to increase the range of molecular species which are soluble in a single supercritical fluid, to alter selectivity by modification of the stationary phase, and to allow operation in a more favorable temperature range [17]. Although not widely applied to date, since the variable solvating power obtained by control of density is so easily utilized, gradient elution methods provide an additional option to modify and enhance fluid solvating properties for SFC [18—22].

The applicability of SFC is largely defined by the availability of suitable supercritical fluid solvents for the analytes of interest. The solubility of analytes typically increases with density, and a maximum rate of increase in solubility with pressure is typically observed near the critical pressure, where the rate of increase in density with pressure is greatest [4]. This results from the fact that there is often a nearly linear relationship between log(solubility) and density for dilute solutions of nonvolatile compounds (up to concentrations where solute—solute interactions become important). In contrast, where volatility is low and at densities less than or near the critical density, increasing temperature typically decreases solubility [4]. However, "solubility" may increase at sufficiently high temperatures, where the solute vapor pressure can also become significant. Thus, while the highest supercritical fluid densities at a given pressure are obtained near the critical temperature, the greatest solubilities within experimental pressure limitations may be obtained at somewhat lower densities but higher temperatures.

As with liquids, polar solutes are most soluble in polar supercritical fluids, although nominally nonpolar fluids can be remarkably good solvents for many moderately polar compounds [4]. Carbon dioxide, for example, at higher pressures can exhibit solvating properties intermediate between those of pentane and methylene

TABLE 1   Typical Supercritical Fluid Mobile Phases

Compound	Boiling point (°C)	Critical temperature (°C)	Critical pressure (bar)	Critical density (g/cm^3)
CO_2 [a]	−78.5 [a]	31.3	72.9	0.448
NH_3	−33.4	132.4	112.5	0.235
N_2O	−88.6	36.5	71.7	0.45
SF_6	−63.8 [a]	45.7	37.1	0.752
CF_3Cl	−31.2	28	38.7	0.579
Ethane	−88.6	32.3	48.1	0.203
Ethylene	−103.7	9.2	49.7	0.218
Propane	−42.1	96.7	41.9	0.217
Pentane	36.1	196.6	33.3	0.232
Benzene	80.1	288.9	48.3	0.302
Methanol	64.7	240.5	78.9	0.272
Ethanol	78.5	243.0	63.0	0.276
Isopropanol	82.5	235.3	47.0	0.273

[a] Sublimes.

chloride, and solvatochromic studies of the solvent environment around solute probe molecules confirm their variable solvent properties [17]. At normal operating pressures, half to several times the critical pressure, solubility typically increases with pressure under isothermal conditions. Under conditions of constant density, solubility generally increases with temperature [4]. However, a temperature increase under isobaric conditions generally results in decreased solubility at pressures less than several times the critical pressure, but at higher pressures increased solubility is typically obtained.

Chromatography with Supercritical Fluids

The recent interest in SFC has been due in large part to the limitations in both chromatographic efficiency and detection methods with HPLC and the introduction of fused silica capillary columns with nonextractable stationary phases for SFC [23,24]. The lower viscosities and higher diffusion coefficients relative to liquids result in significantly enhanced chromatographic efficiency compared to HPLC. In SFC, the mobile phase is maintained at a temperature somewhat above its critical point (often at reduced temperatures of 1.02 to 1.4). The density of the supercritical phase is usually several hundred times greater than that of the gas, but less than that of the liquid at typical SFC pressures (25–500 bar). The mild thermal conditions (determined by the choice of mobile phase) allow application to labile compounds. The use of open-tubular capillary columns results in negligible pressure drops for normal linear velocities and column dimensions. This is an important consideration if the pressure-programming capability of SFC is to be fully exploited. Pressure programming in SFC provides many of the advantages of gradient elution in HPLC. As illustrated later, this also provides the basis for very rapid pressure programming to effect high-speed capillary SFC separations [25].

With the development of small-diameter (<75 μm) fused silica capillary columns coated with cross-linked and nonextractable stationary phases, high-resolution separations with efficiencies approaching those of conventional capillary gas chromatography have been obtained [9]. Studies under isobaric conditions have demonstrated that more than 3000 and 12,000 theoretical plates/m can readily be obtained with 50- and 25-μm (i.d.) columns, respectively [26]. Although this measure of efficiency is not applicable under pressure-programmed conditions, it provides the basis for the high-quality separations obtained in this mode.

Packed Versus Capillary Columns for SFC—MS

Nearly all current SFC research utilizes either columns packed with
5—10 µm particles prepared for HPLC or wall-coated (0.1—1 µm film
thickness) open-tubular fused silica capillaries of 25 to 100 µm i.d.
Simple plate height arguments [27—29] suggest that columns packed
with 5-µm particles should have plate heights somewhat less than
50-µm-i.d. capillaries, resulting in the potential for either faster
separations or generation of greater numbers of effective plates
within a certain time constraint. Packed columns of typical diameter
(1—5 mm i.d.) provide flow rates suitable for conventional liquid-
phase (e.g., HPLC) detectors and generally allow relatively large
sample loadings. The large flow rates and the necessity to remove
the liquid phase, however, make interfacing to most gas-phase detectors somewhat more difficult, particularly for mass spectrometry [5].
The most practical approaches for packed column SFC—MS currently
involve the use of fluid expansions in higher pressure regions, such as
a mediated "thermospray" LC—MS interface. Such a high-flow rate
interface is described later in this chapter. Capillary columns have
demonstrated higher numbers of effective plates than packed columns
because they have much greater permeabilities. Thus, considerably
longer columns can be used before an excessive pressure drop occurs
through the column. The pressure drop is directly proportional to
linear velocity in both packed and capillary columns, and very fast
separations with conventional 25-cm microparticle packed columns result in quite large pressure drops. Capillary columns allow pressure
programming over a wider range of pressures than packed columns
(with the precise comparison depending on column length, packing,
linear velocity, and fluid viscosity). The low flow rates with capillary
columns provide easier interfacing to mass spectrometry [4,7], but
application of such columns also requires strict attention to the elimination of dead volume. In addition, small injection volumes (<50 nl)
are typically required with capillary columns, which are most readily
obtained by flow splitting. Although lower detection limits are obtained with capillary columns, the injected sample size is proportional-
ly smaller, and a larger analytical loading range can often be obtained
with packed columns.

   The above considerations are generally understood, but often a
number of more important considerations regarding the relative merits
of the two column technologies are less widely recognized. These considerations are derived from the large pressure drops incurred with
packed columns and the greater importance of active surface sites for
typical packed columns, as opposed to the relative inertness of deactivated capillary columns. The later limitation is certain to be addressed, but the former represents a fundamental property of densely
packed columns and needs further discussion.

   Figure 1 illustrates the wide variation in chromatographic performance which can be obtained by variation of pressure and flow
rate using a packed column. In these examples, a 1 mm × 25 cm,

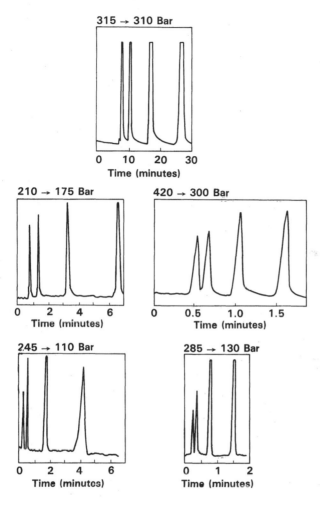

Fig. 1 SFC separations for a mixture of benzene, naphthalene, phenanthrene, and pyrene using CO_2 at 75°C at various linear velocities. Inlet and outlet pressures are indicated for each separation.

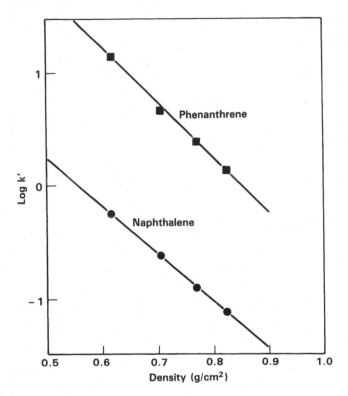

Fig. 2   Capacity ratio versus density plot for phenanthrene and naphthalene obtained under isobaric conditions for the 5-μm C_{18} microbore column used for the examples shown in Fig. 1.

5-μm particle C_{18} "microbore" column was used with a supercritical carbon dioxide mobile phase at 75°C and a 1-μl injection volume. The pressure was monitored at both the column inlet and outlet in separate experiments. The sample was a four-component mixture of benzene, naphthalene, phenanthrene, and pyrene in hexane. A UV detector was used at 254 nm with a 3-μl cell volume. The individual chromatograms show the inlet and exit pressures and suggest a modest increase in efficiency at the lowest linear velocities. It should be noted that all the faster separations resulted in large pressure drops. This can be avoided only by use of shorter columns, lower linear velocities, or a larger particle diameter of packing material.

The pressure or density drop across packed columns is their major drawback, resulting in a reduced ability to utilize the variable solvating power of supercritical fluids at different densities. Figure 2 shows the dependence of the capacity factor (k') for naphthalene and phenanthrene at 75°C as a function of density for the

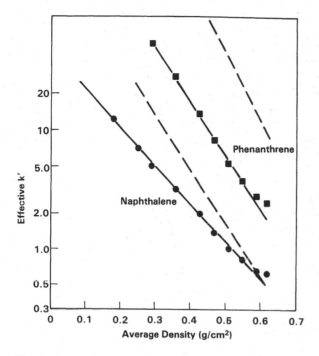

Fig. 3 Comparison of experimental and calculated (dashed lines) effective capacity ratios (k') as a function of average column density under high-pressure-drop conditions for the column used for the data shown in Figs. 1 and 2.

25-cm microbore column. The capacity factor measurements were obtained at very low linear velocities, where the pressure drop was minimal (<5 bar). The nearly linear relationship between log(k') and density seen here is typically observed [25]. For measurements made at high pressure drop conditions a linear relationship between log(k') and the calculated average density is also observed, as shown in Fig. 3. However, the experimental capacity ratios measured under large pressure drop conditions for naphthalene and phenanthrene were up to two orders of magnitude smaller than predicted by numerical calculations based on isobaric retention data and density drop corrections [25]. The primary origin of this discrepancy was that the fluid temperature was not effectively regulated. The actual fluid temperature was less than the oven temperature at the column inlet and probably remained below the oven temperature through much of the column because of the high flow rate and adiabatic cooling from the pressure drop. The net effect of the cooling was a higher fluid density than calculated, which resulted in lower retention. A secondary effect was slightly lower chromatographic

efficiencies due to lower diffusion rates. Although this problem can
be alleviated by using a thermal equilibration cell before the injector,
this approach can severely limit injector operation and choice of sample solvent, depending on the operating temperature. Thus, retention measurements made under high pressure drop conditions, while
representing a practical mode of operation, do not necessarily reflect
actual capacity factors anticipated from isobaric conditions.

Capillary columns coated with bonded and cross-linked stationary
phases provide significantly greater flexibility for regulating retention
through variation of either pressure or temperature. Capacity ratios
obtained by capillary SFC for phenanthrene in supercritical carbon
dioxide, as a function of fluid density at 50, 100, 150, and 200°C, are
shown in Fig. 4. Figure 5 shows similar data for benzo[e]pyrene.
Studies have shown that the pressure drops for 50-μm-i.d. capillary
columns are negligible under typical conditions (linear velocities <10
cm/s and columns <20 m in length). In addition, the large surface

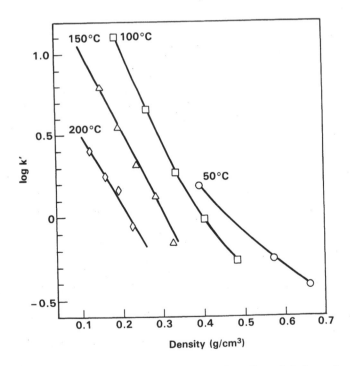

Fig. 4 Capacity ratio versus density plot for phenanthrene in
CO_2 at various temperatures using a 50-μm-i.d. capillary column
with a 0.25-μm cross-linked film of SE-54 as the stationary phase.

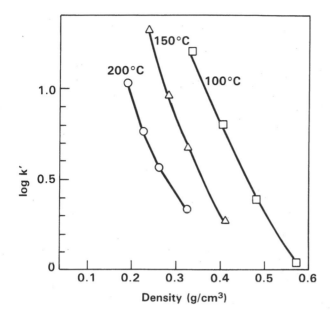

Fig. 5 Capacity ratio versus density plot for benzo[e]pyrene in CO_2 at various temperatures for the capillary column used to obtain the data shown in Fig. 4.

area and small volumetric flow rates typically provide thermal equilibration within a few seconds. The dependence of log(k') on density shows nearly the expected linear behavior, but with increasing curvature noted at the lowest temperature. The observed behavior of k' as a function of temperature at constant pressure is more complex, as shown in Fig. 6. Here the competing roles of density and volatility are most evident at lower pressures (e.g., 150 bar), where retention passes through a maximum with temperature. Although temperature provides a valuable adjustment parameter (not available in practice with the more labile stationary phases often used for packed columns), temperature programming has attracted relatively little attention because of this behavior. Figure 7 illustrates the complex dependence of the "effective capacity ratio" (k_e') at 100 bar using a short 50 μm i.d. × 2 m capillary column at a linear velocity of ∿2 cm/s for temperature ramp rates up to 20°C/min. The three compounds showed quite different retention dependences on temperature ramp rates. Retention was also dependent on the initial temperature and the selected pressure. Thus, temperature programming by itself is not generally useful in SFC. However, simultaneous temperature and pressure programming to either increase density or maintain constant density can offer

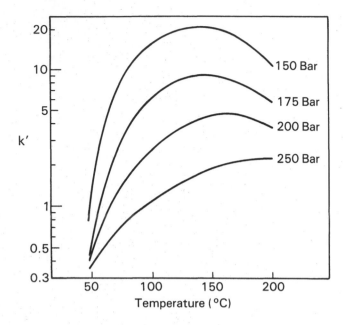

Fig. 6   Capacity ratio versus temperature for benzo[e]pyrene in CO_2 at various pressures for the capillary column used to obtain the data shown in Figs. 4 and 5.

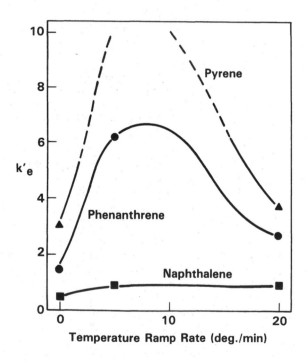

some advantages [30]. In general, however, pressure programming at the highest temperature compatible with the separation offers the advantages of simplicity, improved reproducibility, and maximum chromatographic efficiency (efficiencies are improved at operation at higher temperatures and lower densities because of more favorable diffusion rates).

Columns packed with $3-5$ μm particles can generate $10-10^2$ as many plates per second as 50-μm-i.d. capillary columns [28,28] but are more limited in the range of flow rates or column length because of the large pressure drop. The pressure drop for 5-μm packings at optimum linear velocities is $\sim 3 \times 10^4$ greater per plate than for 50-μm-i.d. open-tubular capillary columns [27]. This difference in column permeability limits effective high-speed application to very short packed columns. Separations requiring large numbers of plates using packed columns are ultimately limited by the rapid increase in plate height at lower than optimum linear velocities [27]. The maximum efficiency in such separations depends on the particular conditions and the maximum pressure drop which can be tolerated during a separation (which is directly proportional to linear velocity). Capillary columns are less subject to this limitation and maximum possible efficiencies (given the same pressure drop) are about 10^2 to 10^3 greater than for packed columns.

The reduced pressure drop in capillary columns is also of practical importance in pressure programming, where much greater flexibility exists than for packed columns. Although 5-μm microparticle packed columns generate $10-10^2$ more plates per second than 50-μm-i.d. capillary columns, the lower pressure drop with capillaries allows greater numbers of plates to be obtained in conjunction with pressure programming. A pressure drop does not intrinsically prevent pressure programming, but it does result in more subtle difficulties. For example, the low pressure drop in capillaries allows the programmed pressure rise to take place across the entire column length almost instantaneously (because the pressure rise is transmitted at near the speed of sound). As the pressure drop increases, the time to reach steady-state conditions also increases; thus changes in k' are much more difficult to predict with packed columns. The pressure drop also limits the minimum pressure at the column entrance and thus the range of pressure and density which can be used to vary retention.

The different retention characteristics of commercially available packed columns and SFC capillary columns are also an important

---

Fig. 7   Effective capacity ratios observed under various temperature ramp rates with CO_2 at 100 bar with a 50 μm × 2 m capillary column and a linear velocity of 2 cm/s.

consideration. For example, comparison of Figs. 2 and 4 shows that
capacity ratios for phenanthrene in carbon dioxide with the 5-μm
C_{18} microbore column are approximately 10^2 greater than with the
SE-54 capillary column (0.25-μm film thickness) at the same density. Greater capacity ratios are observed for packed columns with
carbon dioxide in all cases. As the interest in SFC is for analysis
of more polar and higher-molecular-weight compounds, which often
show greater retention and can require relatively high fluid densities, capillary columns are usually more appropriate. Similar limitations arise from the greater adsorptive activity of packed column
stationary phases. The above considerations combine to make pressure programming for packed columns far less attractive than for
capillaries. These factors have resulted in greater interest in using
solvent modifiers and solvent gradient methods to reduce retention
in packed column applications.

The practical advantages of pressure programming with capillary
SFC are demonstrated in Fig. 8, which shows the separation of C_{12}

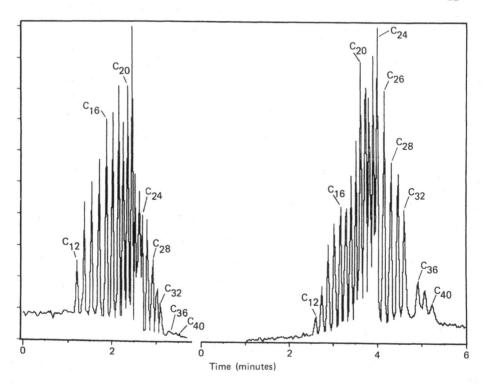

Fig. 8 SFC—MS total-ion chromtograms obtained at two different
linear velocities for the separation of an *n*-alkane mixture at 100°C
using a 50 μm i.d. × 2 m column and 50 bar/min pressure ramp beginning at 75 bar.

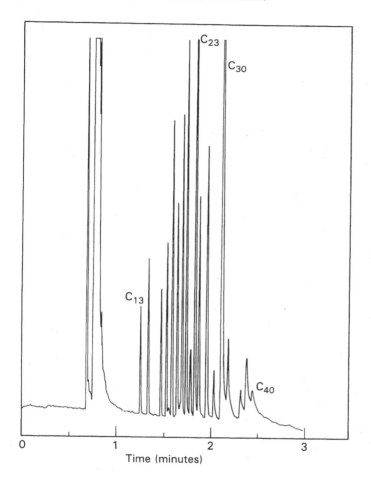

Fig. 9   Fast SFC chromatogram for the n-alkane mixture using a
1.5 m × 25 μm column at 100°C with a 100 bar/min pressure program rate.

to C_{40} n-alkanes obtained using a short 2.0 m, 50-μm-i.d. capillary
column.   The figure shows SFC−MS total ion chromatograms obtained
at 100°C at two different mobile phase linear velocities.   The mobile
phase was pressure ramped at 50 bar/min from an initial pressure
of 75 bar.   Figure 9 shows the improvement in separation efficiency
which can be obtained using a smaller 25-μm-i.d. capillary column
and a higher pressure programming rate of 100 bar/min.   The
theory and practical considerations relevant to pressure-programmed
capillary SFC, with particular emphasis on rapid separations, have
been discussed in detail elsewhere [25].   In practice, higher linear
velocities and higher pressure ramp rates provide better separations

when measured as separation number or Trenzahl number per minute [25]. These performance characteristics allow capillary columns
to be competitive with packed columns for fast separations.

With all the advantages of capillary columns, there are still situations where the use of packed columns is advantageous. These
applications include (a) very high speed separations where very few
effective plates are required, (b) alteration of selectivity using the
wider range of stationary phases currently available, (c) situations
where higher flow rates are required, (d) situations where fluid-
compatible capillary phases do not exist, and (e) situations where
greater sample loading is necessary.

Because of the considerable importance of restrictors in SFC—MS,
particularly with capillary columns, various aspects of restrictor design, selection, and operating conditions are described in the next
section. It is shown that higher SFC flow rates often allow improved performance with gas-phase detectors (e.g., FID or MS).
In this laboratory, SFC—MS instrumentation has been developed for
both capillary and packed column operation so that the distinctive
advantages of each can be exploited.

RESTRICTORS FOR SFC AND SFC—MS

Restrictor performance constitutes the most vital component of an
SFC—MS interface. At present, nearly all capillary SFC utilizes
some form of capillary restrictor, with the dimensions empirically
selected to give the desired linear velocity for the experimental temperature and pressure range. The porous "frit" restrictor is also becoming increasingly popular in some applications due to a decreased
tendency for plugging. The capillary restrictor can be short with a
relatively small inner diameter or longer with a larger i.d.; the latter
is more desirable for situations in which the restrictor follows the detector, as is the case with on-column UV and fluorescence detectors.
The ideal restrictor for capillary SFC and SFC—MS would have the following characteristics: (a) provide uniform pulse-free flow, (b) provide immunity from plugging, (c) be easily replaced or provide for
variation of flow rate, and (d) provide for complete transfer of labile
or nonvolatile solutes to the detector without pyrolysis or formation of
analyte particles.

In this section, the fundamental aspects of restrictor performance relevant to SFC—MS are considered and the methods currently
used to approach the ideal situation noted above are described.

Expansion from Supercritical Conditions;
Thermodynamic Considerations

The flow properties of a restrictor, and to a significant degree
the transfer of less volatile solutes to the gas phase during the

expansion, depend on the state of the fluid prior to expansion, the dimensions of the restrictor, restrictor heating (and heat transfer properties of the restrictor), and the state of the fluid during and after the expansion. Figure 10 schematically illustrates an adiabatic process where the enthalpy (H) of the fluid is the same before and after the expansion process. Expansions relevant to SFC occur under conditions where the Joule-Thomson coefficient, $(\partial T/\partial p)_H$, is generally positive and net cooling results [31]. If the restrictor is very short the process may be considered adiabatic. In this case the equilibrium state of the fluid after expansion (if also isolated from the surroundings) can be predicted directly from thermodynamic data. However, the state of the fluid during and shortly after expansion depends on the physical processes related to the expansion which are kinetically controlled and inherently nonequilibrium.

For a flow process at steady state, the energy balance can be written

$$\Delta H + \frac{\Delta u_b^2}{2g_c} + \Delta z \frac{g}{g_c} = Q - W \tag{1}$$

where ΔH is the enthalpy change between the inlet and outlet streams of the system, u_b is the bulk velocity under a turbulent flow, g_c is the gravitational constant, g is the gravitational acceleration, z is the vertical position, Q is the heat added, and W is the work extracted from the system. The flow of a supercritical fluid through a sufficiently short capillary can be approximated as adiabatic. For horizontal tubes the potential energy term is zero and Eq. (1) can be written as

$$\Delta H = \frac{\Delta u_b^2}{2g_c} \tag{2}$$

When the fluid is accelerated to the speed of sound (as is usually the case), the kinetic energy term is typically <10% of the initial enthalpy. This is usually a small correction and does not apply upon viscous dissipation of the jet after expansion.

Enthalpy data as a function of pressure and temperature are given in Fig. 11 for carbon dioxide [32]. The figure shows the two-phase region where vapor and condensed phase exist in equilibrium. Flow through a short restrictor will be nearly isenthalpic until the end of the restrictor, where isentropic expansion to supersonic velocities occurs. Shock phenomena and interactions with the background gas in the expansion region largely convert the kinetic energy of the gas jet back to internal energy [4,6]. For CO_2 an initial enthalpy of less than 175 cal/g results in the expanding fluid

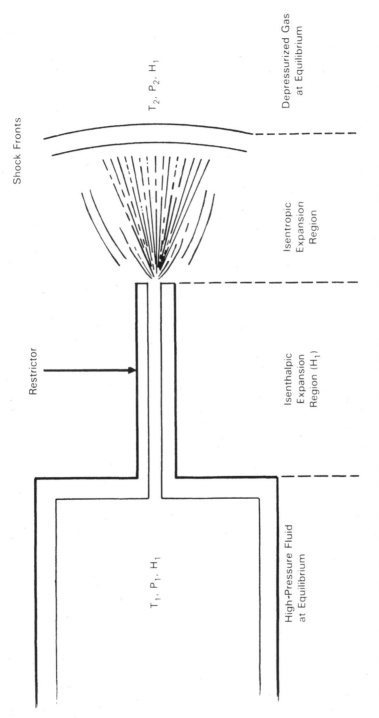

Fig. 10 Schematic illustration of the supercritical fluid expansion for an adiabatic process where the initial and final states of the fluid are known.

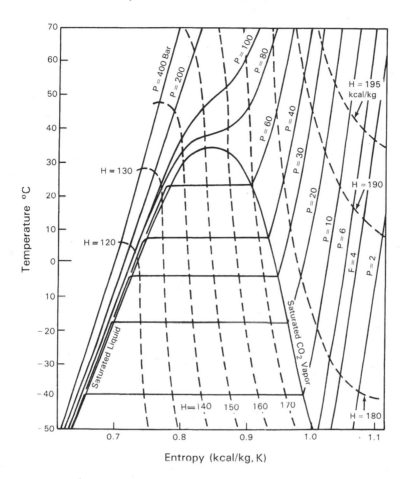

Fig. 11   Temperature-entropy diagram for CO_2. Dashed lines show constant-enthalpy (H) paths for a reversible adiabatic expansion.

transversing the two-phase region.   For a fluid temperature of 70°C, a pressure of >100 bar would result in passing through a two-phase region during an adiabatic expansion.   In order to obtain good flow characteristics and avoid a two-phase region during CO_2 expansion, a fluid temperature of at least 100°C before entering the restrictor should be maintained.   When an expansion occurs from pressures and temperatures which result in two phases, the assumption of adiabatic conditions allows the fraction of fluid in each phase to be calculated.   Figure 12 gives the fraction of CO_2 in the condensed phase after expansion to low pressure as a function of reduced

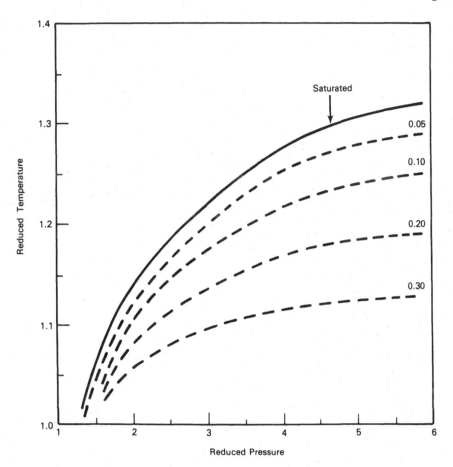

Fig. 12   Reduced temperature—reduced pressure diagram for a
supercritical fluid (CO_2) showing lines of constant vapor-liquid
composition after expansion to low pressures.

pressure and reduced temperature of the supercritical fluid. The
dashed lines give the minimum reduced pressure and temperature
required to avoid passage through a two-phase region for any stage
of an isenthalpic process. The reduced conditions necessary to
avoid a two-phase system given in this figure also provide rough
guidance for other fluid systems.

      Expansion of supercritical water through a restrictor is shown
in Fig. 13. Supercritical water has potential analytical applications,
including controlled pyrolysis-mass spectrometry, high-temperature
inorganic SFC, and introduction for various plasma emission analysis

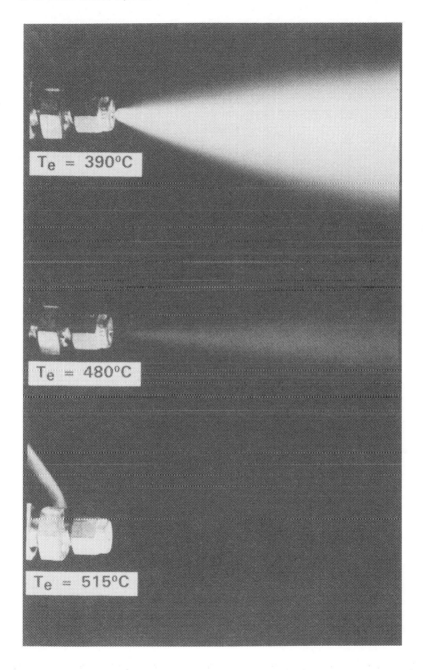

Fig. 13   Photographs of water expansion from 600 bar at various
pre-expansion temperatures (T_e) showing the liquid droplet size
decrease rapidly as temperature is increased.

methods and other spectroscopic methods. In this experiment, water at 600 bar was expanded through a stainless steel restrictor (10 mm long by 0.1 mm i.d.) at various temperatures (T_e). The result is a gas or steam jet with an average droplet size which decreases as the fluid temperature increases, causing the steam jet to become less visible. At the highest temperature, just above the region where a two-phase product is anticipated, only the slightest hint of a visible jet exists even through the flow rate is about 20 ml/min (liquid).

Estimation of Flow Rates Through SFC Restrictors

In addition to providing efficient analyte transfer to the detector, a primary requirement for capillary SFC restrictors is to provide acceptable flow rates or linear velocities over the range of pressures relevant to a given separation. For short restrictors the flow rate during an adiabatic expansion cannot exceed the speed of sound until within a few nozzle diameters of the end of the capillary [33, 34].

A convenient method of estimating flow rates of compressible fluids through straight tubes is based on friction losses from the Fanning friction factor:

$$F = \frac{2fL}{R}\left(\frac{G^2}{2g_c\rho^2}\right) \tag{3}$$

where F is the friction loss, f is the Fanning friction factor, G is the flow rate, ρ is the fluid density, L is the tube length, and R is the radius [33]. For a compressible fluid, the Fanning friction factor has been used successfully in differential form [34]. The differential form of the mechanical energy balance then becomes

$$V\,dP + \frac{u\,du}{g_c} = \left(\frac{g}{g_c} - \frac{fu^2}{g_c D}\right)dx \tag{4}$$

where V is the volume, P the pressure, D the restrictor diameter, and u the instantaneous velocity. A graphical method of integrating this equation is given by Lapple [34] and can be used to estimate flow rates in the adiabatic and isothermal limiting cases. This method accounts for the speed-of-sound limitation at the capillary exit tube and assumes ideal gas behavior so that the V dP term can be rewritten as a function of V only. This does not introduce large errors ($<30\%$) in most practical applications where the two-phase region is avoided. A more exact solution of Eq. (4) for super-critical fluids could be obtained by an integration using an appropriate equation of state.

An important term used by Lapple to nondimensionalize the result is

$$g_{max} = P_1 \left(\frac{g_c}{2.718}\right)\left(\frac{M}{RT_1}\right)^{1/2} \tag{5}$$

where P_1 is the inlet pressure, T_1 is the inlet temperature, M is the molecular weight, and g_{max} is the maximum mass flow ($g/cm^2 \cdot s$) attainable through an infinity short restrictor.

Equation (5) can be evaluated to yield a simple relationship for estimating mass or volumetric flow (Q) of a fluid through a capillary restrictor:

$$Q \ (ml/min) = 31.3 F_r \ \frac{P_1 D^2}{\rho}\left(\frac{M}{T_1}\right)^{1/2} \tag{6}$$

where P_1 is pressure in bar, D is capillary diameter in mm, T is temperature in K, and F_r is a flow reduction parameter from Lapple [34] based on the gas and the "equivalent capillary length." Figure 14 gives values for F_r in terms of the capillary length-to-diameter ratio (L/D) applicable to fluids such as CO_2 and H_2O when sonic conditions exist at the end of the restrictor. (The calculations assume an abrupt edge entrance to the capillary and f = 0.0047.) Also shown in Fig. 14 is the maximum capillary outlet/inlet pressure ratio (P_2/P_1) for sonic velocities at the capillary exit. When P_2/P_1 is below the minimum for sonic flow (as is usually the case), the pressure at the capillary exit (P_{exit}) can be estimated since the ratio becomes P_{exit}/P_1. For such an adiabatic expansion the fluid temperature at the capillary exit (T_{exit}) is given by [34]

$$\frac{T_{exit}}{T_1} = \frac{2}{1 + \gamma} \tag{7}$$

where $\gamma = C_p/C_v$, the ratio of heat capacities at constant pressure and constant volume. Note that when $\gamma = 1.4$, the approximate value for CO_2 and H_2O, then $T_{exit}/T_1 = 0.83$. Thus, the fluid density at the capillary exit can be estimated as

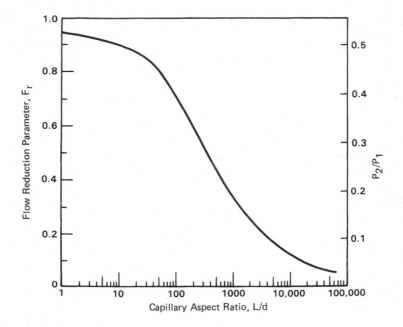

Fig. 14 Flow reduction parameter from Eq. (6) as a function of capillary aspect ratio (L/D) for an adiabatic expansion to sonic velocities at the capillary exit. Also given is the maximum pressure drop ratio (P_2/P_1) resulting in sonic conditions at the exit.

$$\frac{\rho_{exit}}{\rho_1} = \left(\frac{1 + \gamma}{2}\right) \frac{P_{exit}}{P_1} = 1.2 \frac{P_{exit}}{P_1} \qquad \text{for } \gamma = 1.4 \qquad (8)$$

Figure 15 shows flow rate measurements obtained for supercritical pentane $(T_c = 196.6°C)$ at 250°C through 0.64-cm-long restrictors, with 25, 50, and 75 μm i.d., for decompression to atmospheric pressure. Estimated flow rates were calculated from Eq. (6), where $F_r = 0.66$, and show reasonable agreement with experiment. It is interesting to note that the density at the end of the capillary restrictor (to within a few capillary diameters) can still be quite large. For example, at 240 bar ($\rho = 0.63$ g/cm^3) for the 50-μm-i.d. capillary, the pentane density at the capillary exit was approximately 0.19 g/cm^3. These results show that good estimates of the flow rate can be made, providing guidance in selection of restrictor dimensions.

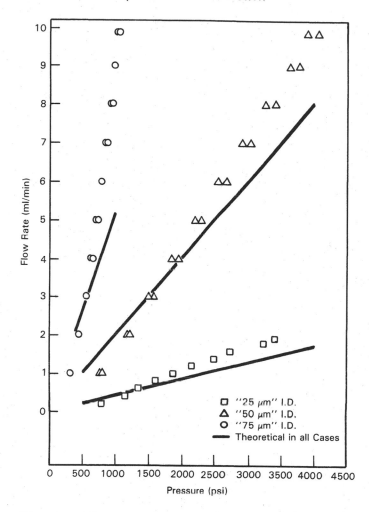

Fig. 15   Flow rate data as a function of pressure for supercritical pentane at 250°C for various restrictor diameters and predictions calculated from Eq. (6).

Practical Restrictors for Capillary SFC and SFC—MS

The flow rates reported in Fig. 15 are approximately three orders
of magnitude larger than desirable for capillary SFC—MS and are
more appropriate for packed HPLC columns. Lower flow rates require much longer or smaller-i.d. restrictors. For very long restrictors the assumption of an adiabatic process becomes invalid,
although the speed-of-sound limitation to maximum velocity still ap-
plies. The analyte will precipitate, and ultimately plug a long restrictor, if it is not heated to vaporize the analyte or cooled to
maintain sufficient fluid density for solubility.

  One effect of increased restrictor temperature is to reduce the
mass flux, as predicted by Eq. (5), if the fluid is a gas. This behavior reflects the temperature dependence of the fluid viscosity.
Figure 16 shows typical linear velocity data for various SFC temperatures using FID with the restrictor heater at 275°C, well above
the fluid critical temperature of 31°C.

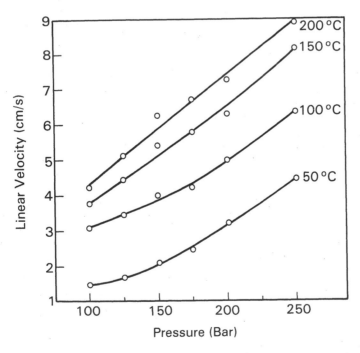

Fig. 16   SFC linear velocity (v) as a function of pressure at various
temperatures for a 5 cm × 8 μm i.d. restrictor at 275°C.

If the SFC column and restrictor are independently thermostated, SFC mass flux (linear velocity times density, $v \cdot \rho$) should be independent of column temperature. Figure 17 gives the SFC mass flux at six pressures as a function of column temperature for a restrictor heater temperature of 275°C. At lower pressures the expected behavior was observed and mass flux was nearly independent of SFC temperature. At higher pressures an increasingly significant decrease in mass flux was observed for higher SFC temperatures. This implies that the effective restrictor temperature was being increased by the greater mass flux. This is consistent with data obtained at a higher restrictor heater temperature (400°C), shown normalized to the lower-temperature data in Fig. 17 (dashed lines). It appears that the lower mass flux and the higher restrictor heater temperature reduced the heating effect of the fluid. These results support the earlier assumption that heating can be inefficient for short restrictors because of the high mass flux and poor thermal conductivity of fused silica.

These relationships also resolve previously unexplained observations of linear velocity either increasing or decreasing during pressure programming. A restrictor heated to well above the fluid's critical temperature results in a nearly linear dependence of mass flux on pressure (Fig. 17). However, if the SFC temperature is near the critical point, the rate of increase in fluid density in the SFC column with pressure can be much greater than the increase in mass flux through the restrictor. Under such conditions linear velocity decreases as pressure is increased. It is therefore possible to select restrictor temperatures (as far as compatible with detection) to control linear velocity during pressure-programmed operation.

Transport of Nonvolatile Compounds to Gas-Phase Detectors

A vital property of SFC restrictors is their effectiveness for transporting "nonvolatile" compounds to the FID or MS ion source. Success depends on the restrictor design, the fluid pressure and temperature, the particular demands of the detector, and, under some conditions, the volatility or melting point of the "nonvolatile" compounds.

It has been observed by several capillary SFC workers [35–37] that FID detection of less volatile or higher-molecular-weight compounds can cause large spikes with a frequency on the order of $0.1-10$ s^{-1}. Similar phenomena are observed with MS detection. Lee and co-workers [35] suggested an origin due to "clustering" and showed that incorporation of a suitably long time constant (a few seconds) can produce a reasonable chromatogram. Chester [38] also observed that increased fluid temperature delays the onset of

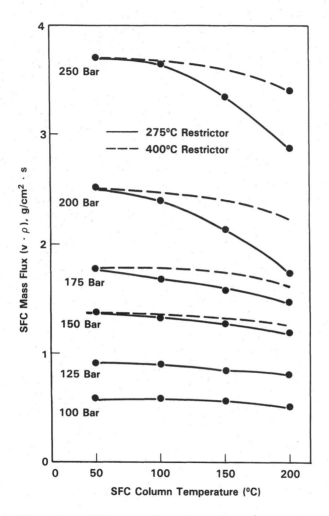

Fig. 17   SFC mass flux at various temperatures as a function of column temperature for restrictor heater temperatures of 275 and 400°C.

the spiking phenomena to later-eluting (and typically less volatile) compounds and suggested a "spiking delivery rate threshold" which varies with solute volatility. Heating the fluid through the relatively long restrictors (L/D > 1000) with FID is also effective [37] and can be done using fused silica restrictors drawn from large tubing to achieve better heat transfer through thinner capillary walls.

For truly nonvolatile compounds two distinct modes of operation appear feasible, the first leading to spiking and the second operating in a "nucleation" regime where it is unlikely. In the spiking mode, the nonvolatile analyte tends to collect on restrictor walls and (if the local temperature is above the analyte melting point) flows toward the end of the restrictor. The liquid collects at the end of the restrictor and is periodically entrained in the high-shear gas flow and transported to the detector. This is a complex process with details which likely depend on the liquid viscosity and restrictor geometry.

The fluid expansion process has been directly observed and various materials expanded from supercritical fluid solutions through 5- to 75-μm-i.d. fused silica capillaries have been collected. In one set of experiments, polycarbosilane, which has a mean molecular weight of approximately 1430 dalton, a melting point of ~240°C, and is not volatile but decomposes to yield silicon carbide at >900°C [39], was studied. The molten polycarbosilane (Dow Corning) solute could be seen to collect at the fused silica capillary exit and be periodically entrained in the gas flow. For polycarbosilane in pentane at 350°C and 100 bar, short fibers were observed (Fig. 18a). The high shear forces can elongate polymer droplets to yield fibers. The particle formation rate was directly proportional to solute concentration. Particle size (either fibers or spheres—depending on the solute, fluid, and restrictor conditions) was found to increase with capillary diameter. These observations are consistent with the observed FID spiking phenomena and suggest a process involving the restrictor walls and solute melting point. The average particle size formed with 5- to 10-μm restrictors (typical of capillary SFC) is in the 2-10 μm range. Assuming an average of 1 ng of a nonvolatile solute in capillary SFC, on the order of 1 to 10^2 particles might be found during elution of the compound. The particle production rates based on elution over 10 s are entirely consistent with observed spiking rates.

A mode of operation producing much smaller particles occurs when analyte nucleation is delayed to near the end of the restrictor. Figure 18b shows polycarbosilane collected under such a set of conditions (30 ppm pentane solution at 240 bar, 250°C) with a 6 mm × 25 μm i.d. restrictor. In this case large quantities of nearly monodisperse particles with an average diameter of ~0.02 μm

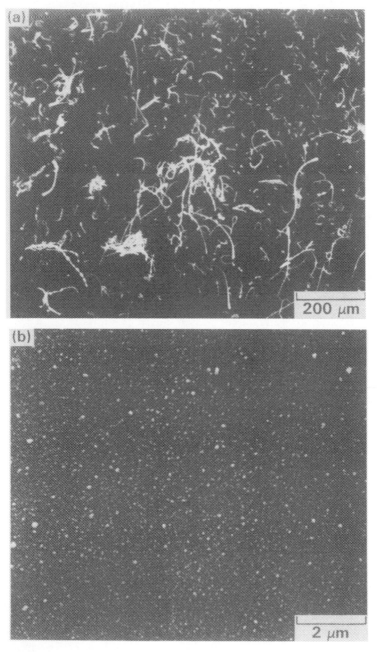

Fig. 18   Photographs showing (a) fibers collected from a polycarbosilane solution expanded from 350°C and 100 bar and (b) ultra-
fine polycarbosilane powder expanded from 250° and 250 bar.

were collected. A single nanogram of analyte will produce on the order of 10^8 particles of this size.

The major requirement for operation in this "nucleation mode" is to maintain solvating conditions to near the end of the restrictor, which is facilitated by the use of very short restrictors and fluid conditions that enhance solubility. It is important to recognize that this does not necessarily correspond to elevated restrictor temperatures. For example, the polycarbosilane particles shown in Fig. 18a were formed at 350°C and 100 bar, while the very fine, nearly monodisperse, particles were produced at 240 bar and 250°C. The solvating power of pentane for polycarbosilane is much greater under the latter conditions. In addition, the particles were formed at gas temperatures below their melting point (due to cooling on expansion) and where adhesion to the capillary walls was apparently insignificant. Under these conditions a very long restrictor (LD $\gg$ 10,000) would plug rapidly but would produce particles (as in Fig. 18a) if heated to >300°C.

Increased pressure always enhances the transport of nonvolatile analytes, but in practice maximum pressure is limited by the SFC conditions needed to yield sufficient retention or the maximum pressure for the system. The effect of temperature on fluid phase solubility is more difficult to predict, but some guidance is provided by analyte vapor pressures. Typically, the more volatile a compound, the greater the solubility increase with temperature [4].

The improved performance obtained from drawn restrictors compared to "linear" restrictors is, at least in part, explained by these observations. In some cases the improved heat transfer improves fluid phase solubilities and allows improved operation. In other cases the role of restrictor heating is more direct. Very few of the compounds that can be solvated by fluids such as CO_2 are truly nonvolatile at the restrictor temperatures used (100—400°C). This effect is illustrated in Fig. 19, which shows SFC chromatograms for polybutadiene (average molecular weight of 1000 daltons) obtained with two different restrictor heater temperatures. A significant improvement in transmission of higher-molecular-weight oligomers was obtained at the higher temperature. In the fast gas flow, compounds having relatively low volatilities can readily be transported, and one would expect to find few compounds separated by SFC with CO_2 which could not, for example, be volatilized from a heated direct-inlet probe at 400°C for mass spectrometric analysis. (Even very labile compounds would show little decomposition by the brief exposure at elevated temperatures.)

It is also apparent that very short restrictors suitable for capillary SFC or SFC—MS are difficult to fabricate and the small orifice size makes them somewhat subject to plugging. This is particularly a problem at fluid conditions leading to saturation during expansion (i.e., lower pressures and temperatures), where frozen or condensed solvent can disrupt flow and cause plugging. At higher flow rates,

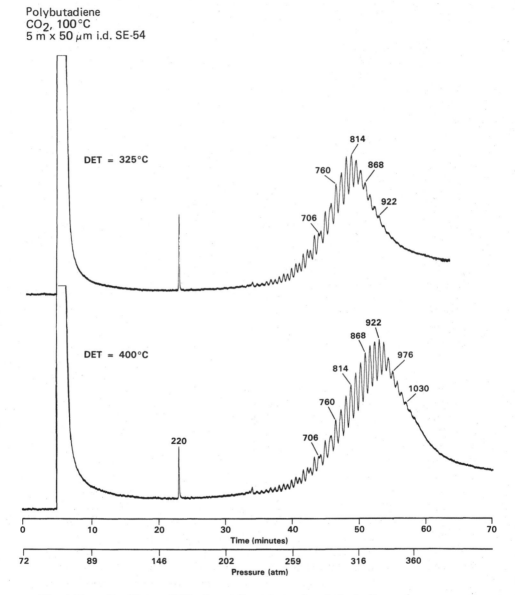

Polybutadiene
CO_2, 100°C
5 m x 50 μm i.d. SE-54

DET = 325°C

814
760   868
922
706

DET = 400°C

922
868
976
814
1030
760
220   706

Fig. 19   Capillary SFC chromatograms of polybutadiene (average
MW 1000) with restrictor heater temperatures of 275 and 400°C.
Peak labels indicate molecular weight as determined by SFC—MS
analysis.   See text for additional chromatographic conditions.

such as that used to produce the fine polycarbosilane powder (Fig. 18a), the requirements are relatively easily met, and the major difficulty is the large gas flow rate to the detector. Under conditions where flow rates are >100 µl/min (liquid), extended operation (>weeks) has been observed for nonvolatile compounds in the nucleation mode for systems as diverse as supercritical water with silica or germanium dioxide solutes, and pentane with solutes such as polycarbosilane.

At present, the most widely used fluid for capillary SFC is carbon dioxide. Because of the limitations of solvating properties of this fluid, few compounds can be separated which cannot be readily detected by FID using a drawn (tapered) or polished (integral) fused silica capillary restrictor with effective heating. However, most fluids with greater solvating power are incompatible with FID. The utilization of the mass spectrometer as a detector removes the constraints imposed by the FID. The more polar fluid systems which are compatible with SFC—MS allow application to more polar and less volatile analytes.

SUPERCRITICAL FLUID CHROMATOGRAPHY— MASS SPECTROMETRY (SFC—MS)

Molecular Beam Mass Spectrometric Detection for SFC

The advantages of coupling a chromatographic technique with mass spectrometry are considerable, as evidenced by the major role of GC—MS in mixture analysis. However, in contrast to GC, where the effluent is compatible with classical gas-phase ionization methods, the ideal approach to interfacing SFC or HPLC with mass spectrometry is less obvious. Currently, the most promising LC—MS interfacing approach "thermospray" ionization [40], is most applicable to highly polar, easily ionized compounds. These compound classes, which generally include materials soluble in aqueous systems, are also the most difficult to address by SFC. Thus, SFC—MS and thermospray LC—MS may be viewed as complementary techniques.

The most direct (and earliest) approach to SFC—MS interfacing involves expansion of the fluid through a nozzle or restrictor and production of a collimated beam, using molecular beam skimmers, which is then ionized and analyzed using conventional electron impact ionization (EI) mass spectrometry. This approach was first proposed in the literature by Milne [41], who discussed the potential problems of the approach and suggested that the high pressures could be handled with a sufficiently small-diameter nozzle orifice and a multistage, differentially pumped vacuum system. Subsequently, Giddings et al.

[42] presented a more quantitative evaluation of the molecular beam approach for "the gas phase isolation of nonvolatile molecules by high pressure jets for mass spectrometry." This work provided the basis for the studies of Randall and Wahrhaftig [43–45], which provided a detailed examination of the molecular beam approach to SFC–MS interfacing.

The instrumentation constructed by Randall and Wahrhaftig utilized a three-stage vacuum system. Their work demonstrated that solutes of low volatility could be injected into a low-pressure environment. Randall and Wahrhaftig produced results that not only indicated solute–solute clustering was negligible, but also solute–solvent clustering was much less than expected. The significant complexity of the molecular beam approach involving the multistage vacuum system and operational difficulties (such as beam alignment) prevented serious application. More important, however, was that this approach provided relatively poor sensitivities (microgram range), which may have been indirectly related to the frequent nozzle "plugging" problem noted by these workers [44].

SFC–MS Interfaces Based on Direct Fluid Injection

The development of fused silica capillary columns with stabilized stationary phases suitable for SFC presented the opportunity to eliminate the disadvantages of the molecular beam approach. Typical fluid flow rates for 50-μm-i.d. columns are in the range of 1 to 5 μl/min (as liquid); this range can be handled by conventional mass spectrometers with two-stage pumping systems (as typically configured for chemical ionization capability) without the use of the liquid nitrogen cryopumps required for commercial direct light introduction LC–MS interfaces [46].

In the first capillary SFC–MS interface, supercritical conditions were maintained in the column to a region immediately adjacent to the ion source, where the fluid was rapidly decompressed by expansion through an orifice. The orifice was made by either crudely pinching platinum-iridium tubing or utilizing a laser-drilled orifice, such as that used in the direct liquid introduction LC–MS approach [6]. The lower fluid flow rates with capillary columns avoided the complex multistage pumping systems and alignment problems characteristic of the molecular beam approach. Rather than attempting to sample the unperturbed expanding gas by conventional electron impact ionization, this direct fluid injection (DFI) method utilized the shock fronts and rapid collisional processes in the expanding jet to disrupt solvent clusters, as described in the last section. Thus, the DFI method not only greatly simplified instrumental design but also significantly reduced the two major drawbacks of the molecular beam approach: clustering phenomena and inadequate sensitivity.

An improved understanding of the DFI process can be gained by consideration of cluster formation during expansion of high-pressure jet through a nozzle, as illustrated earlier in Fig. 10. When an expansion occurs in a chamber with a finite background pressure, the expanding gas interacts with the background gas, producing a shock wave system. This includes barrel and reflected shock waves as well as a shock wave perpendicular to the jet axis (the Mach disk). The Mach disk serves to heat and break up solvent and solute—solvent clusters formed during the expansion process. The breakup of clusters is further aided by the rapid collisional processes in the chemical ionization source, where the average molecule undergoes on the order of 10^5 collisions before escaping, assuming no wall losses. Since the solute concentration for a large peak in capillary SFC is typically <10 ppm, it is clear that solute clustering of neutral molecules in the ion source will usually be negligible. (Similar arguments apply for clustering of the ionized species, since a typical average number of ion-molecule collisions in a chemical ionization source is in the range of 10^4 to 10^6.)

In reality, an ideal pinhole-type orifice is impractical for capillary SFC—MS interfaces, as discussed in the last section. In the initial work with 100 to 200-μm-i.d. capillary columns, this concept was viable with $1-3$ μm holes [6]. However, even in this case the minimum substrate thickness was ~15 μm (and readily deformed if not precisely supported) and had a conical channel due to the nature of the laser drilling process. Although, greater success was achieved with orifices drilled in material 50 to 100 μm thick, this typically required multiple laser "shots" and was less reproducible. In addition, this approach presented significant mechanical difficulties associated with producing a high-pressure, low dead volume seal and the necessary precise alignment with smaller capillaries.

A practical compromise initially involved the use of short lengths ($1-3$ cm) of small inner diameter ($5-20$ μm) fused silica tubing or, more favorably, tapered restrictors made by drawing fused silica tubing. The compromises necessary in this approach were discussed in the last section, and it was shown that at very low flow rates heating may be advantageous for transfer or "less volatile" analytes to the ion source, but is less appropriate for truly nonvolatile compounds.

The distance from the orifice to the Mach disk may be crudely estimated from experimental work [47] as $0.67D(P_1/P_2)^{1/2}$, where D is the orifice diameter and P_1 is the fluid pressure. For a capillary restrictor the fluid pressure is more correctly described by the fluid pressure at the end of the capillary (P_{exit}), which can be estimated as described in the previous section, and represents typically 5 to 20% of the SFC column pressure. Thus, if $P_f = 60$ bar, $P_v = 1$ torr, and $D = 5$ μm, the distance to the Mach disk is

∿0.7 mm. The extent of cluster formation is related to the fluid
pressure, temperature, and orifice dimensions. Since initial cluster
formation involves volatile solvent molecules, heat applied in the
later stage of this expansion reduces solvent clustering and can also
facilitate transfer of solute molecules. However, as already demonstrated, as the capillary restriction becomes longer, solute clustering can become significant, leading to precipitation of the solute.
This can be manifested as "spikes" in the mass spectrometer signal or
plugging of the restrictor. While nucleation processes will ultimately
result in solute clusters, this typically will not prevent application for
sufficiently short restrictors and SFC analyte concentrations that avoid
fluid conditions yielding a two-phase system on expansion. Alternative
restrictor designs for SFC—MS have been discussed in greater detail
elsewhere [53].

Capillary SFC—MS Instrumentation

The instrumentation has continued to evolve to provide more effective application to less volatile analytes and increased compatibility
with 25- to 50-μm-i.d. capillary columns. A simplified schematic
illustration of one of the three SFC—MS configurations developed in
this laboratory is given in Fig. 20. A high-pressure, programmable
syringe pump is used to generate a pulse-free flow of a high-purity
fluid. Injection utilizes a high-pressure valve with volumes from
0.06 to 0.2 μl and flow splitting at (typically) ambient temperatures.
The split occurs as a subcritical liquid, so discrimination between
sample components is expected to be minimized. Split ratios range
from as little as 1:3 for conventional separations on long (>15 m)
50-μm-i.d. columns, to as high as 1:80 for fast separations on
25-μm-i.d. columns. (The need for injection splitting is indicated
by consideration of the fact that the total volume of a 1 m × 25 μm
column is only 0.5 μl.) The column is mounted in a constant-
temperature oven, which also serves to heat the air circulated
through the transfer probe. A zero dead volume union is used to
connect the column to the flow restrictor. The restrictor and probe
tip are independently heated to optimize analyte transport to the
ionization region.

The mass spectrometer ion source and required pumping speeds
are similar to those utilized for GC—MS. Figure 21 shows a detailed
view of the capillary interface and the design utilized for restrictor
temperature control. The restrictor is located in a stainless steel
capillary, which can be regulated to over 600°C by direct electrical
heating.

Both electron impact and chemical ionization mass spectrometry
have been evaluated for SFC detection. Methane, isobutane, and
ammonia are the most frequently used chemical ionization reagent

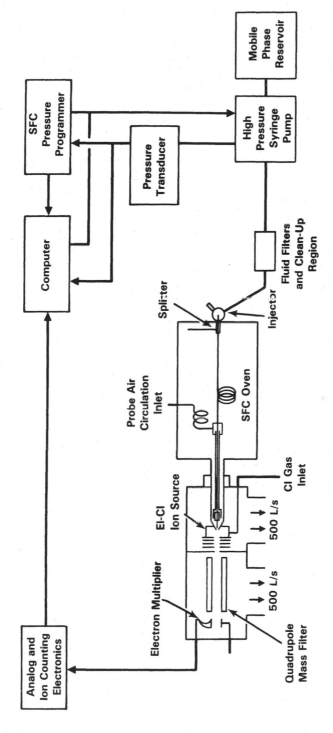

Fig. 20   Schematic diagram of capillary SFC—MS instrumentation.

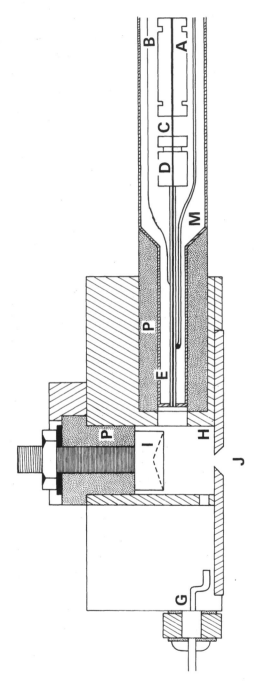

Fig. 21   Detailed schematic of the capillary SFC–MS interface and ion source for one instrument constructed in this laboratory.   (A) Capillary union, (B) electrical lead for heater, (C) restrictor, (D) vacuum seal to stainless steel capillary heater, (E) probe tip, (G) ionizer filament, (H) CI ionization volume, (I) repeller, (J) ion source exit, (M) thermocouple leads, (P) ceramic insulation.

gases. An advantage of SFC—MS with small-diameter capillary columns is that the flow rates are low enough that any CI reagent gas may be used. Typical detection limits range from 0.1 to 10 pg, depending on the compound, analysis time, separation efficiency, and CI reagent gas.

Figure 22 gives a schematic illustration of the SFC—MS interface and ion source developed for electron ionization. The SFC column was coupled to the mass spectrometer through an oven air-heated probe. The fused silica capillary restrictor allowed injection of the fluid into a heated expansion region (1 cm × 0.14 cm i.d. with a 0.1-cm orifice to the ionization volume), which provided the higher pressure (0.1—1 torr) necessary for cluster breakup prior to the ionization region. The temperature of this region was typically 50 to 150°C higher than the mobile phase critical temperature to avoid a two-phase region during the expansion (see Figs. 11 and 12). When the two-phase region was avoided there was no evidence of solvent clustering. However, at temperatures which resulted in a

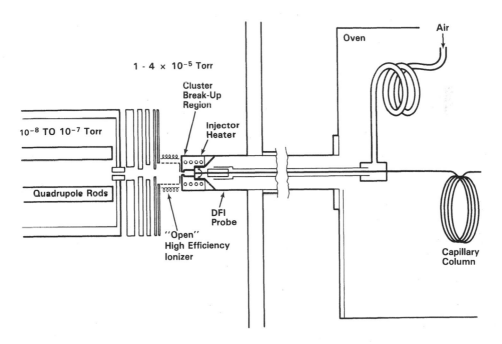

Fig. 22   Detailed schematic of the SFC—MS interface for electron impact ionization.

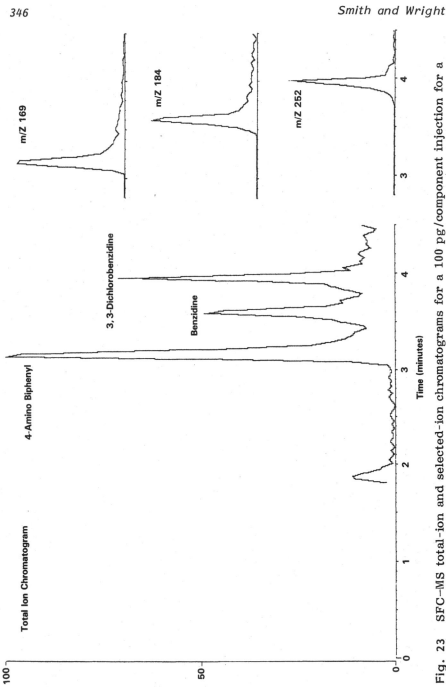

Fig. 23 SFC—MS total-ion and selected-ion chromatograms for a 100 pg/component injection for a mixture of 4-aminobiphenyl, benzidine, and 3,3'-dichlorobenzidine.

two-phase system during expansion (see Fig. 12), extensive solvent clustering was observed. This two-phase region can be avoided by higher SFC temperatures, restrictor heating, or heating of the expansion region.

The expansion region efficiently directs the SFC effluent into the ionization volume of a high-efficiency ionizer having an open design. Ion source chamber pressure was in the range $1-5 \times 10^{-5}$ torr and the mass spectrometer chamber pressure was $<10^{-6}$ torr, allowing conventional 70 eV EI. Unlike some higher pressure systems which yield CO_2 charge exchange CI spectra, this approach provides true EI spectra and is thus useful for mixed fluid systems. Although the sensitivity is somewhat less than that obtained with CI, good EI mass spectra are usually obtained with 100-pg injections [8]. The flexibility in selection of the ionization method and the ability to use the existing EI spectral libraries provide additional advantages for SFC—MS relative to most LC—MS methods.

The sensitivity of this approach is demonstrated in Fig. 23, which shows a total-ion chromatogram obtained for an injection of 100 pg each of a three-component mixture of 4-aminobiphenyl, benzidine, and 3,3'-dichlorobenzidine. These results were obtained by using a short 1.75 m × 50 μm column and mass spectral scans over the range of m/s 90—350. Nearly linear calibration curves were obtained for injections of 50 pg to 5 ng; larger injections showed evidence of column overloading. The single-ion plots in Fig. 23 show a signal-to-noise ratio of >50 for aminobiphenyl, and a detection limit of approximately 5 pg was determined for this compound.

Microbore SFC—MS Instrumentation

The capillary SFC—MS instrumentation cannot be used directly for SFC—MS with conventional HPLC packed columns unless linear velocities far below optimum or flow splitting is used. The pumping speed of conventional two-stage vacuum systems used for chemical ionization limits maximum flow rates to <100 μl/min (liquid) unless exceptional measures are taken.

Crowther and Henion [29] have described a simple approach for SFC—MS using equipment developed for direct liquid injection (DLI) mass spectrometry, which utilizes flow splitting and a laser-drilled orifice for sample introduction. Although this approach provides for favorable sample introduction and was initially explored without flow splitting for capillary SFC—MS [6], it suffers drawbacks related to the pressure limitations of laser-drilled orifices and decreased sensitivity due to flow splitting.

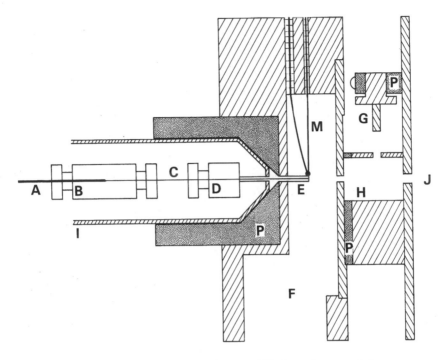

Fig. 24   Detailed schematic of SFC—MS interface and ion source
for "microbore" or high flow rate operation. (A) Uncoated capillary
union to microbore column, (B) zero dead volume union, (C) capillary restrictor, (F) port for mechanical pumping, (G) ionization
filament mount, (H) CI source volume, (I) probe sheath, (J) ion
source exit aperture, (M) thermocouple and capillary heater leads,
(P) ceramic insulation.

An alternative approach which is compatible with the capillary
SFC—MS interface developed in this laboratory is based on modification of the mass spectrometer ion source. This new approach provides direct pumping to a region immediately adjacent to the ionization region using a relatively low speed and inexpensive mechanical
pump, which is effective at higher pressures. A schematic drawing
of the interface and ion source region is shown in Fig. 24. Fused
silica tubing (100 μm i.d.) is connected to the outlet of the microbore column and through an air-heated probe to a zero dead volume

union which makes the high-pressure connection to the restrictor. Another "half-union" provides a vacuum seal to a stainless steel capillary, which serves as a holder for the restrictor. When the probe is introduced into the vacuum chamber it slides the restrictor termination into a relatively snug stainless steel capillary sheath, which can be electrically heated to over 600°C. The expansion occurs into a mechanically pumped chamber behind the ion source at pressures in the range of 5 to 50 torr. The exit orifice is aligned with the capillary restrictor and provides flow into the chemical ionization region, where 300-eV electrons are used as the ionization source for CI at pressures of 0.2 to 1 torr. The plate dividing the two regions also serves as the CI repeller electrode. Otherwise, the CI source is of conventional design and operation and can be utilized with any CI reagent gas.

The new ion source design has several advantages compared to alternative approaches. These include

1. Higher SFC flow rates than for other direct inlet designs
2. Closer to optimum CI conditions compared to arrangements which provide pumping directly to the CI source volume (as in thermospray)
3. Possible enrichment of higher-molecular-weight species (as in jet separators for GC—MS)
4. Easier coupling to high-voltage ion sources through easing of the "discharge" phenomena

Although both the sophistication and performance of SFC—MS interface designs have advanced significantly in the past few years, it is unlikely that any of the present methods represents an optimum approach. The detailed understanding now being gained for the fluid expansion process should serve to guide further improvements in both sensitivity and application to truly nonvolatile compounds.

SFC—MS APPLICATIONS

The interest in SFC—MS is centered around application to labile and less volatile samples not easily amenable to GC—MS. SFC—MS can usually provide superior results compared to LC—MS methods because of greater chromatographic efficiencies and detection limits. Currently, the limited knowledge of fluid phase solvating properties and limited practical experience with alternative (more polar) SFC mobile and stationary phases constitute the primary restraints on much wider application.

One of the primary advantages of SFC—MS is illustrated in Fig. 25, which shows the chromatographic resolution obtainable during

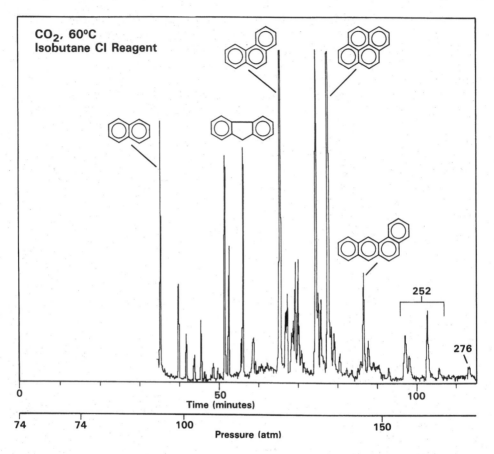

Fig. 25   Capillary SFC—MS separation of a coal tar extract.   See text for chromatographic conditions.

the separation of a polycyclic aromatic hydrocarbon (PAH) mixture using a 10 m × 50 μm i.d. column.   The figure shows the total-ion chromatogram with the base peaks, usually $(M)^+$ ions, for the major PAH components obtained with isobutane CI.

The capability for complex mixture analysis is further illustrated in Fig. 26 with the separation of a supercritical fluid extract of a solid hazardous waste sample.   This analysis was obtained with a 12 m × 50 μm i.d. column coated with a 50% phenyl polymethyl-phenylsiloxane stationary phase.   Supercritical pentane at 230°C was used for the mobile phase, which was pressure-programmed from 25 bar at rates of 0.5, 1.0, and 2.0 bar/min sequentially during the analysis.   The molecular weights for many of the

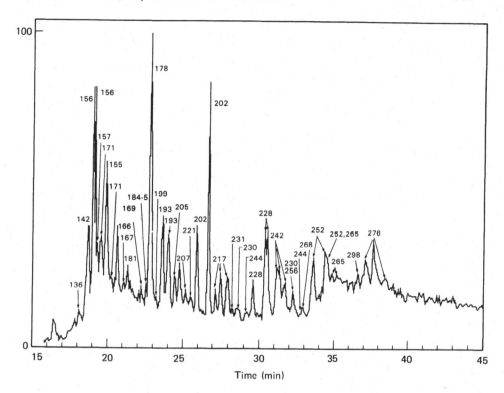

Fig. 26   Capillary SFC−MS separation of a solid waste sample. Probable molecular weights for the major components are given.   See text for chromatographic conditions.

components in the mixture are noted on the total-ion chromatogram (TIC).   Many of the components have been tentatively identified as polycyclic aromatic compounds.

An important application area for SFC−MS is the analysis of higher-molecular-weight components that exhibit limited volatility so that gas chromatography is either impossible or impractical.   In many instances the lack of a sufficiently sensitive detector prevents successful liquid chromatographic analysis.   Such an example is shown in Fig. 27.   This SFC−MS separation of a Triton X-100 polymer (average molecular weight ∿600) was obtained with a 10 m × 50 μm i.d. column coated with a 5% phenyl polymethylphenylsiloxane stationary phase and a mixed carbon dioxide-isopropanol (5%, v/v) fluid at 110°C for the mobile phase.   Pressure program rates of 2.5 and 2.0 bar/min beginning at 140 and 230 bar, respectively, were used.   A portion of the chemical ionization mass spectrum from one

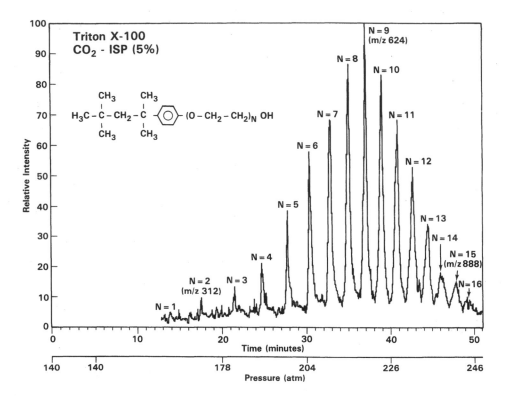

Fig. 27   Total-ion chromatogram obtained from the SFC—MS analysis
of a Triton X-100 polymer sample.   See text for chromatographic
conditions.

of the later-eluting oligomers (N = 10, m/Z 668) is shown in Fig.
28.   The ammonia chemical ionization yielded a gentle fragmentation
process that produced a dominate ammonium adduct ion, $(M + 18)^+$,
for this oligomer.

Another example of the SFC—MS separation of higher-molecular-
weight compounds is shown in Fig. 29.   In this separation three
triacylglycerols (trimyristoylglycerol, tripalmitoylglycerol, and
tristearoylglycerol) were separated in approximately 15 min using a
5 m × 50 μm i.d. column coated with a 5% phenyl polymethylsiloxane
stationary phase.   Carbon dioxide at 100°C was used for the mobile
phase with a pressure ramp of 10 bar/min beginning at 130 bar.
The methane chemical ionization mass spectrum for tripalmitoylglycerol is shown in Fig. 30.   This detection mode provided sufficiently "soft" ionization that protonated molecules, $(M + 1)^+$, were

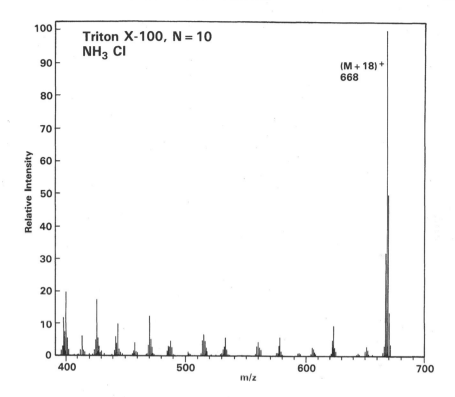

Fig. 28   Ammonia chemical ionization mass spectrum of the N = 10
(m/Z 668) oligomer obtained from the SFC—MS analysis shown in
Fig. 27.

formed, yet still provided characteristic fragmentation (loss of the
acid group R—COOH) that would be useful for identification.

Rapid separations are clearly desirable where sufficient selectivity exists for the particular components of interest.   Other advantages include the potential for enhanced sensitivity and reduction of adsorptive and reactive losses of highly polar or labile compounds.   The low pressure drop across capillary SFC columns allows
the solvating properties of the fluid to be manipulated rapidly [25].

Figure 31 shows a rapid capillary SFC—MS separation of diacetoxyscirpenol (DAS) and four macrocyclic trichothecene mycotoxins at 100°C [48].   The separation was obtained on a 1.7 ×
50 μm i.d. column.   The mass spectrometer was scanned through
the range of m/z 300 to 600; thus the solvent which eluted at ∿2
min is not seen in the total-ion chromatogram.   This separation
used a linear pressure ramp of 50 bar/min from an initial pressure

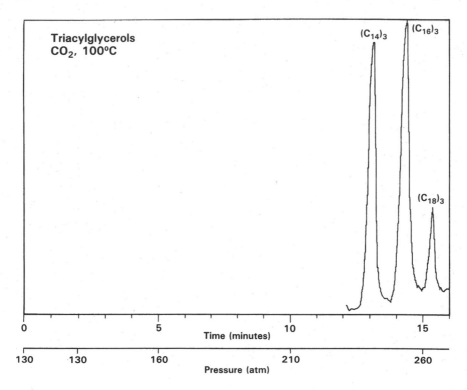

Fig. 29   Capillary SFC—MS separation of a mixture of three tri-
acylglycerols.   See text for chromatographic conditions.

of 75 bar at the time of injection to a maximum pressure of 300 bar
at 4.5 min.   The DAS is easily separated from the solvent and the
other macrocyclic trichothecenes, while the sets of macrocyclic
roridin and verrucarin compounds are not chromatographically resolved in the total-ion chromatogram.   However, ions representative
of the molecular species show no interference and the reconstructed
single-ion chromatograms clearly show the contributions of the individual components [48].

    Ammonia CI mass spectra provide excellent sensitivity but few
structural data for either confirmation or identification of unknown
components [49].   Isobutane CI can be used, as shown in Fig. 32,
to provide abundant fragmentation due to the more energetic proton
transfer process.   Isobutane CI still provides molecular ions,
$(M + 1)^+$, whereas more energetic methane CI provides excessive
fragmentation and much poorer detection limits.   The major fragment

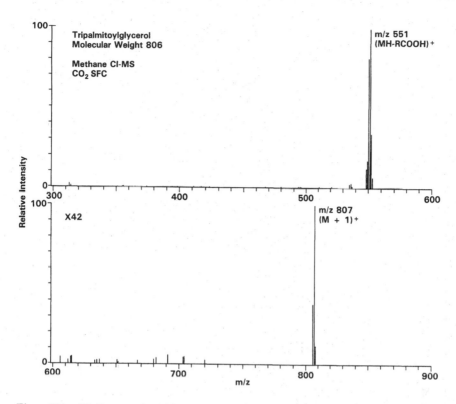

Fig. 30 Methane chemical ionization mass spectrum of tripalmitoy-glycerol obtained from the SFC—MS analysis shown in Fig. 29.

ions for methane CI are at m/z < 150, where background contributions are also greater.

The high proton affinity of ammonia prevents ionization of many potentially interfering species, either chromatographically separated or in the mass spectrometer "background," providing an additional element of selectivity. This advantage is particularly significant because a supercritical carbon dioxide mobile phase will not solvate many of the polar compounds which have higher proton affinities than ammonia. Care must be taken with ammonia CI in the selection of ion source parameters, however, because ammonia ion clusters, $(NH_3)_nNH_4^+$, act as different CI reagents and their relative abundances can influence the resulting mass spectra.

Several alternatives exist if the combination of chromatographic resolution and selective chemical ionization is insufficient. Complete

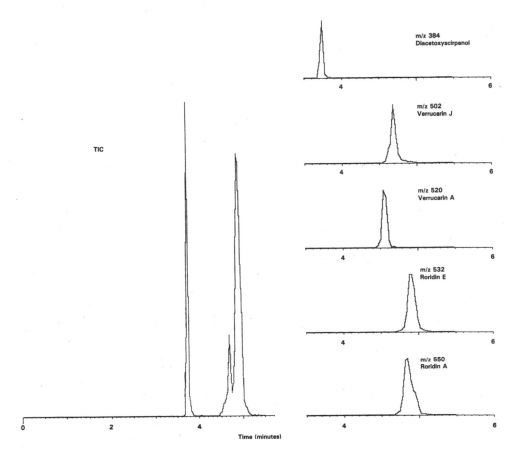

Fig. 31 Rapid capillary SFC–MS separation of a standard mixture of DAS and the macrocyclic trichothecenes verrucarin A, verracurin J, roridin A, and roridin E. The total-ion chromatogram (left) and selected-ion chromatograms for each component in the mixture are shown. See text for chromatographic conditions.

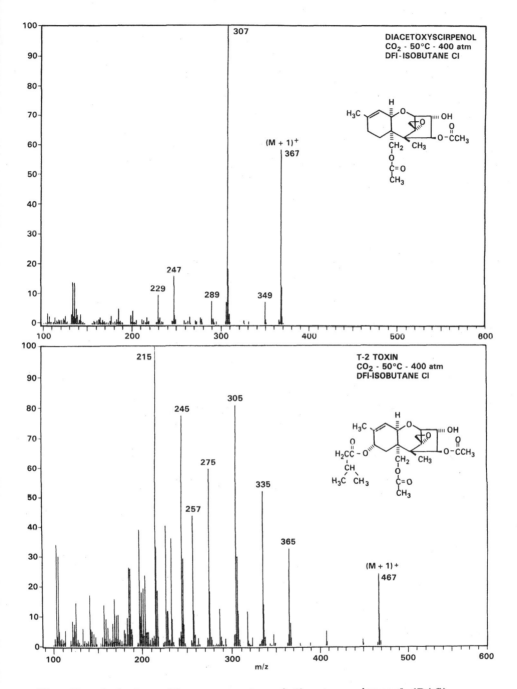

Fig. 32 Isobutane CI mass spectra of diacetoxyscirpenol (DAS) and T-2 toxin obtained with supercritical CO_2 at 50°C.

spectra or several selected ion profiles may be used for confirmation with more energetic CI reagents (such as isobutane) with a corresponding moderate drop in sensitivity. Also, if even more selectivity is required, tandem mass spectrometric methods (e.g., MS/MS) utilizing collision-induced dissociation (CID) of a selected ion can be used [4,49]. Figure 33 illustrates the CID spectrum obtained from the $(M + 1)^+$ peak from isobutane CI of DAS. Nearly any ion or combination of ions can be used, depending on the confidence level required and the complexity of the sample.

The application of these methods to an actual complex sample matrix is shown in Fig. 34 [50]. In this example, a wheat extract contaminated with approximately 5 ppm of DAS and T-2 toxin was analyzed using a 4 m × 50 μm i.d. column with CO_2 at 100°C and ammonia CI detection. The relatively uncomplicated mass spectrum obtained for the m/z 200–500 scan range, due to the selectivity of ammonia CI, is evident in the total-ion chromatogram. The single-ion chromatograms for m/z 384 and m/z 484 show peaks obtained at the same retention as for authentic standards, and the mycotoxins

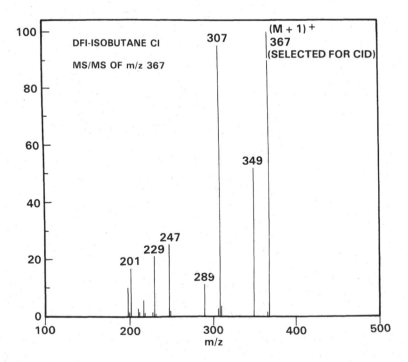

Fig. 33   MS/MS spectrum obtained for the dissociation of the $(M + 1)^+$ ion from the isobutane CI of diacetoxyscirpenol.

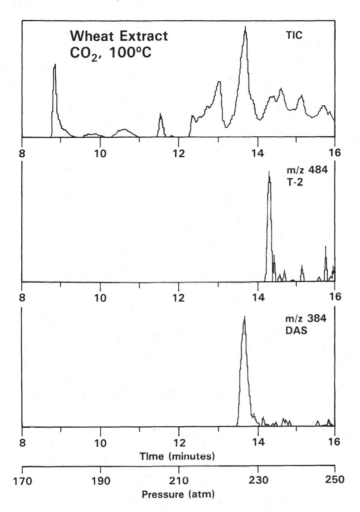

Fig. 34   Total-ion chromatogram and selected-ion chromatograms for m/z 484 and 384 showing detection of diacetoxyscirpenol (DAS) and T-2 toxin from a 5 ppm wheat extract.   See text for chromatographic conditions.

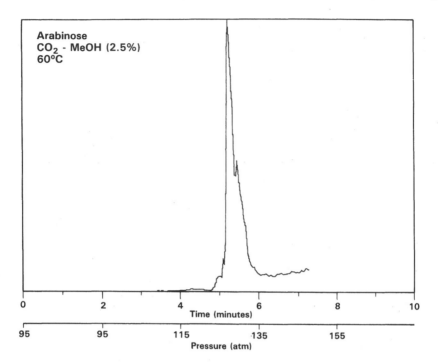

Fig. 35 Total-ion chromatogram obtained from the capillary SFC—MS analysis of arabinose. See text for chromatographic conditions.

can be identified on this basis with reasonable certainty. The identification has been confirmed using MS/MS methods [50]. These results suggest feasible detection limits for trichothecenes of <100 ppb while scanning and perhaps 10 ppb using single-ion monitoring.

More polar thermally labile molecules can also be analyzed using SFC—MS methods. The total-ion chromatogram of the simple sugar arabinose obtained without derivatization is shown in Fig. 35. This analysis was obtained using a 2 m × 50 µm i.d. column coated with a 5% phenyl polymethylphenylsiloxane stationary phase [51]. A methanol-modified carbon dioxide (2.5 wt. % methanol) mobile phase was employed to provide the necessary selectivity. A temperature of 60°C and an initial pressure of 95 bar with a pressure ramp of 10 bar/min were utilized to ensure that the fluid mixture was above the critical point (in a single-phase region) and to elute the compound. The mass spectrum obtained by ammonia CI during the SFC—MS analysis is shown in Fig. 36. A dominant $(M + 18)^+$ molecular ion was observed. However, the m/Z 150 ion, which corresponds to a molecular ion, is probably the loss of water from an ammonium adduct ion, e.g., [(M +

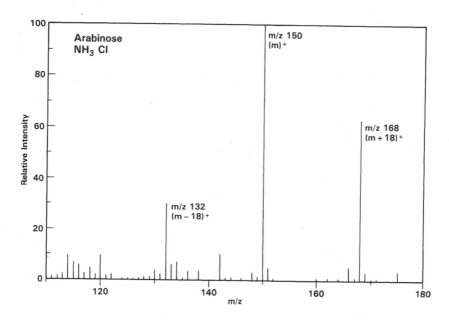

Fig. 36   Ammonia chemical ionization mass spectrum of arabinose obtained during the SFC–MS analysis shown in Fig. 35.

18) − 18)]$^+$. As expected from the structure of a sugar, additional losses of water were also evident.

For the reasons described earlier, packed column SFC–MS is somewhat more limited in application due to the greater activity of the stationary phase, particularly if restricted to pure carbon dioxide as the mobile phase. Figure 37 illustrates that limitation, showing an SFC–MS separation of a C_{12} to C_{40} n-alkane mixture using a 5-μm particle C_{18} microbore column with the instrumentation described earlier. The separation used carbon dioxide at 50°C with an inlet pressure of 400 bar and an outlet pressure of ∿355 bar with isobutane CI mass spectrometric detection. Comparison of the total-ion chromatogram with the earlier n-alkane separations on short 50-μm-i.d. capillaries (Fig. 8) shows that pressure programming allows roughly comparable separations to be obtained between 5 and 10 times faster, although the two columns have similar plate numbers at their optimum linear velocities. Another interesting observation based on Fig. 37 is the much greater retention for the larger alkanes (C_{36} elutes at ∿50 min and is evident only in the selected-ion chromatogram). However, these separations have the advantage that sample loading constraints are eased and detection dynamic range is well over an order

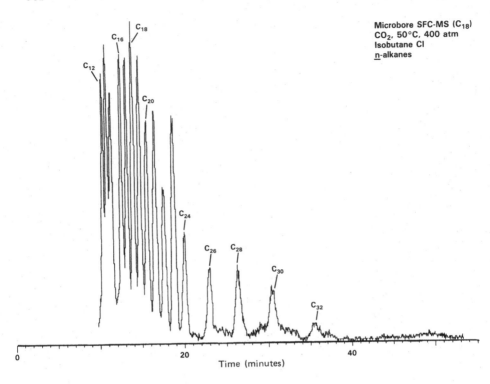

Fig. 37   Microbore SFC—MS total-ion chromatogram of an *n*-alkane mixture.   See text for chromatographic conditions.

of magnitude greater than with the capillary column. In addition, split injection is not required, which makes quantitation more straight-forward. An improved approach utilized pressure programming with packed columns for SFC—MS, as recently demonstrated for organo phosphorous pesticides [54].

The use of solvent modifiers considerably extends the utility of available packed column phases. Figure 38 shows a separation of eight pesticides using the C_{18} microbore column described above, but with addition of a 1% (by weight) methanol modifier. The sep-aration was obtained at 50°C and 450 bar inlet pressure using iso-butane CI detection. Selected-ion chromatograms given in Fig. 39 show peaks corresponding to each of the pesticides, clearly show-ing the contributions of the individual components where sufficient chromatographic resolution does not exist for their separation (e.g., propachlor and propoxur). Although the quality of the separation is clearly less than that obtained with pressure-programmed capillary methods, the results suggest that application of improved (less active) stationary phases, gradient methods, and pressure pro-gramming to packed column separations may provide significant

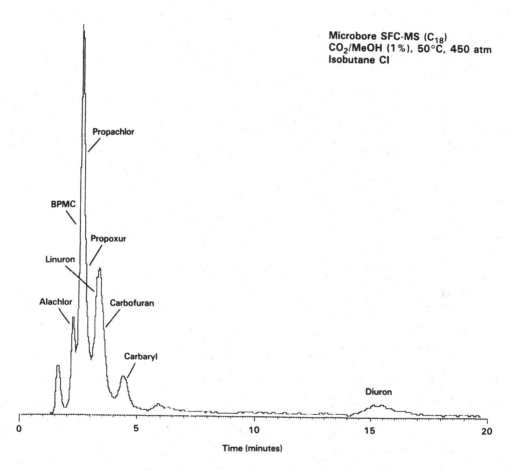

Microbore SFC-MS (C$_{18}$)
CO$_2$/MeOH (1%), 50°C, 450 atm
Isobutane CI

Fig. 38   Microbore SFC—MS total-ion chromatogram of a pesticide mixture.   See text for chromatographic conditions.

Pesticides

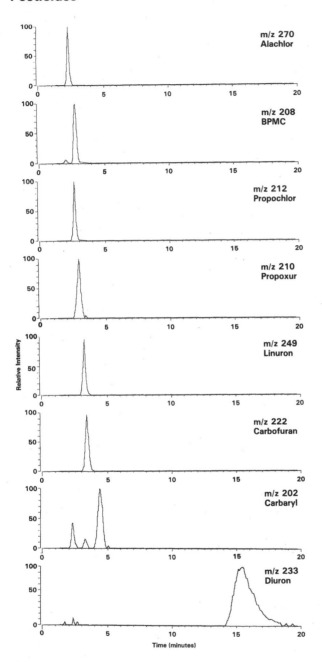

Fig. 39 Selected ion chromatograms for components in the pesticide separation shown in Fig. 38.

improvements in performance [52]. The advantages of the microbore
SFC—MS interface described earlier are also evident, but improvements in column lifetime, methods for solvent gradient production
for highly compressible fluids, and alternative solvent modifiers for
SFC—MS are required.

PRESENT LIMITATIONS OF SFC—MS AND RESEARCH NEEDS

The recent progress in capillary SFC—MS has demonstrated significant potential for a unique role intermediate between GC—MS and
LC—MS and as a favorable alternative to LC—MS, where applicable.
The capillary SFC instrumentation has improved rapidly and the
practice of SFC with fluids such as CO_2 is approaching a stage
amenable to routine laboratory application. The improved chromatographic resolution, sensitivity, and interface operation possible with
SFC—MS compared to LC—MS make this approach attractive.

Major applications of SFC and SFC—MS will undoubtedly exist in
the analysis of labile and high-molecular-weight polymer and complex
organic mixtures. The potential for separation of high-molecular-
weight polymers has been well established with spectroscopic detectors. While the potential exists for rapid growth in SFC—MS applications, problems remain to be solved. Improved sample injection
techniques are required, in particular, on-column injection techniques
which remove the sample solvent prior to analysis and reduce constraints due to low sample solubility or limited injection volumes.
The limitations for more polar fluid systems are mostly unexplored.
A large fraction of the compounds presently separated by HPLC
can, in principle, be analyzed using SFC. In fact, it is not un-
reasonable to predict that essentially any compound soluble in an
organic solvent should be amenable to SFC. Many higher polar materials are soluble in supercritical fluids such as ammonia [4,49].
At present, the most likely classes of compounds to be inaccessible
by SFC—MS are those soluble only in aqueous solutions and those
for which specific chemical interactions (such as hydrogen bonding)
are required for solubility, although derivatization methods may
render many of these cases accessible. The recent discovery [55] and
application of reverse micelle phases for SFC [56] suggests one possible approach to this extension.

The current limitations of SFC—MS technology result largely
from the limited knowledge of supercritical fluid solvating properties and interactions with condensed phases. The demands imposed
by supercritical fluids are different from those encountered in GC
and are related primarily to the necessity for chemical and physical
stability of the stationary phase. At present, higher temperatures
and more polar fluids often lead to degradation of stationary phases

which are perfectly suitable for GC at similar temperatures. Similarly, improved methods for column deactivation and formation of homogeneous stationary phase films of varying polarity in smaller-diameter (<40 μm) capillary columns would be useful in the separation of polar analytes.

The mass spectrometric interface for SFC also needs further development to improve sensitivity and performance with higher-molecular-weight compounds. Fundamental studies are required to understand nucleation processes relevant to the transport and detection of truly nonvolatile analytes, although our understanding of these phenomena has advanced rapidly [53]. The current status of SFC—MS also reflects the limitations of the mass spectrometer as well as that of the interface. Developments in extended-range mass spectrometry (m/z 5000—50,000) could provide a unique role for SFC—MS, in that it would provide the only mass spectrometer introduction method for operation in this molecular weight range not utilizing some form of desorption ionization [40].

The recent progress in capillary SFC—MS methods has demonstrated significant potential for many important applications. The flexibility offered by the various mass spectrometry ionization methods coupled with the wide variety of SFC analysis methods that are available provides tremendous analytical potential. This potential will continue to grow as SFC techniques and instrumentation advances are extended by the introduction of new mobile and stationary phases and as improved mass spectrometric detectors for higher-molecular-weight compounds (>1000 dalton) become available.

ACKNOWLEDGMENTS

We gratefully acknowledge the support of the U.S. Department of Energy, Office of Health and Environmental Research, through Contract DE-AC06-76RL0 1830. We also thank H. T. Kalinoski, H. R. Udseth, and A. J. Kopriva for technical contributions to this work.

REFERENCES

1. F. W. McLafferty, R. Knutti, R. Venkataraghavan, P. J. Arpino, and B. G. Dawkins, Anal. Chem. 47, 1503 (1975).
2. W. H. McFadden, H. L. Schwartz, and S. J. Evans, J. Chromatogr. 122, 389 (1976).
3. C. R. Blakley and M. L. Vestal, Anal. Chem. 55, 750 (1983).
4. R. D. Smith and H. R. Udseth, Anal. Chem. 55, 2266 (1983).
5. R. D. Smith, W. D. Felix, J. C. Fjeldsted, and M. L. Lee, Anal. Chem. 54, 1883 (1982).

6. R. D. Smith, J. C. Fjeldsted, and M. L. Lee, *J. Chromatogr.* *247*, 231 (1982).

7. R. D. Smith, H. T. Kalinowski, H. R. Udseth, and B. W. Wright, *Anal. Chem. 56*, 2476 (1984).

8. R. D. Smith, H. R. Udseth, and H. T. Kalinoski, *Anal. Chem. 56*, 2971 (1984).

9. B. W. Wright, H. R. Udseth, R. D. Smith, and R. N. Hazlett, *J. Chromatogr. 314*, 253 (1984).

10. C. R. Yonker, B. W. Wright, H. R. Udseth, and R. D. Smith, *Ber. Bunseges, Phys. Chem. 88*, 908 (1984).

11. L. McLaren, N. Myers, and J. C. Giddings, *Science 159*, 197 (1968).

12. G. M. Schneider, E. Stahl, and G. Wilke, eds., *Extraction with Supercritical Gases, Verlag Chemie*, Deerfield Beach, Fla. (1980).

13. T. H. Gouw and R. E. Jentoft, *Adv. Chromatogr. 13*, 1 (1975).

14. U. Van Wassen, I. Swaid, and G. M. Schneider, *Angew. Chem. Int. Ed. Engl. 19*, 575 (1980).

15. C. R. Yonker, S. L. Frye, D. R. Kalkwarf, and R. D. Smith, *J. Phys. Chem., 90*, 3022 (1986).

16. R. C. Petersen, D. W. Matson, and R. D. Smith, *J. Am. Chem. Soc., 108*, 2100 (1986).

17. C. R. Yonker and R. D. Smith, *J. Chromatogr., 361*, 25 (1986).

18. F. P. Schmitz and E. Klesper, *Makromol. Chem. Rapid Commun. 2*, 735 (1981).

19. F. P. Schmitz and E. Klesper, *Makromol. Chem. Rapid Commun. 3*, 959 (1982).

20. F. P. Schmitz and E. Klesper, *Polymer 24*, 142 (1983).

21. F. P. Schmitz, H. Hilgers, and E. Klesper, *J. Chromatogr. 267*, 267 (1983).

22. A. L. Blilie and T. Greibrokk, *J. Chromatogr. 349*, 317 (1985).

23. P. A. Peaden, B. W. Wright, and M. L. Lee, *Chromatographia 15*, 335 (1982).

24. B. W. Wright, P. A. Peaden, M. L. Lee, and T. J. Stark, *J. Chromatogr. 248*, 17 (1982).

25. R. D. Smith, E. G. Chapman, and B. W. Wright, *Anal. Chem. 57*, 2829 (1985).

26. B. W. Wright and R. D. Smith, *J. High Resolut. Chromatogr. Chromatogr. Commun., 9*, 73 (1986).

27. G. Guiochon and H. Colin, *J. Chromatogr. Lib. 28*, 1 (1984).

28. L. G. Randall, in *Ultrahigh Resolution Chromatography*, Am. Chem. Soc. Symp. Ser. 252, 135 (1984).

29. J. B. Crowther and J. D. Henion, *Anal. Chem. 57*, 2711 (1985).

30. S. M. Fields and M. L. Lee, *J. Chromatogr. 349*, 305 (1984).
31. A. B. Cambel and B. H. Jennings, *Gas Dynamics*, McGraw-Hill, New York (1958).
32. *Gas Encyclopedia*, Elsevier, North-Holland, New York (1976).
33. H. W. Liepmann and A. Roshko, *Elements of Gas Dynamics*, Wiley, New York (1957).
34. C. E. Lapple, *Trans. Am. Inst. Chem. Eng. 39*, 385 (1943).
35. J. C. Fjeldsted, R. C. Kong, and M. L. Lee, *J. Chromatogr. 279*, 449 (1983).
36. B. E. Richter, *J. High Resolut. Chromatogr. Chromatogr. Commun. 8*, 279 (1985).
37. T. L. Chester, D. P. Innis, and G. D. Ownens, *Anal. Chem. 57*, 2243 (1985).
38. T. L. Chester, *J. Chromatogr. 299*, 424 (1984).
39. S. Yajima, Y. Hagegawa, J. Hayashi, and M. Iimura, *J. Mater. Sci. 13*, 2569 (1978).
40. M. L. Vestel, *Mass Spectrom. Rev. 2*, 447 (1983).
41. T. A. Milne, *Int. J. Mass Spectrom. Ion Phys. 3*, 153 (1969).
42. J. C. Giddings, M. N. Myers, and A. L. Wahrhaftig, *Int. J. Mass Spectrom. Ion Phys. 4*, 9 (1970).
43. L. G. Randall and A. L. Wahrhaftig, *Anal. Chem. 50*, 1705 (1978).
44. L. G. Randall, Ph.D. Thesis, University of Utah (1979).
45. L. G. Randall and A. L. Wahrhaftig, *Ref. Sci. Instrum. 52*, 1283 (1981).
46. J. Ashkenas and F. S. Sherman, in *Proceedings of the 4th International Symposium on Rarefied Gas Dynamics* (J. H. deLeevw, ed.), Academic Press, New York (1966).
47. O. F. Hagena and W. Obert, *J. Chem. Phys. 56*, 1793 (1972).
48. R. D. Smith, H. R. Udseth, and B. W. Wright, *J. Chromatogr. Sci. 23*, 192 (1985).
49. R. D. Smith, H. R. Udseth, and R. N. Hazlett, *Fuel 64*, 810 (1985).
50. H. T. Kalinoski, B. W. Wright, and R. D. Smith, *Anal. Chem. 58*, 2421 (1986).
51. B. W. Wright and R. D. Smith, *J. Chromatogr. 355*, 367 (1986).
52. R. D. Smith and H. R. Udseth, *Anal. Chem. 59*, 13 (1987).
53. R. D. Smith, H. T. Kalinoski, and H. R. Udseth, *Mass Spectrom. Rev. 6*, 445 (1977).
54. H. T. Kalinoski and R. D. Smith, *Anal. Chem. 60*, 529 (1988).
55. R. W. Gale, J. L. Fulton, and R. D. Smith, *J. Amer. Chem. Soc. 109*, 920 (1987).
56. R. W. Gale, J. L. Fulton, and R. D. Smith, *Anal. Chem. 59*, 1977 (1987).

11

Practice and Applications of Supercritical Fluid Chromatography in the Analysis of Industrial Samples

THOMAS L. CHESTER / The Procter & Gamble Company, Cincinnati, Ohio

INTRODUCTION

The use of a supercritical fluid as a chromatographic mobile phase was first reported in 1962 by Klesper et al. [1]. They used supercritical Freons to separate metal porphyrins on a short packed column. Developments in packed column supercritical fluid chromatography (SFC) continued throughout the 1960s and 1970s to the present. The work prior to 1968 is summarized in a review by Giddings et al. [2]. More recent packed column developments were reviewed by Gouw and Jentoft in 1972 [3] and by Gere in 1983 [4]. The first report of open-tubular, or capillary, SFC was made by Novotny et al. in 1981 [5]. Significant new growth in research and application of capillary SFC occurred following its introduction [6–29].

Supercritical fluid chromatography, especially when used with open-tubular columns, has the potential for solving a major fraction of industrial analytical separation problems currently going unsolved [30]. The practical, problem-solving, and cost-effective strength of capillary SFC lines in its combination of a relatively high-efficiency separation using a low-temperature solvating mobile phase of programmable elution strength and its compatibility with a universal detector

for organic compounds—the flame ionization detector (FID). The
next few pages develop a perspective on this unique capability, followed by a discussion of the hardware necessary for one approach
toward FID usage. The final paragraphs focus on successful separations of industrially important problems that have, so far, proved
difficult or impossible to solve by conventional chromatography.

Since supercritical fluids have physical properties between those
of ordinary gases and liquids, it is not surprising that, depending
on conditions, SFC can behave like gas chromatography (GC), like
liquid chromatography (LC), or have properties of both [24]. Using
open-tubular column chromatography as an example, we will explore
some of the operating characteristics that distinguish supercritical
fluids from other mobile phases.

The height equivalent to a theoretical plate, H, of an open-
tubular column is given by the Golay equation,

$$H = \frac{B}{v} + Cv = \frac{2D_m}{v} + \frac{(1 + 6k + 11k^2)d_c^2 v}{96(1 + k^2)D_m} \tag{1}$$

where D_m is the solute diffusion coefficient in the mobile phase, v
is the average mobile phase velocity, k is the solute capacity factor
$[= (t_r - t_0)/t_0$, where t_r and t_0 are retention times of the solute
and unretained material, respectively], and d_c is the inside diameter (i.d.) of the column. Factors for pressure drop effects and
the term for stationary phase effects have been omitted as they do
not affect the present discussion or the conclusions reached.

The optimum mobile phase velocity, v_{opt}, resulting in the maximum column efficiency is

$$v_{opt} = \sqrt{\frac{B}{C}} = 4.2 \frac{D_m}{d_c} \tag{2}$$

for lightly loaded columns with well-retained peaks. Because solute
diffusion coefficients are always larger in supercritical fluids than
in liquids (typically $10^{-4}-10^{-3}$, and 10^{-5} cm^2/s, respectively),
optimum mobile phase velocities are always higher for SFC than for
LC for any single column and for a solute capable of being conveniently eluted by both mobile phases. This can be seen in Fig. 1,
where the Golay equation has been solved and plotted for a hypothetical column operated with both liquid and supercritical fluid
mobile phases. Similarly, GC optimum velocities are higher than
both LC and SFC for a given column because gaseous mobile phases

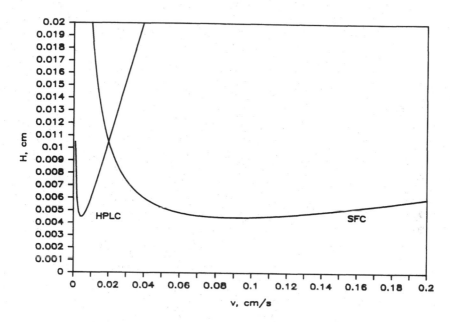

Fig. 1   Height equivalent to a theoretical plate, H, versus mobile phase velocity, v, for a 50-μm-i.d. column with liquid and super-critical fluid mobile phases (D_m = 5 × 10^{-6} and 10^{-4} cm^2/s, respectively).

have even higher solute diffusion coefficients (about 10^{-1} cm^2/s). The behavior of the same column with a gaseous mobile phase cannot be displayed in Fig. 1 because the GC curve is so far to the right and so flat compared to the LC and SFC curves. However, column behavior is basically the same regardless of mobile phase if relative, not absolute, velocities are considered. This is evident in Fig. 2, a plot similar to Fig. 1 but using a logarithmic velocity scale.

As Figs. 1 and 2 show, whenever optimum velocities are used and the column is not so long that pressure drop becomes an issue, column efficiency is determined by the column, not the mobile phase. For open-tubular columns, the minimum height equivalent to a theoretical plate, H_{min} (obtained at v_{opt}), is

$$H_{min} \approx 0.9d_c \tag{3}$$

for a well-retained peak regardless of the mobile phase used.

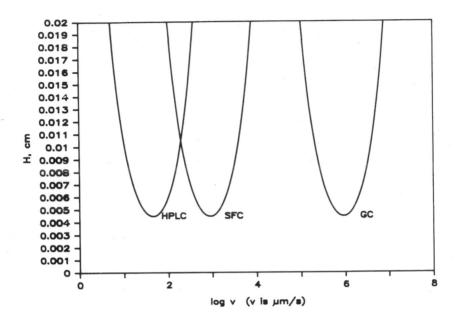

Fig. 2 Height equivalent to a theoretical plate, H, versus log mobile phase velocity, v. Column and mobile phase assumptions phases. The gas $D_m = 10^{-1}$ cm^2/s. (Reproduced from Ref. 30 by permission of Preston Publications, Inc.)

In practice, when a maximum limit is placed on the analysis time we are willing to tolerate and higher-than-optimum velocities are chosen to reduce analysis time, the efficiency of any column is directly proportional to the solute diffusion coefficient in the mobile phase. Under these conditions, a well-retained peak would produce a height equivalent to a theoretical plate of

$$H = 0.11 \frac{d_c^2 v}{D_m} \tag{4}$$

This simple analysis suggests that GC is the method of choice whenever it will work: Comparing the three types of mobile phases at optimum efficiency for a given column, Eq. (2) predicts analysis times to vary as

GC < SFC < LC

(fastest)   (slowest)

This relationship also holds true when comparing the three mobile phases at equal, but not optimum, column efficiencies. Now, with a maximum limit placed on time we are willing to spend on a separation, column efficiency would vary as

GC > SFC > LC

(most   (least
efficient)   efficient)

Of course, this analysis assumes that the solute in question can be eluted with a reasonable capacity factor by any of the mobile phases. When this is the case, no other separation technique known can match GC for efficiency or speed of analysis. It's enough to make one wonder why we'd ever need anything else.

SCOPE OF INDUSTRIAL PROBLEMS— THE NEED FOR CAPILLARY SFC—FID [30]

When separation efficiency, speed of analysis, and instrumental simplicity are all considered in choosing a separation method, open-tubular GC should always be the first choice of an analyst whenever it will work. No other separation method known today can match it on these combined merits. Especially noteworthy is the large variety of detectors available. They range from a superb universal detector, namely the FID, to a variety of highly selective or highly sensitive detectors, such as thermionic detectors and the electron capture detector. The FID is the detector most often used in industrial problem solving, however, because of the combination of its high sensitivity (giving rise to subnanogram detection limits for narrow peaks) and its universal applicability for detecting organic compounds. Thus, GC—FID is the first choice for a separation-detection combination in general. Unfortunately, GC doesn't work on every problem.

There are two basic limitations associated with GC: solute volatility and thermal stability. Because mass transport through the column is via the gas phase, some vapor pressure is necessary for any solute to be eluted. Elevated temperatures are used to raise vapor pressures, but this introduces the thermal stability requirement. Gouw and Jentoft [3] pointed out that only about 15% of the

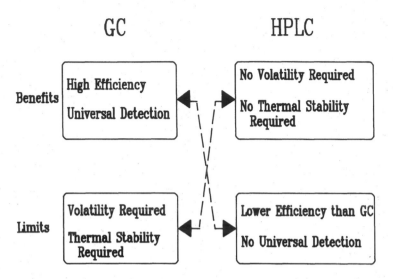

Fig. 3   Complementary benefits and limits of GC and HPLC. (Reproduced from Ref. 30 by permission of Preston Publications, Inc.)

compounds known can be volatilized without being thermally degraded. Chemical derivatization is often used as a means of increasing volatility, reducing polarity, or improving thermal stability of
analytes, thus making GC possible in many additional situations. If
derivatization approximately doubles the scope of GC, then about a
third of the separation problems faced in industry can be addressed
by GC.

   High-performance liquid chromatography (HPLC) is an ideal complement to GC (see Fig. 3). As HPLC is usually run at ambient
(or at least not highly elevated) temperature, it has no significant
thermal stability requirement. And volatility is not necessary because mass transport occurs in the liquid phase. However, there
are two new limitations. First, ordinary HPLC (that is, with packed
columns) is not nearly as efficient as ordinary GC (because capillary
columns are commonly used). Practically speaking, 25,000 theoretical plates is considered quite good for a commercial packed HPLC
column, whereas commercial capillary GC columns are readily
available that generate over 200,000 theoretical plates. Thus, HPLC
is somewhat limited compared to GC in the analysis of complex mixtures requiring high column efficiency. The other limitation of
HPLC is detection. Sensitive detection in HPLC requires particular
functional groups to be present on the solutes. Ultraviolet absorbance and fluorescence detectors are often used, but they require a chromophore or fluorophore to be present on the solute

molecule. Electrochemical detectors can be very sensitive, but only for electroactive solutes. There is no simple, sensitive, universal detector for HPLC (yet). Instead, when no available selective detector is suitable for a particular problem, a bulk property detector, such as the refractive index detector, is used. These relatively insensitive detectors further limit HPLC by imposing a requirement of isocratic elution only, the primary result being a limit on the range of solute polarity or molecular weight that can be determined with a single injection. This translates into lost analysis efficiency in a real-time sense. Of course, as in GC, derivatization can be used to increase the scope of HPLC, this time by adding detectability (as well as by reducing polarity to improve peak shapes in some cases). Despite these limitations, HPLC is extremely successful in solving separation problems—more than half of the industrial separation and detection problems faced by chemists are solved using HPLC.

However, a "hole" is left between GC and HPLC containing problems that cannot be solved by either (see Fig. 4). These are problems where the analyte either is not volatile or is thermally unstable *and* where it either is present in a very complicated mixture or has no convenient means of detection in organic (or organically modified) mobile phases. In other words, both GC and HPLC have serious difficulties with these chemical analysis problems. Unfortunately, about a fourth of the separation problems faced by industrial

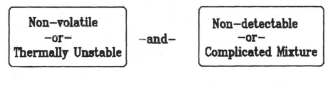

When the solute is

| Non-volatile
—or—
Thermally Unstable | —and— | Non-detectable
—or—
Complicated Mixture |

then

GC won't work   —and—   HPLC won't work

To fill the hole we need:
- a high efficiency separation
- a low-temperature, solvating mobile phase
- universal detection

Fig. 4   The hole existing between GC and HPLC analysis capabilities. (Reproduced from Ref. 30 by permission of Preston Publications, Inc.)

chemists fall into this hole (partially because GC and HPLC overlap somewhat for problems that can be solved by both of these techniques).

The ideal technique to fill the hole would be one with a low-temperature solvating mobile phase of programmable strength, high chromatographic efficiency, and universal detection. SFC, when used with CO_2, N_2O, and several other mobile phases, is compatible with the universal FID. Thus, it fills the hole between GC and HPLC and has the potential for solving a large fraction of industrial problems currently going unsolved.

However, SFC—FID has some limitations. Solutes must have some minimum solubility in an FID-compatible mobile phase. The mobile phases used most often in SFC (CO_2, N_2O, pentane, etc.) are relatively nonpolar. High-polarity solutes, such as carbohydrates (unless derivatized), do not have enough mobile phase

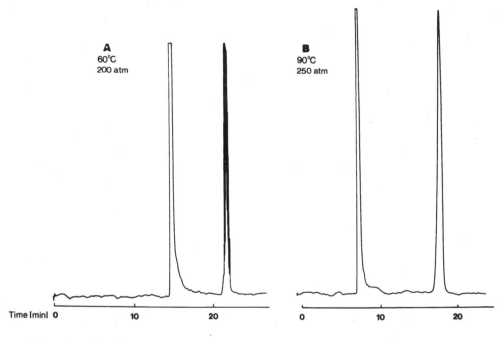

Fig. 5 Glycerol tristearate chromatograms under conditions giving detector spiking (A, 60°C, 200 atm) and spike-free (B, 90°C, 250 atm). Conditions: CO_2 mobile phase, methyl silicone column, 15 m × 100 μm i.d., 0.25-μm film. (From Ref. 14.)

solubility for SFC—FID analysis. Similarly, no reports of analyses of ionic solutes by capillary SFC have appeared in the literature. Finally, a practical molecular weight limit is imposed by the FID— solutes must be delivered to the FID in a gas stream without being allowed to condense appreciably during the decompression step. If significant condensation occurs, the chromatographic peaks will be littered with sharp spikes (see Fig. 5) as solute particles enter the flame and are ionized [2]. This condition renders the chromatography useless as spikes often cannot be distinguished from genuine peaks.

MINIMIZATION OF SPIKING IN THE FID OUTPUT SIGNAL

General Considerations—The Root of the Problem

Solvent strength of a supercritical fluid depends on its density and, of course, on the externally applied pressure. That pressure must be maintained to the largest extent possible over the entire length of the chromatographic column. Any pressure drop would result in continuously increasing retention and a concomitant increase in the average pressure required to elute a solute peak. In isobaric (constant head pressure) separations, an on-column pressure drop is manifested as a loss in the selectivity between solutes [9]. Thus, there is a basic incompatibility between an SFC column and an atmospheric-pressure detector. The only method proved successful for interfacing the two has been to use a low-volume flow restrictor to maintain pressure over the entire column and then drop the pressure abruptly just as the eluent enters the detector.

Decompression of a high-pressure fluid phase results in its cooling via the Joule-Thomson effect. In addition, as there is no volatility requirement for solutes separated by SFC, it is reasonable to assume that relatively nonvolatile solutes will be eluted most of the time (otherwise, GC would be a better choice if the analytes could stand the heat). This places even greater requirements on the column-to-detector interface. The design and performance of this interface are crucial in making SFC useful for the applications where it is needed most.

The FID spiking can be controlled fairly successfully by making use of whatever small amount of volatility exists in a solute. This is accomplished by postponing the decompression as long as possible (to allow less time for nucleation) while providing heat to the mobile phase, especially at the restrictor outlet, where the cooling effect of decompression is greatest. Thermal stability is not necessary, as the separation is already accomplished before the solutes experience

heating within the restrictor tubing and any pyrolysates formed
would be detected with about the same sensitivity as their parents.

Tapered Restrictors

Small-diameter fused silica tubing can be used to interface a capillary
column to an FID. The length of the restrictor tubing can be conveniently short if the diameter is small enough. Typically, 10−40
cm lengths of 5−10 μm i.d. tubing are used to interface 50−100 μm
i.d. columns. However, straight-walled restrictors usually do not
work very well for solutes of much more than about a thousand mass
units: the decompression occurs over too long a period of time and
distance, and the walls are generally quite thick, insulating the
mobile phase from the detector heat source. Pinched metal tubes
have also been used as interfaces by various workers (for example,
Ref. 12). They conduct heat better than fused silica tubes but,
since all of the restriction is provided by the pinch rather than the
length of the tube, the limiting orifice must be very small, typically less than 1 μm for capillary column flow rates. Thus, they tend
to plug easily.

A very convenient compromise is possible by using thin-walled,
tapered, fused silica capillary tubes as restrictors (Fig. 6). The
taper postpones decompression effectively while allowing wall thickness to be reduced, but does not require an extremely small outlet
orifice. Also, the taper can be prepared from larger-i.d. tubing
than would be practical for a straight-walled restrictor. This allows a relatively low mobile phase velocity through the bulk of the
detector. Thus, greater heating time for the mobile phase is provided before decompression occurs. And the wall thickness at the
outlet can be quite small (<10 μm), effectively minimizing the insulation where most of the cooling occurs.

These thin-walled restrictors are tapered by heating the tubing
in a flame and pulling. However, they are nearly impossible to prepare by hand because of the delicacy of the tubing. The best restrictors have been made from tubing of only about 150 μm o.d. and
25 μm i.d.

Restrictors can be prepared reproducibly from delicate tubing
with an automated system. The first of these is described in Ref.
25 and is based on a laboratory robot. It is illustrated in Fig. 7.
This system is capable of reproducing taper dimensions to within
several micrometers on a particular lot of tubing, but it still requires microscopic grading and sizing of individual restrictors and
custom cutting of each outlet to make acceptable restrictors with
the desired orifice diameter. The outlet must be cut as square as
possible. This can be done with a fiber-optic cleaver of sufficient
hardness (such as sapphire) or with a broken edge of a silicon

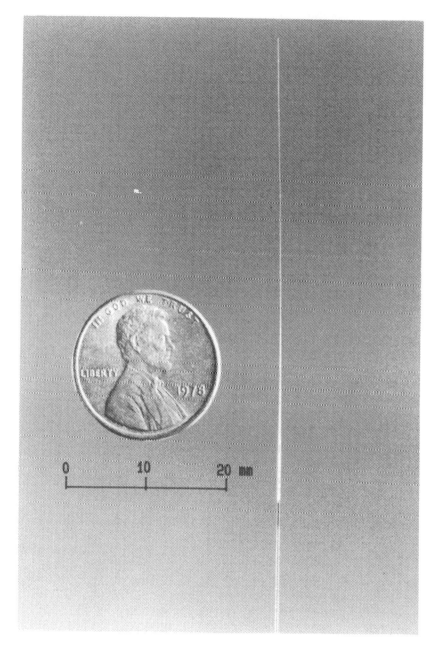

Fig. 6 A hand-pulled, tapered capillary restrictor, made from 14-μm-i.d. fused silica with a 250-μm o.d. (The smaller, robot-pulled capillaries, typically made from 25-μm-i.d., 150-μm-o.d. fused silica, are more difficult to photograph.)

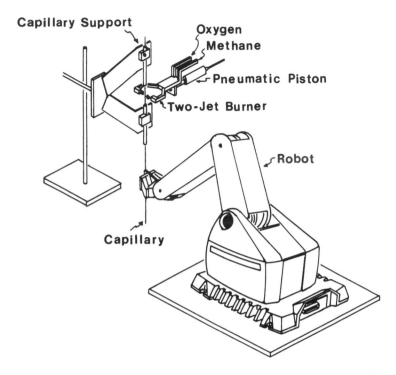

Fig. 7 Robot-based automated capillary puller. (Reprinted with permission from Ref. 25. Copyright 1985, American Chemical Society.)

wafer. The restrictor is placed against a smooth surface, such as a glass microscope slide, scored, and broken.

Polyimide-coated fused silica tubing is used as the starting material. But since the coating is burned away from the taper during the pulling operation, the tapered section is more delicate than an ordinary piece of fused silica chromatography tubing. However, it is surprisingly rugged as long as the bare silica surface is not scratched. Installation into the FID is best done by inserting the inlet end through the outlet of the flame jet. Once the restrictor inlet is visible in the oven, the connection is made to the column using an appropriate capillary union. This is done so that pressure can be applied to the system to check for flow and leaks and to form a small ball of ice on the end of the restrictor. The ice makes the outlet visible as the restrictor is lowered into final position, usually flush with the jet tip or as much as a millimeter below the jet tip. In any case, care must be taken not to leave the restrictor protruding from the jet, where it will be within the FID flame. When this happens the restrictor usually clogs with carbon after only a few injections.

As suggested by Springston [21], it is possible to prepare a restrictor directly on the end of the chromatography column. The principal advantage is in avoiding the column-to-restrictor union and any associated dead volume. One potential disadvantage is the necessity of heating the mobile phase in the presence of stationary phase as the column enters the detector block. The temperature increase will cause a local decrease in the mobile phase density and solvating power and may locally increase the solute capacity factor. This may adversely affect the separation. This is not a problem when separate restrictors are made from blank tubing since no stationary phase is present at the point where heat is applied—although the mobile phase experiences the same change in strength on heating, solute remains in the mobile phase. The necessary connection to the column, while potentially introducing postcolumn band broadening, offers an advantage—the column is easily demounted from the detector and can be changed in minutes without disturbing a working restrictor. Chromatograms obtained with this approach suggest that postcolumn broadening is not a problem: very narrow peaks can be obtained early in the chromatogram (with low pressure, small solute molecular weights, and high D_m) with broader peaks following (with increased pressure, larger solute molecular weights, and lower D_m), suggesting that the broadening is occurring mainly on the column. A detailed study of both approaches and their effects on band broadening would be a welcome addition to the literature.

The use of tapered restrictors is admittedly a difficult approach to SFC—FID for workers without access to some kind of robot or other automated system for the pulling operation and a quality microscope for sizing and inspecting the finished restrictors. Those just getting started in SFC, or limiting their applications to solutes with rather low molecular weights (perhaps up to 1500), have several other alternatives. Porous frit restrictors [31] and integral restrictors [32] are commercially available from several sources. Straight-walled capillaries can be used with limited success, especially with elevated oven temperatures or extra heating in the detector block, as demonstrated by Richter [23]. Both of these arrangements heat the mobile phase prior to decompression, with the latter being preferred in order to maintain flexibility in the separation via oven temperature selection. Pinched metal tubes can be used but, as mentioned earlier, clog more frequently and require much more maintenance on the system than capillary restrictors. (Tapered capillaries have been used as long as 6 months with no noticeable changes in their characteristics.) It is possible to hand-pull fused silica restrictors from 300–400 μm o.d. tubing (with inside diameters ranging from 10 to about 25 μm). Good technique, combined with some trial and error, is required. But usable restrictors can be made that allow spike-free detection of solutes with molecular weights well over 1000 [14].

Finally, some degree of spiking can be tolerated by data systems, or simply filtered out of the detector output signal as shown by Fjeldsted et al. [10]. This approach works as long as the average frequency of spike occurrence is low enough for the system to recover before the next spike. Several spikes per second can be handled this way. But as spiking worsens, the data system will eventually be overwhelmed. Minimizing spiking at the source is the first step in maximizing the molecular weight range. Electronic filtering is the last step.

The other instrumental requirements for capillary SFC−FID (pumps, injectors, etc.) have been covered in several references [e.g., 7,15] and are relatively easy to implement.

SELECTIVITY CONSIDERATIONS

Stationary phase selection in capillary SFC follows the same basic rules as in capillary GC. The major difference is that there is not yet a universally accepted method for indexing stationary phases as is done in GC. For a first approximation in selecting a stationary phase for SFC, GC-derived McReynolds constants can be used. They are fairly predictable of interactions between solute functional groups and stationary phases under SFC conditions, but the user must keep in mind the possibility of additional interactions between the solute and the mobile phase that could change the selectivity. Also, any stationary phase used in SFC should be bonded or cross-linked in some way and must be nonextractable.

Since CO_2 is nearly universally used as the mobile phase for SFC−FID, mobile phase selection has not yet been a major issue. In a few applications N_2O has been used, but it is very similar to CO_2. Unlike CO_2, N_2O does have a permanent dipole moment. However, in practice, these two mobile phases are nearly identical in strength and tend to elute most solutes with about the same pressures (or densities).

The more important difference between the two mobile phases is chemical reactivity; N_2O is less acidic than CO_2 and can be used to elute some bases that would react with CO_2 to form ionized (and noneluting) products. Other chemical differences between these two materials that affect their performance as mobile phases are unexplored and unpublished at this time. However, practically speaking, the selectivity differences between CO_2 and N_2O, if significant at all, are small for most solutes. Also, N_2O gives a larger background signal from an FID and a more steeply sloping baseline with pressure programming than does CO_2.

Several other mobile phases are possible in combination with the FID. Xenon works well but would be ridiculously expensive for use with an FID since there would be no major benefit over more traditional mobile phases. In contrast, it is an ideal mobile phase for spectrometric detectors, particularly for Fourier transform infrared absorbance detection [29]. Sulfur hexafluoride and sulfur dioxide are possible mobile phases for use with the FID. Ammonia was used by Giddings et al. [2,31] and is particularly well suited for separating more polar solutes. However, its critical temperature is 132°C, and thus it is unsuitable for applications involving thermally labile solutes. It also dissolves polyimide (Vespel) and quickly corrodes many copper-containing metals, especially brass. Finally, it is a nuisance to use because of its odor and toxicity coupled with the certainty of having small leaks from time to time in any high-pressure system.

Another selectivity mechanism belongs uniquely to SFC: temperature selection. SFC can behave like GC, like LC, or somewhere between the two for any particular solute, as shown in Fig. 8. The

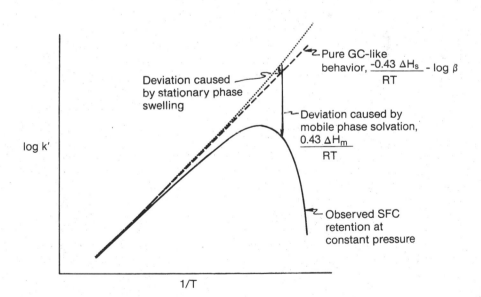

Fig. 8 Dual nature of SFC retention characteristics versus temperature. The positive slope region to the left is GC-like, and the negative slope region to the right is LC-like. (From Ref. 24.)

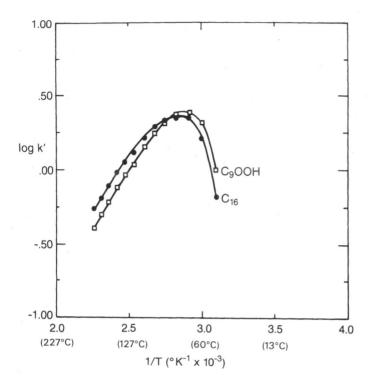

Fig. 9 Selectivity change with temperature for hexadecane and nonanoic acid, eluted with CO_2 from a DB-1 column. (From Ref. 24.)

extent of GC versus LC behavior can vary from one solute to the next in a sample. Thus, some solutes can be retarded while others are advanced or remain unchanged (with respect to their relative elution positions) as the temperature is changed. In fact, pressure (or density) programming at constant temperature is the preferred means of programming elution strength in SFC, rather than temperature programming at constant pressure, because of the often unpredictable nature of the latter. However, with only a little knowledge of the retention behavior of solutes with respect to temperature, a separation can be improved with the right temperature selection. Figure 9 shows an example of this selectivity behavior where a temperature shift was able to reverse the elution order of the two solutes. In practice, any separation under development that looks promising, but not quite good enough, should be run again at one or two different temperatures before changing the column or mobile phase. And any process to optimize a separation for maximum resolution or minimum analysis time should include a study of temperature effects. This should usually be considered a fine-tuning of the selectivity rather than a means of effecting a large selectivity change. For larger changes, the stationary phase selection should be reconsidered.

The unique feature of this selectivity mechanism is the fact that the oven temperature is an instrumental parameter, controlled electrically. Thus, this selectivity control can be accomplished without making a direct chemical change in the chromatographic system. This is a mechanism with no counterpart in GC or LC systems, as only SFC can have partitioning affected simultaneously by both volatility and mobile phase solvation.

APPLICATIONS

Capillary SFC is capable of eluting a surprising variety of solutes, including many with polar functional groups. For example, free hydroxyl groups can be handled reasonably well, as can carboxylic acids and some amines. Esters and amides are even easier. In addition, several polar functional groups can be present on the same solute molecule and good chromatographic behavior can still result.

From the limited number of applications of SFC to polar solutes published to date, the key to elution, at least with CO_2 mobile phase, seems to be in avoiding solute ionization. SFC separations of solutes that react with CO_2 to form ionized products, such as low-molecular-weight diamines, which are reasonably strong bases in CO_2, and low-molecular-weight dicarboxylic acids, which are relatively strong acids in CO_2, have not been reported. Similarly, SFC separations of solutes capable of reacting with themselves to form inter- or intramolecular ions, such as phospholipids and amino acids, respectively, have not been reported (at least not in CO_2 mobile phase). McLaren et al. [33] showed that ammonia is capable of dissolving polar solutes such as amino acids and sugars, whereas CO_2 is better for lower-polarity solutes.

Chemical derivatization can be used to reduce the polarity and increase the CO_2 solubility of highly polar solutes. Fortunately, the same chemistry, already highly developed for GC applications, is directly applicable to SFC. Furthermore, if several kinds of functional groups are present on a solute that cannot be eluted as is, it may not be necessary to derivatize all of the functional groups to make a separation possible—the easiest-to-derivatize functional groups can be treated first with an attempted separation following every derivatization. Thus, multistep derivatizations for polyfunctional solutes can be shortened, compared to GC, with a corresponding savings in work for any resulting routine analyses.

The following examples were chosen to show some of the types of industrial applications where SFC—FID has been used. In many of these examples, SFC—FID is the only separation-detection combination capable of solving the problems. Unfortunately, the most exciting, challenging, and rewarding work done in industry is usually held proprietary and is not often reported in print (except possibly the patent literature). So these examples should be taken as only an indication of larger possibilities.

Fats

There are two advantages of SFC over GC for the determination of glycerides. First, SFC can elute mono- and diglycerides without derivatization. Second, SFC is free of on-column thermal decomposition, whereas GC usually requires column temperatures in excess of 300°C for triglycerides. Figure 10 shows the ability of SFC to elute mono-, di-, and triglycerides in a single analysis with pressure programming. In fact, it is possible to include free fatty acids in the chromatogram by injecting at a lower pressure and adding time to the analysis. The best example of triglyceride analysis reported to date is the work of White and Houck [22]. They were able to elute compounds as heavy as triervonin ($C_{75}H_{140}O_6$) at 90°C and pressures not exceeding 300 atmospheres using a 19 m × 100 μm DB-5 column with 0.25-μm film thickness. In addition, they demonstrated that a more polar stationary phase, DB-225, was capable of separating four triglycerides of the same carbon number but differing in the number of double bonds present. These compounds coeluted on the DB-5 column.

Fatty acids and alcohols are trivial compared to triglycerides. However, care must be taken in column selection in order to have a stationary phase with sufficient capacity. It is relatively easy to overload a nonpolar phase, like polydimethylsiloxane, with acids or alcohols. Usually, a medium-polarity stationary phase (such as OV-17, OV-1701, or the equivalent, and cross-linked) is best. A highly polar stationary phase (Silar 10 or the equivalent, for example) will also tend to overload because of the influence of the nonpolar hydrocarbon end of fatty compounds.

Some artificial fats have been separated by SFC [25]. In this case, the molecular weight range covered by the pressure-programmed run went to nearly 2500. Flame ionization provided universal detection of these compounds, too heavy for GC and with no convenient detection mechanism by HPLC.

Polyglycerol esters are fats made by esterification of polyglycerol. Molecular weights can easily exceed those of common

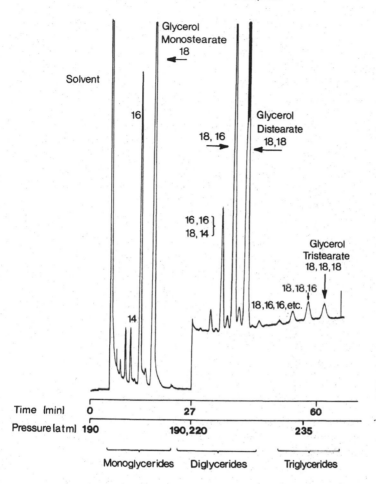

Fig. 10 Commercial glycerol monostearate sample, containing mono-, di- and triglycerides. Separation was by pressure-programmed SFC—FID using a 12 m × 100 μm BP-10 column (from SGE) with CO_2 mobile phase at 90°C. (From Ref. 14.)

triglycerides, but several free hydroxyl groups can be present on the molecule as well. Again, much information is available using SFC without derivatization, but the quality of the chromatograms and the amount of information available go down as the number of free hydroxyl groups increases [27]. Depending on the information sought, considerable gains can be accomplished by silylating the free hydroxyl groups to improve the chromatogram (Figs. 11 and 12). However, it is important to realize that detailed information is not needed in every industrial application and that broad, merged peaks, fully understood and available with no sample preparation, may be preferable to beautiful, but labor-intensive, separations. For example, diglycerol monoesters can be distinguished as a class from triglycerol monoesters without derivatization.

Surfactants

Nonionic surfactants can be separated relatively easily by SFC. They behave very much like fats. Ethoxylated alcohols have been convenient for testing new chromatographs and columns. Triton X-100 is a popular test material because of its good chromatographic behavior and because it contains an aromatic ring and can be detected by UV absorbance. Significant improvements have been made in Triton X-100 separations in just the past several years. These are attributable mainly to improvements in columns and injection procedures. For example, Fig. 13 is the first published capillary SFC chromatogram of Triton X-100 [14]. Figure 14 is a more recent chromatogram produced with a 10 m × 50 μm i.d. column of DB-17 stationary phase (J&W Scientific) combined with solute focusing on the head of the column using a "retention gap" type of technique, similar to that pioneered by Grob for GC [34].

Other simple ethoxylated alcohols work equally well and can be analyzed without derivatization to molecular weights well beyond 1000. The alkyl portion of the molecule really is not a requirement for good separation. Polyethylene glycols and polypropylene glycols have been separated over similar molecular weight ranges by SFC without derivatization [23,28].

Silicones

Silicones are almost anomalously well behaved in SFC. Silicone oil (that is, polydimethylsiloxane) chromatograms have been published by several workers, demonstrating the ability to elute more than 40 oligomers [15]. This good behavior tends to persist as substitution of the methyl groups is done. The only obvious restriction is that ionic or strongly ionizing functional groups cannot be present since, as with other examples, they increase retention tremendously.

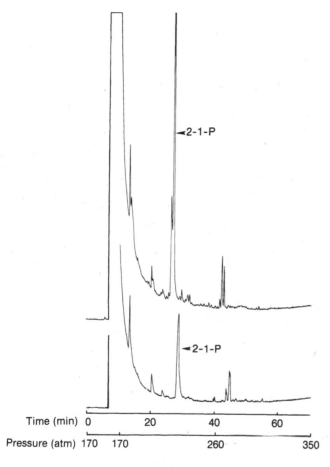

Time (min) 0   20   40   60

Pressure (atm) 170   170   260   350

Fig. 11 Chromatograms of derivatized (upper trace) and underivatized (lower trace) polyglycerol ester containing predominantly diglycerolmonopalmitate (2-1-P). A DB-1 column (J&W Scientific), 10 m × 50 μm i.d., was used with CO_2 mobile phase at 140°C. (From Ref. 27.)

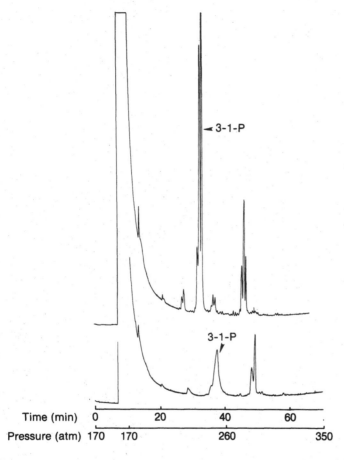

Fig. 12 Chromatograms of derivatized (upper trace) and underivatized (lower trace) polyglycerol ester sample containing predominantly triglycerolmonopalmitate (3-1-P). Conditions are the same as in Fig. 11. (From Ref. 27.)

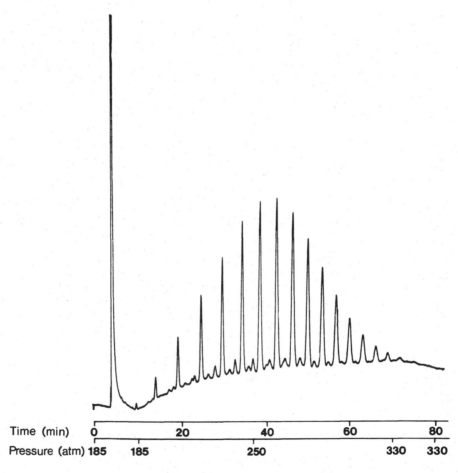

Fig. 13 Chromatogram of Triton X-100 from Ref. 14 (with permission, work done in 1983). A 9 m × 100 μm i.d. BP-10 column was used with a CO_2 mobile phase at 110°C.

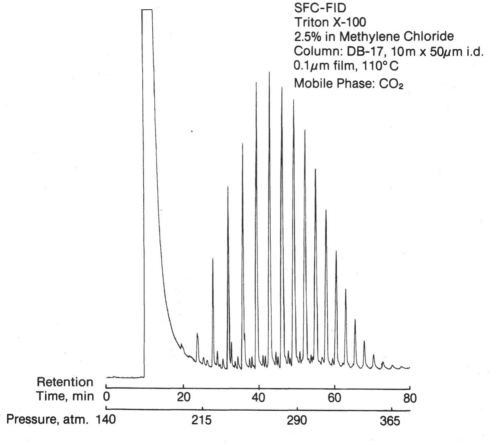

SFC-FID
Triton X-100
2.5% in Methylene Chloride
Column: DB-17, 10m x 50µm i.d.
0.1µm film, 110°C
Mobile Phase: CO_2

Retention
Time, min 0   20   40   60   80

Pressure, atm. 140   215   290   365

Fig. 14   Chromatogram of Triton X-100 using a 10 m × 50 µm i.d.
DB-17 column and a retention gap-style injection. (See Ref. 25 for
injector description.) CO_2 mobile phase was used with a column
temperature of 110°C.

Figure 15 demonstrates the difference in informing power of size-
exclusion chromatography and SFC in analyzing a low-molecular-
weight silicone. Two different terminal functional groups were
present in the oligomers of this sample, giving rise to the two distinct series of peaks clearly visible by SFC.
      Silicones are so well behaved that they could be used as mark-
ers for a retention index system for SFC. Properly chosen silicones would work better than hydrocarbon markers for two reasons. First, because the molecular weight range of SFC is so

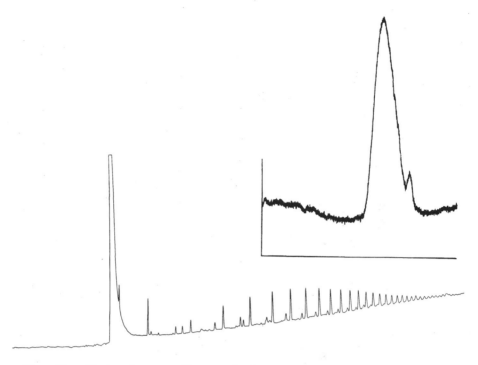

Fig. 15 Upper trace: Size-exclusion chromatogram of a low-molecular-weight polyfunctional silicone. Refractive index detection was used. Lower trace: Capillary SFC–FID separation of the same silicone. Two chemically different sets of oligomer peaks are clearly visible.

much higher than that of GC, hydrocarbon markers used in a Kováts index fashion would be too close together above about C$_{40}$. Second, n-hydrocarbons are not very well behaved on the polar stationary phases. The ideal retention marker would be of medium polarity, would contain oligomers covering the molecular weight range of interest, and would have a convenient molecular weight difference between oligomers. It would also be easily detectable with a variety of detectors. A material like poly(50%-phenyl)phenylmethyl-siloxane, if sufficiently purified and regular in phenyl substitution, would be perfect.

Carbohydrates and Highly Polar Solutes

So far we have seen many situations where derivatization has not been necessary and borderline cases where improvement was made

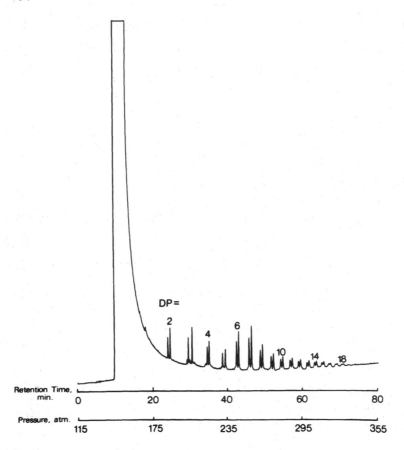

Fig. 16 Capillary SFC—FID separation of silylated oligo- and poly-saccharides in a corn syrup solid. The numbers represent the degree of polymerization, that is, the number of glucose units present in the peaks. (From Ref. 26.)

with derivatization but chromatograms could be obtained on underiv-atized samples. However, carbohydrates are a class of compounds where derivatization is absolutely necessary if low-temperature SFC separations are desired. Sugars are easily derivatized using conventional techniques and produce excellent chromatograms [26]. The powerful combination of pressure programming and universal detection is extremely attractive here, since it allows the programmed elution of polysaccharides over a broad molecular weight range in a single analysis. Figure 16 is an example of what can be done on such a sample. The molecular weight range of the undervatized polysaccharides represented by the peaks in Fig. 16 is over 2900, and the molecular weight range of the actual peaks is about 7000. GC can achieve better resolution of the lower-molecular-weight sugars (through the hexoses) but cannot come close to the molecular weight range provided by either SFC or HPLC. And HPLC methods generally fall well short of the resolving power shown by SFC for polysaccharides.

Other Applications

Much work has been done in applying SFC to the analysis of fossil fuel-derived samples and combustion products [13,16,17]. Flame ionization detection was used in many of these applications and allows the determination of aliphatic and aromatic components with approximately equal sensitivity. Fluorescense, UV absorbance, and mass spectrometric detection systems were also used in many of these applications to provide detection selectivity and aid in peak identification.

SFC is being applied to other areas, such as pesticides, drugs, and biologically active materials [19,20,23]. The limited availability of commercial equipment for capillary SFC initially prevented its proliferation into industrial laboratories, where it is often not considered cost effective to build instruments. As commercial instruments have become available, we have seen an explosion in the application of SFC which will eventullay bring it much closer to GC in numbers of industrial problems solved.

FUTURE TRENDS

Two areas stand out in terms of providing new problem-solving capability when teamed with SFC. The first is the development of more selective detectors for use with SFC. Despite the advantages of FID cited earlier, there are always times when selectivity in detection is necessary. A full array of detectors, including UV absorbance, fluorescense, flame photometric, thermionic, electron capture, and radioactivity monitors, need to be developed commercially and made available to chromatographers using SFC. As in any other form of chromatography, selective detectors like these will be of immense benefit whenever it is desirable to limit response to a particular solute class.

The second stand-out area is SFC-mass spectrometry. Tremendous benefits of combining SFC and mass spectrometry have already been demonstrated in numerous publications [see Refs. 16 and 20, for example]. However, only the surface has been scratched—today's quadrupole mass spectrometers simply do not match the mass range of SFC. Magnetic sector instruments with larger mass ranges are often considered too costly for use as dedicated chromatography detectors. And even many of these instruments fall short of the mass range required by applications already amenable to SFC. Perhaps the answer lies in a cheap, low-resolution instrument optimized to cover a larger mass range. At present, virtually nothing is known of the ionization and fragmentation problems associated with solutes in the 1000–10,000 dalton range delivered to an ionization source in a supercritical fluid. And with further improvements in columns, pumping systems, and injectors for capillary SFC, the 10,000 dalton range may soon be exceeded.

For the chemist, tremendous opportunities lie just ahead, both for research in SFC and for the application of its unique capabilities to the solution of industrial problems.

ACKNOWLEDGMENTS

The contributions of D. P. Innis are gratefully acknowledged. Also, thanks are due G. D. Owens, J. D. Pinkston, and P. A. Rodriguez for their support.

REFERENCES

1. E. Klesper, A. H. Corwin, and D. A. Turner, *J. Org. Chem.* 27, 700–701 (1962).
2. J. C. Giddings, M. N. Meyers, L. McLaren, and R. A. Keller, *Science 162*, 67–73 (1968).
3. T. H. Gouw and R. E. Jentoft, *J. Chromatogr.* 68, 303–323 (1972).
4. D. R. Gere, *Science 222*, 253 (1983).
5. M. Novotny, S. R. Springston, P. A. Peadon, J. C. Fjeldsted, and M. L. Lee, *Anal. Chem.* 53, 407A–414A (1981).
6. S. R. Springston and M. Novotny, *Chromatographia 14*, 679–684 (1981).
7. P. A. Peadon, J. C. Fjeldsted, M. L. Lee, S. R. Springston, and M. Novotny, *Anal. Chem.* 54, 1090–1093 (1982).
8. J. C. Fjeldsted, W. P. Jackson, P. A. Peadon, and M. L. Lee, *J. Chromatogr. Sci.* 21, 222–225 (1983).
9. P. A. Peadon and M. L. Lee, *J. Chromatogr.* 259, 1–16 (1983).
10. J. C. Fjeldsted, R. C. Kong, and M. L. Lee, *J. Chromatogr.* 279, 449 (1983).
11. M. Novotny and S. R. Springston, *J. Chromatogr.* 279, 417–422 (1983).

12. K. Grob, *J. High Resolut. Chromatogr. Chromatogr. Commun.* 6, 178—184 (1983).
13. M. L. Lee, *Separation of Asphaltenes Using High Resolution Supercritical Fluid Chromatography—Final Report September 1981—August 1983*, U.S. Department of Energy, Fossil Energy Program, Grant No. DE-FG22-81PC40809 (1983).
14. T. L. Chester, *J. Chromatogr.* 299, 424—431 (1984).
15. J. C. Fjeldsted and M. L. Lee, *Anal. Chem.* 56, 619A—628A (1984).
16. B. W. Wright, H. R. Udseth, R. D. Smith, and R. N. Hazlett, *J. Chromatogr.* 314, 253—262 (1984).
17. B. W. Wright and R. D. Smith, *Chromatographia* 18, 542—545 (1984).
18. M. Novotny, *J. Chromatogr. Lib.* 30, 105—120 (1985).
19. B. W. Wright and R. D. Smith, *J. High Resolut. Chromatogr. Chromatogr. Commun.* 8, 8—11 (1985).
20. R. D. Smith, H. R. Udseth, and B. W. Wright, *J. Chromatogr. Sci.* 23, 192—199 (1985).
21. S. R. Springston, Ph.D. Thesis, Indiana University (1984).
22. C. M. White and R. K. Houck, *J. High Resolut. Chromatogr. Chromatogr. Commun.* 8, 293—296 (1985).
23. B. E. Richter, *J. High Resolut. Chromatogr. Chromatogr. Commun.* 8, 297—300 (1985).
24. T. L. Chester and D. P. Innis, *J. High Resolut Chromatogr. Chromatogr. Commun.* 8, 561—566 (1985).
25. T. L. Chester, D. P. Innis, and G. D. Owens, *Anal. Chem.* 57, 2243—2247 (1985).
26. T. L. Chester and D. P. Innis, *J. High Resolut. Chromatogr. Chromatogr. Commun.* 9, 209—212 (1986).
27. T. L. Chester and D. P. Innis, *J. High Resolut. Chromatogr. Chromatogr. Commun.* 9, 178—181 (1986).
28. R. K. Houck, 190th American Chemical Society National Meeting, Chicago, September 8—13 (1985).
29. S. V. Olesik, S. B. French, and M. Novotny, *Chromatographia* 18, 489—495 (1984).
30. T. L. Chester, *J. Chromatogr. Sci.* 24, 226—229 (1986).
31. B. E. Richter, The Pittsburgh Conference and Exposition, March 10—14, 1986, Paper No. 514, Atlantic City, NJ, USA.
32. E. J. Guthrie and H. E. Schwartz, *J. Chromatogr. Sci.* 24, 236—241 (1986).
33. L. McLaren, M. N. Meyers, and J. C. Giddings, *Science 15*, 197—199 (1968).
34. K. Grob, Jr., G. Karrer, and M.-L. Riekkola, *J. Chromatogr. Sci.* 24, 236—241 (1986).

Index

Printed and bound by CPI Group (UK) Ltd, Croydon, CR0 4YY

23/10/2024

01778237-0016